Stem Cell Biology and Regenerative Medicine

Volume 68

Series Editor
Kursad Turksen
Ottawa Hospital Research Institute
Ottawa, ON, Canada

Our understanding of stem cells has grown rapidly over the last decade. While the apparently tremendous therapeutic potential of stem cells has not yet been realized, their routine use in regeneration and restoration of tissue and organ function is greatly anticipated. To this end, many investigators continue to push the boundaries in areas such as the reprogramming, the stem cell niche, nanotechnology, biomimetics and 3D bioprinting, to name just a few. The objective of the volumes in the Stem Cell Biology and Regenerative Medicine series is to capture and consolidate these developments in a timely way. Each volume is thought-provoking in identifying problems, offering solutions, and providing ideas to excite further innovation in the stem cell and regenerative medicine fields.

More information about this series at http://www.springer.com/series/7896

Alexander Birbrair
Editor

Biology of Pericytes – Recent Advances

Humana Press

Editor
Alexander Birbrair
Columbia Univ Med Cent, Dept. Radiology
Univ Federal de Minas Gerais, Dept Patho
Belo Horizonte, Minas Gerais, Brazil

ISSN 2196-8985 ISSN 2196-8993 (electronic)
Stem Cell Biology and Regenerative Medicine
ISBN 978-3-030-62128-5 ISBN 978-3-030-62129-2 (eBook)
https://doi.org/10.1007/978-3-030-62129-2

This Humana imprint is published by the registered company Springer Nature Switzerland AG
The registered company address is: Gewerbestrasse 11, 6330 Cham, Switzerland

This book is dedicated to my mother, Marina Sobolevsky, of blessed memory, who passed away during the creation of this volume. Professor of Mathematics at the State University of Ceará (UECE), she was loved by her colleagues and students, whom she inspired by her unique manner of teaching. All success in my career and personal life I owe to her.

My beloved mom Marina Sobolevsky of blessed memory (July 28, 1959 - June 3, 2020)

Preface

This book's is covering the novel aspects of pericyte biology in distinct pathophysiological conditions. The continuous great interest in this topic is leading to exciting research which brings new discoveries every year. The reading of this volume can be combined with the revision of other three volumes that we published earlier: "Pericyte Biology – Novel Concepts"; "Pericyte Biology in Different Organs"; and "Pericyte Biology in Disease".

This book "Pericyte Biology - Recent Advances" presents contributions by expert researchers and clinicians in the multidisciplinary areas of medical and biological research. The chapters provide timely detailed overviews of recent advances in the field. This book describes the major contributions of pericytes to different organs biology in physiological and pathological conditions. Further insights into the biology of pericytes will have important implications for our understanding of organ development, homeostasis and disease. The authors focus on the modern methodologies and the leading-edge concepts in the field of cell biology. In recent years, remarkable progress has been made in the identification and characterization of pericytes in several tissues using state-of-art techniques. These advantages facilitated identification of pericyte subpopulations and definition of the molecular basis of pericytes role within different organs. Thus, the present book is an attempt to describe the most recent developments in the area of pericyte behavior which is one of the emergent hot topics in the field of molecular and cellular biology today. Here, we present a selected collection of detailed chapters on what we know so far about the pericytes. Eleven chapters written by experts in the field summarize the present knowledge about novel roles of pericytes in different circumstances.

Andy Y. Shih and colleagues from University of Washington discuss in vivo optical imaging and manipulation of brain pericytes. Brad A. Sutherland from University of Tasmania describes the complex and integral roles of pericytes within the neurovascular unit in health and disease. Yoichi Morofujia and colleagues from Nagasaki University compile our understanding of the role of pericytes in brain metastasis. Donald J. Alcendor from Meharry Medical College updates us with what we know about the effects of Cytomegalovirus on pericytes. Turgay Dalkara and colleagues from Hacettepe University summarise current knowledge on pericytes in retinal ischemia. Yasufumi Kataoka and colleagues from Fukuoka University address the importance of inflammatory mediators released by brain pericytes as sensors and effectors in blood-brain barrier dysfunction. Alessandra Stasi and colleagues from University of Foggia focus on TLR-4 signaling in pericytes. Jun-ichi Kawabe from Asahikawa Medical University introduces our current knowledge about the roles of EphA7+ multipotent pericytes. Marni D. Boppart and colleagues from University of Illinois at Urbana-Champaign introduce what we know so far about skeletal muscle-resident pericyte responses to conditions of use and disuse. Nikolaos G Frangogiannis and colleagues from Albert Einstein College of Medicine talk about pericytes in myocardial diseases. Finally, Paolo Madeddu and colleagues from University of Bristol give an overview of adventitial and skeletal muscle pericytes in health and in ischemic tissue regeneration.

It is hoped that the articles published in this book will become a source of reference and inspiration for future research ideas. I would like to express my deep gratitude to my wife Veranika Ushakova, and Dr. Gonzalo Cordova from Springer Nature, who helped at every step of the execution of this project.

Belo Horiz, Minas Gerais, Brazil Alexander Birbrair

Contents

About the Editor

Alexander Birbrair Dr. Alexander Birbrair received his bachelor's biomedical degree from Santa Cruz State University in Brazil. He completed his PhD in Neuroscience, in the field of stem cell biology, at the Wake Forest School of Medicine under the mentorship of Osvaldo Delbono. Then, he joined as a postdoc in stem cell biology at Paul Frenette's laboratory at Albert Einstein School of Medicine in New York. In 2016, he was appointed faculty at Federal University of Minas Gerais in Brazil, where he started his own lab. His laboratory is interested in understanding how the cellular components of different tissues function and control disease progression. His group explores the roles of specific cell populations in the tissue microenvironment by using state-of-the-art techniques. His research is funded by the Serrapilheira Institute, CNPq, CAPES, and FAPEMIG. In 2018, Alexander was elected affiliate member of the Brazilian Academy of Sciences (ABC), and, in 2019, he was elected member of the Global Young Academy (GYA). He is the Founding Editor and Editor-in-Chief of Current Tissue Microenvironment Reports, and Associate Editor of Molecular Biotechnology. Alexander also serves in the editorial board of several other international journals: Stem Cell Reviews and Reports, Stem Cell Research, Stem Cells and Development, and Histology and Histopathology.

Affiliations and Expertise
Department of Radiology, Columbia University, USA

Department of Pathology, Federal University of Minas Gerais, Belo Horizonte, Brazil

Chapter 1
In Vivo Optical Imaging and Manipulation of Brain Pericytes

Andrée-Anne Berthiaume, Vanessa Coelho-Santos, David A. Hartmann, and Andy Y. Shih

Abstract

Introduction

Pericyte serve myriad roles in the developing and adult brain, but many of their functions remain poorly understood. In less than a decade, the advent of new tools and approaches has revolutionized brain pericyte research by enabling direct visualization of pericyte structure and function in the living mouse brain. In addition to accessing pericytes in a physiological context, *in vivo* imaging has revealed fascinating new biology on how distinct pericyte subtypes influence cerebral blood flow in both health and disease. This chapter highlights novel approaches and provides up-to-date guidance on pericyte research using two-photon microscopy and related optical applications.

Methods

We surveyed literature for studies that have imaged brain pericytes *in vivo* in order to summarize their methodologies and approaches. Keywords included pericyte, mural cell, cerebrovasculature, brain, capillary, blood-brain barrier, cerebral blood-

---

A.-A. Berthiaume
Department of Neuroscience, Medical University of South Carolina, Charleston, SC, USA

Center for Developmental Biology and Regenerative Medicine, Seattle Children's Research Institute, Seattle, WA, USA

V. Coelho-Santos
Center for Developmental Biology and Regenerative Medicine, Seattle Children's Research Institute, Seattle, WA, USA

D. A. Hartmann
Department of Neuroscience, Medical University of South Carolina, Charleston, SC, USA

A. Y. Shih (✉)
Center for Developmental Biology and Regenerative Medicine, Seattle Children's Research Institute, Seattle, WA, USA

Department of Pediatrics, University of Washington, Seattle, WA, USA

Department of Bioengineering, University of Washington, Seattle, WA, USA
e-mail: Andy.Shih@SeattleChildrens.org

A. Birbrair (ed.), *Biology of Pericytes – Recent Advances*, Stem Cell Biology and Regenerative Medicine 68, https://doi.org/10.1007/978-3-030-62129-2_1

flow, two-photon microscopy, cranial window, endothelium, cre recombinase, *in vivo* imaging, microvasculature, calcium imaging, cell ablation.

Results
We discuss the identification of true capillary pericytes and set them apart from transitional phenotypes in murine cerebral cortex. An array of available murine Cre drivers and other approaches to target pericytes and brain mural cells are compared and contrasted. Surgical and experimental parameters to perform rigorous quantification of pericyte influence on capillary flow and integrity are also discussed. Lastly, we move beyond correlative observation to cutting-edge approaches for optical ablation and manipulation of pericytes, concomitant with real-time imaging of cerebrovascular function.

Conclusion
The information provided can serve as a roadmap to rigorous *in vivo* imaging of pericytes in the healthy and diseased brain.

1.1 Introduction

Deficiency in the function of small brain vessels is a common link between many age-related brain diseases. A recent emergence of new tools and approaches has allowed researchers to observe and modify cerebral blood vessels, as well as their cellular components, in the living brain. In particular, pericytes have taken center stage, and their role in maintaining a functional neurovascular unit is now in a deep phase of exploration. Pericytes find their niche within the basement membrane of all brain capillaries, and are characterized by their long filamentous processes stemming from protruding ovoid cell bodies. Pericytes, along with vascular smooth muscle cells, fall under the broad category of "mural cells," which exhibit diverse morphologies along cerebrovascular networks. The abundance of pericytes along the walls of small cerebral vessels in the brain compared to the vascular beds of other tissues suggests a critical contribution to cerebrovascular health and stability (Armulik et al. 2005). Indeed, pericytes are integral to the formation of the blood-brain barrier (BBB) (Blanchette and Daneman 2015; Daneman et al. 2010; Armulik et al. 2010) and the stabilization of vascular structure and tone (von Tell et al. 2006; Berthiaume et al. 2018a; Hellstrom et al. 2001), and may even play some roles in immune function (Stark et al. 2018; Rustenhoven et al. 2017) and scar formation following injury (Reeves et al. 2019; Zehendner et al. 2015). The congenital loss of pericytes through genetic knockout is often prenatally lethal (Lindahl et al. 1997; Lindblom et al. 2003), and decreased pericyte numbers has been reportedly involved in aging and Alzheimer's disease (Sengillo et al. 2013; Montagne et al. 2015; Halliday et al. 2016; Miners et al. 2019). Loss of pericytes under these circumstances has been related to increased BBB permeability (Miners et al. 2019; Dore-Duffy et al. 2000; Berthiaume et al. 2018b; Arango-Lievano et al. 2018).

Much of what we know about cerebrovascular pericytes has come from important studies in primary pericyte cultures and brain slices (Kamouchi et al. 2004; Mishra et al. 2014; Hall et al. 2014; Nakagawa et al. 2007; Peppiatt et al. 2006;

Orlidge and D'Amore 1987; Sato et al. 1990). These systems provide the benefit of studying pericytes in relative isolation to determine their fundamental attributes and functions. However, there are also clear limitations to an *in vitro* approach. Pericytes function within the context of a complex neurovascular network. There are aspects of the intact brain that cannot be replicated with *in vitro* systems, notably blood flow, blood pressure, and network dynamics of the vascular beds that house pericytes. Furthermore, evidence suggests that cultured pericytes may develop a gene expression profile that differs dramatically from their *in vivo* counterparts, making it difficult to relate in vitro findings to true pericyte physiology (Guimaraes-Camboa et al. 2017). Accordingly, there has been a recent surge in the number of studies that use *in vivo* imaging to study brain pericytes, particularly with respect to blood flow regulation (Hall et al. 2014, 2015; Grutzendler and Nedergaard 2019; Hartmann et al. 2020). This move to *in vivo* preparations was facilitated by the development of several key techniques to target, visualize, and manipulate pericytes within the intact mouse brain. This chapter will focus on methods to study pericyte *in vivo*, including guidance on genetically engineered mouse lines, gaining optical access to the live brain, and *in vivo* two-photon microscopy. In addition, we describe approaches to identify different pericyte subtypes, and to properly characterize and quantify aspects of these cells in their native environment.

Early studies that expressed bright fluorescent reporters in mural cells of the brain revealed their remarkable morphological heterogeneity (Grant et al. 2017; Hartmann et al. 2015) (Fig. 1.1a). These range from the circumferential smooth muscle cells on arteries to the filamentous pericytes of the capillary network, with transitional cells between these classic forms. Venules and veins are also covered by a mixture of stellate shaped mural cells that do not completely encircle the vessel. The function of venous mural cells remains poorly defined, though their depletion can lead to disturbances in venous structure (Lai et al. 2015). All these cells occupy the abluminal surface of the vessel wall, but each has distinct identities in respect to morphology, topological organization, genetic expression, and physiology (Berthiaume et al. 2018a; Vanlandewijck et al. 2018). Given this diversity, there has been some confusion as to what mural cell subtypes should be considered true "pericytes," since mural cell classification has not been consistent across research groups https://pubmed.ncbi.nlm.nih.gov/26661200/. For this reason, care should be taken to properly identify and report exact microvascular location and morphology when studying pericytes *in vivo*.

1.2 Identifying Brain Pericytes

The first challenge of *in vivo* brain pericyte imaging is knowing how to consistently identify them. Several distinct mural cell morphologies appear as cortical penetrating arterioles branch into pre-capillary arterioles and then capillaries. In our work, we categorize these mural cells based on their morphology, topological location, and expression of contractile protein α-smooth muscle actin (α-SMA) (Fig. 1.1a). We refer to any mural cell type with a protruding ovoid cell body and elongated processes as "pericytes." However, we have applied nomenclature to three

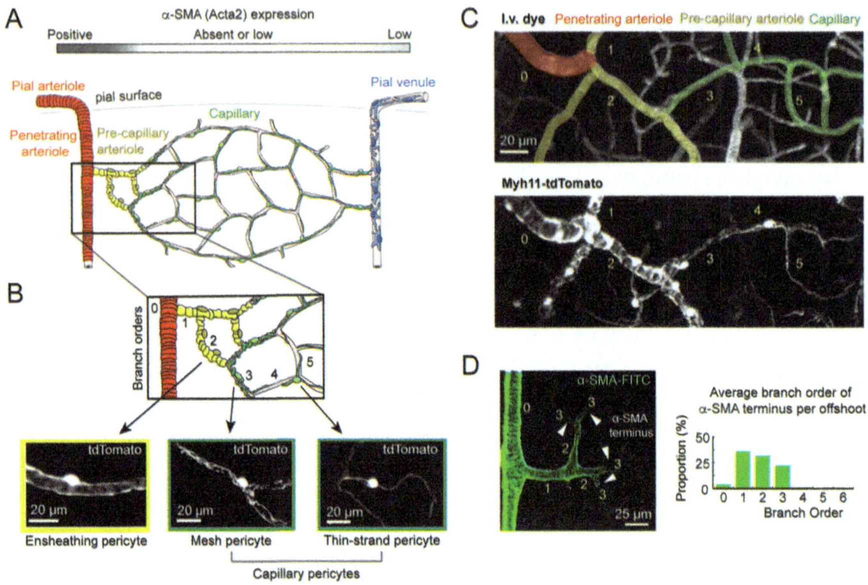

Fig. 1.1 Identifying mural cell subpopulations in the living mouse brain. (**a**) (Top) Relative expression of contractile protein α-SMA across microvascular zones. (Bottom) Schematic of mural cell morphologies residing on vessels of the mouse cerebral cortex. Colors represent distinct mural cell subpopulations. Adapted from Berthiaume et al. 2018 (Berthiaume et al. 2018a). (**b**) (Top) Inset from (**a**), showing a close-up of the vascular network forming the arteriole (red), pre-capillary (yellow), and capillary zones (green). Numbers indicate branch orders from penetrating arteriole, 0, where each bifurcation increases vascular branch order by 1. (Bottom) *In vivo* two-photon images of the three major pericyte subtypes. From left to right, ensheathing pericytes of the precapillary zone, mesh pericytes of the capillary zone, and thin-strand pericytes of the capillary zone. Left image from PDGFRβ-tdTomato mouse, center and right images from Myh11-tdTomato mice. (**c**) *In vivo* i.v. dye image (top) marked with branch order and colors corresponding to colors in (**b**). Corresponding two-photon image of mural appearance (bottom), with labeled branch orders. These images were from a Myh11-tdTomato mouse. (**d**) Immunohistological staining for α-SMA using a FITC-conjugated antibody, showing sharp cutoffs in protein abundance. Quantification of average branch order for α-SMA termini per offshoot from penetrating arterioles, showing undetectable expression after fourth branch order. (Adapted from Grant et al. 2017 (Grant et al. 2017))

morphological sub-classes of pericytes, which we refer to as ensheathing, mesh, and thin-strand pericytes (Grant et al. 2017; Vanlandewijck et al. 2018; Grubb et al. 2020) (Fig. 1.1b).

Ensheathing pericytes are found primarily on pre-capillary arterioles of the mouse cortex. Though the term "pre-capillary arteriole" is not universally used, we have denoted this zone as the first few branch orders that diverge from the penetrating arteriole. Ensheathing pericytes fully cover the microvessel with tightly packed processes running circumferential to the endothelial tube (Fig. 1.1c), suggesting they are well equipped for vascular contraction and dilation (Grubb et al. 2020). Their circumferential processes are reminiscent of vascular smooth muscle cells

(SMCs) on upstream arteries, though they are set apart by their protruding cell bodies and roughly two-fold greater length along the longitudinal capillary axis. Ensheathing pericytes share genetic similarities with SMCs, most notably the high expression of α-SMA (Vanlandewijck and Betsholtz 2018). In contrast, this contractile protein is low or absent in mesh and thin-strand pericytes downstream of the pre-capillary arteriole, reflecting an important functional distinction between ensheathing pericytes and other microvascular pericytes (Grant et al. 2017; Vanlandewijck et al. 2018). While robust α-SMA expression is ideally used to identify ensheathing pericytes, its expression level is not easily ascertained *in vivo* without ACTA2 reporter mice (Hill et al. 2015). However, they can be generally identified based on their circumferential morphology and topological location (i.e. branch order). Several groups have now shown that, in mouse cortex, α-SMA expression on penetrating arteriole offshoots does not surpass the 4th branch order (Grant et al. 2017; Grubb et al. 2020; Wei et al. 2016)(Fig. 1.1d). Thus, examining microvessel branches that are within the 4th order increases the likelihood of studying a region occupied by ensheathing pericytes, but does not guarantee it. However, examining beyond 4th branch order guarantees that one is within the capillary zone, occupied by mesh or thin-strand pericytes.

As one goes distally from ensheathing pericyte territory, α-SMA expression becomes low or absent and pericyte processes begin to lose their ring-like appearance, instead taking on a complex mesh-like shape (Fig. 1.1c). These mesh pericytes also have characteristic protruding cell bodies, but unlike ensheathing pericytes, their processes do not completely cover the underlying endothelium. Beyond the 4th and 5th order branches, pericyte processes with the classic thin-strand morphology become prominent. Thin-strand pericytes are the most abundant pericyte form, as they cover the majority of the capillary bed. These pericytes are easily recognizable with several long, thin processes emerging from the soma, which protrudes from the capillary wall like a "bump on a log." We have grouped mesh and thin-strand pericyte types together into the category "capillary pericytes" because their morphologies are difficult to objectively distinguish. Capillary pericytes form an extensive network along the brain microvasculature, where neighboring pericytes occupy non-overlapping territories (Berthiaume et al. 2018b). Each cell therefore acts as a "link" in a chain of capillary pericytes distributed throughout the capillary bed. These morphological features, as well as their topological location within the cerebrovascular network (beyond 4th branch order), make capillary pericytes identifiable with certainty using *in vivo* optical imaging.

1.3 Distinct Functions of Pericyte Subtypes

The morphological and topological heterogeneity of pericytes translates into functional differences *in vivo*. Ensheathing pericytes and capillaries pericytes show distinct dynamics during neurovascular coupling, and in response to ischemic pathology and small vessel disease. In the normal brain, ensheathing pericytes are first responders to local neural activity, resulting in rapid dilation of the pre-capillary arteriole and influx of blood to the brain's capillary bed (Hall et al. 2014; Cai et al. 2018; Rungta et al. 2018). This dilation outpaces even the dilation speed of upstream SMC-covered arterioles, suggesting that ensheathing pericytes are critical sensors to neural activity and a central locus for blood flow control. In contrast, capillary dilations are absent or small in neurovascular coupling, and appear to be slower if present (Hill et al. 2015; Rungta et al. 2018; Kisler et al. 2017). As expected, intracellular calcium levels are strongly anti-correlated with diameter change in pre-capillary arterioles, suggesting engagement in classic actomyosin contractile machinery, but this relationship is less clear in capillary pericytes (Hill et al. 2015; Rungta et al. 2018). This likely indicates that a separate, slower mechanism of contraction and tone generation exists in capillary pericytes (Hartmann et al. 2020).

The response of these two pericyte subtypes is also strikingly different during brain ischemia. Capillary pericytes are particularly vulnerable to blood flow loss, dying from necrosis within tens of minutes to hours (Hall et al. 2014), undergoing pathological contraction (Yemisci et al. 2009), and producing matrix-metalloproteinase activity, which leaves breaches in the BBB (Underly et al. 2017). In contrast, ensheathing pericytes are more resilient, but still undergo a sustained pathological contraction at the pre-capillary-capillary interface, which impedes flow to the capillary bed even after successful re-cannulation of the occluded artery (Hill et al. 2015). There is also emerging evidence that pericyte subtypes behave differently during chronic pathologies, such as small vessel disease. In a mouse model carrying a mutation in collagen 4 (Col4) relevant to human small vessel disease, spontaneous hemorrhagic stroke occurs in deep brain regions. Interestingly, ensheathing pericytes (referred to as transitional pericytes in this study) proliferate and heighten their expression of contractile proteins in Col4 mutant mice (Ratelade et al. 2020), which does not occur with capillary pericytes or arteriolar SMCs. This is thought to produce hyper-contractility, specifically at the transitional precapillary zone, leading to buildup of intravascular pressure and rupture of upstream arterioles.

Even within the capillary zone, there may be further division of pericyte subtypes to be made, but this remains an open area of research. Under conditions of brain injury, some pericyte populations respond differently than others. For example, NG2-positive capillary pericytes can be either nestin-negative (Type 1) or nestin-positive (Type 2) under basal conditions. Following brain injury, Type 1 pericytes proliferate around the injury site and contribute to scar tissue, unlike Type 2 pericytes (Birbrair et al. 2014a). Type 1 pericytes differ from NG2-positive oligodendrocyte precursor cells, and from PDGFRβ-positive cells that might be perivascular or meningeal fibroblasts, confirming a distinct pericyte type. Further, when Type 1 or Type 2 pericytes are injected into brain tumor masses *in vivo,* Type 2 pericytes are preferentially recruited into the tumor and promote angiogenesis over a 5-week period (Birbrair et al. 2014b). In contrast, Type 1 pericytes do not have this effect, instead decreasing in number over the same time frame. Thus, there is mounting evidence that pericytes are heterogeneous in both the normal and diseased brain. Referring to all pericytes simply as "pericytes" is now insufficient, as there are subdivisions that require further delineation.

1.4 Targeting Brain Pericytes for *In Vivo* Imaging

1.4.1 Cre/Lox System for Genetic Targeting of Pericytes

The Cre/lox system is an effective way to achieve a stable fluorescent labeling of brain mural cells, when bred with Cre-dependent fluorescent reporter mice (Madisen et al. 2010). There are several Cre-driver options, each with advantages and disadvantages. The optimal choice of Cre-lox pairing will vary depending on the experimental requirements (Fig. 1.2b).

1.4.2 PDGFRβ-Cre

For constitutive labeling of all mural cells in brain vasculature, the PDGFRβ-Cre mouse line is a reliable and convenient line (Cuttler et al. 2011) (Fig. 1.2h, i). When these mice are crossed with fluorescent reporters (e.g. Ai14 for tdTomato), mural cell labeling is bright and contiguous throughout the cerebrovasculature, illuminating the entire spectrum of mural cells. Comparing overlap of tdTomato expression with CD13 immunolabeling, an established marker of pericytes, we find that 99% of CD13-positive pericytes are labeled in these mice (Hartmann et al. 2015; Damisah et al. 2017). PDGFRβ-Cre mice are thus very useful for studying the capillary pericyte network as a whole, as well as the interactions between adjacent mural cells under basal conditions and with pericyte manipulation. The PDGFRβ-Cre line is not appropriate for lineage tracing studies, however, because Cre is constitutively active and PDGFRβ expression occurs in non-mural cell types during early development (Guimaraes-Camboa et al. 2017).

PDGFRβ-Cre mice provide highly specific labeling for mural cells, but there is some off-target labeling. Indeed, recent transcriptomic studies have discovered PDGFRβ-expressing fibroblasts that occupy a perivascular location similar to that of mural cells (Vanlandewijck et al. 2018). Perivascular fibroblasts appear mainly on larger vessels, which could allow them to be confused with arteriolar mural cells or ensheathing pericytes. Another consideration in using the PDGFRβ-Cre line for *in vivo* imaging is the high expression of PDGFRβ in meningeal fibroblasts, which seem to increase in abundance following craniotomy, likely indicating a role of these cells in the fibrotic response to invasive surgery. The dense labeling of the meninges can sometimes make it difficult to obtain clean images of mural cells in pial vessels and pericytes of superficial capillaries. Furthermore, caution should be taken if one wishes to breed PDGFRβ-Cre mice to homozygosity because the location of the gene insertion is unclear.

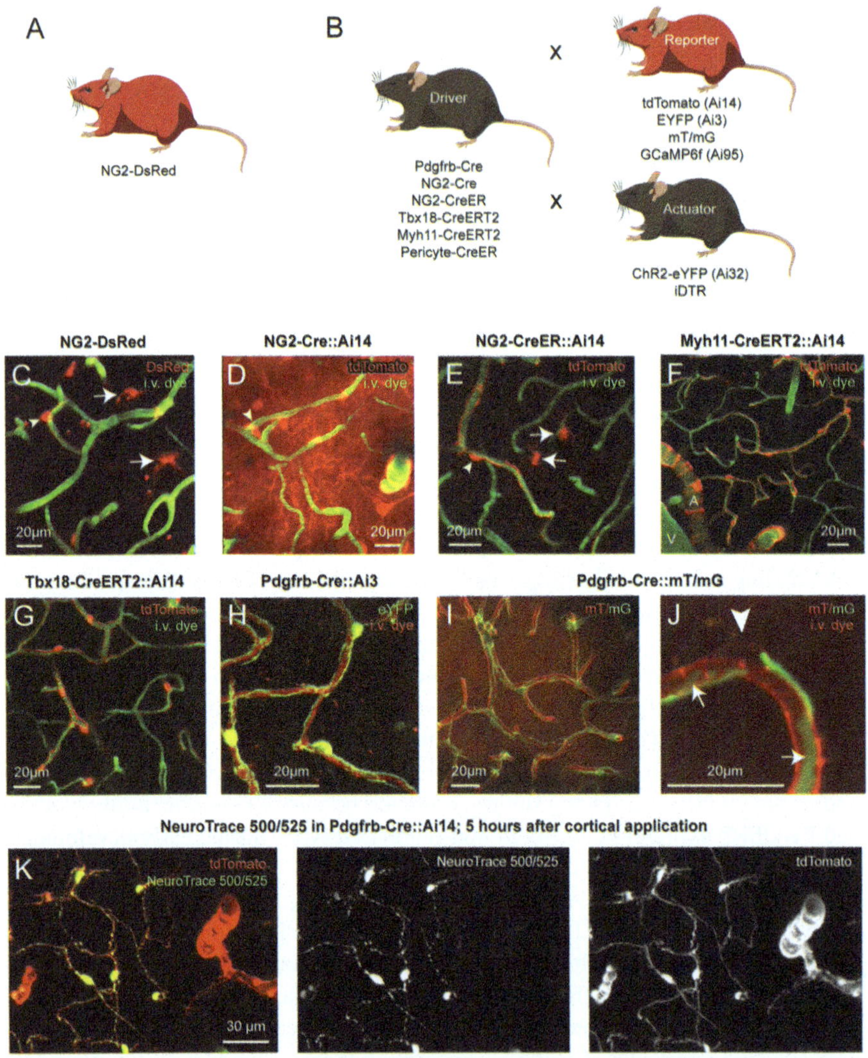

Fig. 1.2 Targeting brain pericytes for *in vivo* optical imaging. (**a**) The NG2-DsRed mouse endogenously expresses DsRed under the NG2 promoter independent of Cre recombination. (**b**) Outline of commonly used Cre/lox breeding schemes for brain pericyte research. (**c**) Constitutive labeling of capillary pericytes (arrowhead) in the NG2-DsRed mouse, with off-target labeling of oligodendrocyte precursor cells (OPCs)(arrows). Pericyte processes are difficult to visualize in this line. (**d**) Constitutive labeling of mural cells (arrowhead) in NG2-Cre::Ai14 mouse. Developmental off-target labeling of OPCs and astrocyte progenitors results in high background fluorescence. (**e**) Tamoxifen-inducible NG2-CreER::Ai14. OPC labeling persists (arrows), but can be spatially distinguished from mural cells (arrowhead). Pericyte processes are more easily visualized with tdTomato. (**f**) Sparse labeling of mural cells with tamoxifen-inducible Myh11-CreERT2::Ai14. Notice extensive arterial smooth muscle cell labeling relative to capillary pericyte and venular mural cell labeling. (**g**) Tamoxifen-inducible sparse labeling of pericytes and other mural cells with Tbx18-CreERT2::Ai14. (**h**) Constitutive mural cell reporter expression in Pdgfrb-Cre::Ai3 mouse.

1.4.3 NG2-Cre, NG2-CreER, and NG2-DsRed

Another widely used promotor to target brain mural cells pericytes is the *Cspg4* promoter, which encodes the NG2 protein. This driver has the advantage of not labeling meningeal or perivascular fibroblasts as strongly as PDGFRβ-Cre (Vanlandewijck et al. 2018). However, along with smooth muscle cells and pericytes, several other cells are NG2-positive in development, notably astrocytes and oligodendrocyte precursor cells (OPCs) (Zhu et al. 2008) (Fig. 1.1d). However, it is possible to visually distinguish astrocytes and OPCs in the adult brain. In mice with constitutive Cre expression under the *Cspg4* promoter (NG2-Cre), most mural cells are labeled. That being said, the background parenchymal fluorescence from neurons and OPCs may be too dense for some applications, such as those seeking to study pericyte morphology. However, one unexpected advantage found when crossing these mice with mice expressing a genetically-encoded calcium indicator was ectopic transgene expression by neurons, providing a convenient way to examine both neuronal and mural cell activity in unison during neurovascular coupling (Hill et al. 2015).

An NG2-CreER driver is available for tamoxifen-inducible labeling of brain mural cells (Huang et al. 2014). By administering tamoxifen at post-developmental timepoint, the problem of astrocyte progenitor labeling is reduced. OPC expression of NG2 occurs throughout the lifetime, so this off-target labeling remains even with the inducible form of the driver (Fig. 1.1e). Interestingly, a recent study showed strong correlation between OPC calcium transients and neuronal activity, again providing a convenient tool for perform neurovascular coupling, where the timing of local neural activity is crucial to measure (Rungta et al. 2018). As with all tamoxifen-inducible labeling schemes, the sparseness of fluorescent cells is titratable based on tamoxifen dose administered. With minimal tamoxifen (i.e. one i.p. injection of 100 mg/kg), extremely sparse labeling of mural cells can be achieved. This strategy allows for the morphologies of individual pericytes to be discerned (Grant et al. 2017). It also allows single mural cells to be tracked over time *in vivo (*Berthiaume et al. 2018b*)*.

A related NG2 mouse line, the NG2-DsRed line, has contributed greatly to the *in vivo* study of brain pericytes (Zhu et al. 2008). They have been used widely for both *in vivo* and *ex vivo* imaging studies on blood flow control (Mishra et al. 2014; Hall et al. 2014; Grubb et al. 2020; Cai et al. 2018). In these BAC transgenic mice,

Fig. 1.2 (continued) (**i**) Membrane-localized tdTomato in all cell membranes and membrane-localized eGFP in capillary pericytes with the Pdgfrb-Cre::mT/mG mouse. No i.v. dye is present in mT/mG images. (J) High resolution image from Pdgfrb-Cre::mT/mG mouse, showing the gap between two adjacent pericyte processes (green), near a faintly visible endothelial cell body (arrowhead). Note how the mGFP label reveals thin sheet-like outcroppings of pericyte processes that are less visible with cytosolic fluorescent proteins (arrows). (**k**) Images from a Pdgfrb-Cre::Ai14 animal, 5 hours after cortical application of the capillary pericyte-specific dye, NeuroTrace 500/525

DsRed protein expression is driven directly by the *Cspg4* promoter, eliminating the need for Cre recombination and increasing the convenience of their use. Mural cells are visible by *in vivo* two-photon microscopy, and pericyte cell bodies can be easily visualized (Fig. 1.2c). However, the DsRed protein appears insufficiently bright to consistently visualize the fine processes of pericytes, compared to bigenic Cre-reporter mice expressing tdTomato. As with NG2-Cre mice, oligodendrocyte precursors are also labeled in NG2-DsRed mice.

1.4.4 Tbx18-CreERT2

The tamoxifen-inducible Tbx18-CreERT2 mouse line is an excellent tool for sparse-labeling of both smooth muscle cells and pericytes. This mouse line has been successfully utilized for lineage-tracing studies of pericytes in the adult brain. For example, one recent study used the mouse line to show that pericytes of the adult brain do not transdifferentiate into other cell types during brain injury (Guimaraes-Camboa et al. 2017). The *tbx18* gene may be highly expressed in certain fibroblast populations, according to the RNA sequencing databank created by the Betsholtz group (http://betsholtzlab.org/VascularSingleCells/database.html), so any cell under study must be positively identified as pericytes through detailed morphological analysis. This also leads to fluorescence expression in some meningeal and perivascular fibroblast populations, though the labeling is not as dense as with the PDGFRβ-Cre mice. Unlike the NG2-CreER inducible labeling system, no parenchymal cells appear to be consistently labeled under this Cre driver. This makes the Tbx18-CreERT2 mouse line a relatively "clean" way to sparsely label pericytes for *in vivo* imaging (Fig. 1.2g).

1.4.5 Myh11-CreERT2

Another inducible mouse Cre line that has been used for brain pericyte labeling is Myh11-CreERT2 (Wirth et al. 2008). This promoter is not expressed in any non-mural cell type in the brain, including fibroblasts, and therefore leads to a highly specific mural cell label with no meningeal labeling. However, there is considerably greater expression levels of the *myh11* promoter in mural cells of arterioles and pre-capillary arterioles compared to capillary pericytes (Betsholtz single cell RNA transcriptome database (Vanlandewijck et al. 2018)). This leads to the strong preferential labeling of mural cells on arterioles with low doses of tamoxifen (Fig. 1.2f). With added doses of tamoxifen, capillary pericytes will also begin expressing the designated reporter. Sparse labeling of capillary pericytes is easily achievable through this method, but the sparse labeling of smooth muscle cells or ensheathing pericytes of pre-capillary arterioles is unlikely due to more complete labeling of these cell types. A disadvantage of the Myh11-CreERT2 mouse line is that the BAC transgene carrying Cre recombinase is inserted within the Y chromosome, which limits reporter expression to only male progeny of crosses. This makes it impossible to run a study with a balanced number of male and female mice if using only this mouse line, which can be a problem when considering sex as a biological variable. Newer versions of the Myh11-CreERT2 line have managed to create a female mouse line to address this limitation, though it has not yet been utilized in brain pericyte research (Liao et al. 2017). In our hands, using the original Y-linked Myh11-CreERT2 mouse line, we have detected some leakiness of Cre activity in the absence of tamoxifen. Constitutive expression is seen specifically in brain smooth muscle cells and ensheathing pericytes when breeding with homozygous Ai14 reporter animals (unpublished observation). This "leaky" Cre is not an issue if the mice are used for cell visualization, but would be problematic if the goal is to modify gene expression in mural cells of the brain in an inducible manner.

1.4.6 Pericyte-CreER

A Cre-driver to specifically target capillary pericytes, while excluding α-SMA positive mural cells, is critical considering the distinct physiological roles of mural cell classes. Recently, a capillary pericyte-specific Cre line was generated by using a combinatorial strategy that requires target cells to express both *Cspg4* and *Pdgfrb* (Nikolakopoulou et al. 2019). While both genes are expressed by smooth muscle cells and ensheathing pericytes, they are expressed at relatively lower levels than in capillary pericytes, reducing probability of Cre expression in α-SMA positive cells. When paired with a fluorescent reporter, this pericyte-CreER line labels capillary pericytes of the brain with remarkable specificity in a tamoxifen-inducible manner. The mural cells of pial/penetrating arterioles, and even pre-capillary arterioles are excluded from expression. As a newly developed mouse line, more characterization will be required from independent groups to fully test the functionality of the line under varying circumstances. In particular, whether pericyte-CreER also excludes mural cells of the veins and venules remains to be seen, as these mural cells express both *Cspg4* and *Pdgfrb* at levels similar to capillary pericytes (Betsholtz single cell RNA transcriptome database (Vanlandewijck et al. 2018)). One application of the pericyte-CreER mouse using a diphtheria toxin strategy to cause mass deletion of brain capillary pericytes has already been implemented, confirming its functionality (Nikolakopoulou et al. 2019; Kisler et al. 2020).

1.4.7 Fluorescent Reporters and Actuators

The most common fluorescent reporters paired with the aforementioned Cre drivers are Ai14 (tdTomato) and Ai3 (eYFP), which are both incorporated into the Rosa26 locus and lead to high expression of their respective fluorescent proteins (Madisen et al. 2010). TdTomato generally performs better than YFP for *in vivo* two-photon microscopy, partially because red light is less easily scattered than yellow/green. However, the Ai3 YFP reporter can be strategically utilized when mice already have a red fluorescent label. Additional reporters can also be paired with any of the discussed Cre drivers to create useful phenotypes. For instance, the mT/mG reporter (Muzumdar et al. 2007) constitutively labels all cellular membranes with membrane-associated tdTomato. With the action of Cre, however, the targeted cell type will switch to expression of a membrane bound eGFP fluorophore. This phenotype is especially useful in cerebrovascular biology, as eGFP-expressing pericytes and tdTomato-expressing capillary endothelial cells can be visualized concurrently (Fig. 1.2i). The endothelial labeling appears bright relative to other parenchymal cells. Endothelial cell somata can also visualized, though with occasional difficultly due to the diffuse tdTomato expression from the brain parenchyma, (Fig. 1.2j). It is also worth noting that fine sheathes of capillary pericyte processes, beyond thin strands, can be resolved with mGFP labeling compared to cytosolic dyes such as tdTomato. Another useful reporter is Ai95, encoding the GCaMP6f protein (Madisen et al. 2010). This reporter has been successfully utilized to study pericyte fast calcium transients in the living mouse brain (Rungta et al. 2018). The Ai96 reporter (GCaMP6s) is also available to detect slow calcium transients in cells (Madisen et al. 2010), though this reporter line has not yet been directly used in a pericyte study. Looking ahead, an extensive array of next-generation fluorescent reporters have been under development in recent years using the permissive TIGRE locus (Daigle et al. 2018). These reporters generate even higher protein expression than the previous set of Rosa26 locus reporters. Finally, Ai32 mice expressing the original variant of ChR2-YFP have been crossbred with mural cell specific Cre drivers. These reporters have been used to query the contractile dynamics of smooth muscle cells and pericyte *in vivo* during two-photon imaging (Hill et al. 2015).

1.4.8 Viral Targeting of Brain Pericytes

Viruses are a valuable tool to target cells *in vivo*. They can be introduced at any time in the lifespan of the mouse, and into any brain region. Genetically engineered viral vectors can lead to high levels of ectopic gene expression while maintaining user safety, as any pathogenic genes of the virus are absent (Wang et al. 2018; Davidson and Breakefield 2003). Several parameters ultimately dictate the level of viral expression, including the volume, titer and penetration of the virus into the brain tissue, the efficiency of its uptake by the targeted cell type, and the activity of its promoter (Packer et al. 2013; Stoica et al. 2013; Lowery and Majewska 2010). The exclusivity of viral uptake to certain cell types is often problematic. Viruses targeting brain pericytes are not exempt from these challenges.

For use with *in vivo* optical imaging, one virus has been demonstrated to target brain mural cells. This virus is the adeno-associated virus 2/5 (AAV2/5) serotype carrying a Pdgfrb promoter for the expression of the GCaMP6s calcium indicator in vascular mural cells (Khennouf et al. 2018). The virus was microinjected near the somatosensory cortex region of mice. Its specificity was not limited to only mural cells, but also revealed punctate GCaMP6s expression throughout the parenchyma of the injection area with unclear specificity. However, positive expression in pericytes could be confirmed using a pericyte-labeling Cre mouse line in conjunction with the virus. This approach will undoubtedly prove useful as it alleviates a significant bottleneck in the genetic modification of pericytes *in vivo* for future mechanistic studies. However, specificity for capillary pericytes still needs to be improved. Other studies have also targeted brain pericytes using a viral approach, though none for *in vivo* optical imaging use. One group selectively manipulated pericytes with a virally expressed shRNA to silence Ephrin-B2 signaling in pericytes, by creating a AAV viral construct with the NG2 promoter (Coucha et al. 2019).

1.4.9 In vivo *Dye Labeling of Brain Pericytes*

Thus far, the methods discussed for targeting brain pericytes *in vivo* have relied on genetic approaches. Furthermore, most of these methods are unable to disentangle between capillary pericytes and other populations of mural cells, with the exception of the new pericyte-Cre driver. The use of a specialized dye is an excellent method to temporarily label capillary pericytes for *in vivo* imaging. Recently, Damisah and colleagues demonstrated that the green fluorescent Nissl dye NeuroTrace 500/525 labeled capillary pericytes with high specificity (Damisah et al. 2017). The preferential uptake of NeuroTrace 500/525 into capillary pericytes requires topical application or intracortical microinjection of the dye into the living mouse brain. Remarkably, only mesh and thin-strand pericytes seem to be labeled by this dye, and mural cells of both pre-capillary arterioles and penetrating arterioles, as well as downstream venules, appear to be excluded. The dye labeling appears several hours after its application, and remains for up to 3 days after. With topical application, dye penetration reaches at most 400 micrometers in cortical depth. Capillary pericytes that internalize the dye exhibit robustly labeled cell bodies and punctate labeling of their processes (Fig. 1.2k). The molecular transporters responsible for selective NeuroTrace 500/525 uptake in capillary pericytes remains unknown.

1.5 Visualization of Brain Pericytes *In Vivo*

1.5.1 Cranial Windows for in vivo *Multi-Photon Imaging*

To gain optical access into the living mouse brain, a cranial window must be implanted in the skull (Fig. 1.3). These windows can be created in adult animals, ranging from approximately 2 to 24+ months, with an increase in bone thickness,

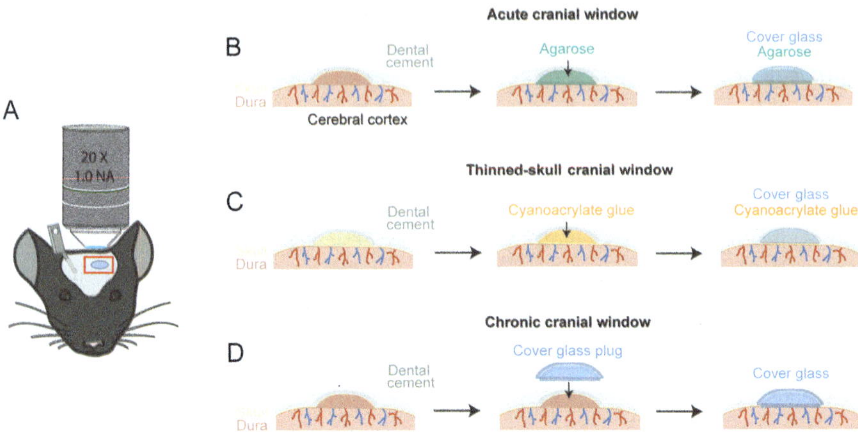

Fig. 1.3 Cranial windows for optical imaging. (**a**) Schematic of a mouse with implanted cranial window and metal flange for head fixation. (**b**) Acute cranial window design, with a layer of agarose separating the coverglass from the cortical surface. (**c**) Thinned-skull window, with cyanoacrylate glue to stabilize and bind the coverglass to the thin remaining layer of skull. (**d**) Chronic cranial window schematic, featuring a double coverglass "plug" to impede bone regrowth over time

brittleness, and porosity (and thereby, surgical difficulty) with increased age. The choice of cranial window depends largely on the goals and timeline of the experiment. For example, some studies require long-term access with high imaging quality. Other studies require minimal perturbation of the intracranial environment. With each cranial window there are advantageous and disadvantages that should be carefully weighed during the development of the study.

1.5.2 Acute Cranial Window

For short experiments that require only a single imaging timepoint, an acute cranial window can be generated. This involves the removal of a circular piece of skull (diameter ~ 3–4 mm) over the desired location of mouse cortex. In experiments where cortical location is not critical, windows can most easily be placed over the mouse somatosensory cortex, which lies well within the bounds of the bregma and lambda sutures. The acute cranial window preparation involves the addition of a layer of transparent agarose to provide a buffer between an overlying cover glass and the brain (Shih et al. 2012a) (Fig. 1.3b). The edges of the cover glass are then sealed into place with dental cement. The agarose limits brain swelling and reduces motion artifacts caused by distension of the brain during breathing or heartbeat. Acute windows are best used within ~4 hours after implantation, as the window quality tends to decline over time, likely due to increasing inflammatory responses. An advantage of the acute window is that skull removal allows for the application of compounds directly to the brain before closing the window. This can be done by dissolving compounds into the agarose that sits atop the pial surface, or by leaving an edge of the window unsealed such that dyes or drugs can be injected with a micropipette. In more sophisticated preparations, cannulas for continuous inflow and outflow of fluid can be created as part of the implantation (Nishimura et al. 2010). The full craniotomy also allows optimal imaging depth and resolution since light scattering by the bone is minimized. A disadvantage of the acute cranial window is that the surgical process is invasive and disruption of the brain could affect the physiological processes being studied.

1.5.3 Thinned-Skull Window

A cranial window type which allows for reasonable optical penetration with reduced impact on the brain environment is the thinned-skull window (Xu et al. 2007), or the similar PORTS window (Shih et al. 2012b) that allows for longer term imaging through the thinned-skull. As the name implies, this style of cranial window does not breach the skull, instead leaving a thin layer of bone between the brain and the glass coverslip (Fig. 1.3c). The skull is thinned to a thickness of ~20 μm using a sharp scalpel blade (acute) or very stable hand-held drill (chronic), such as the Osada, EXL-M40. The amount of thinning required to achieve a ~ 20 μm thickness can be fine-tuned over separate surgeries followed by measurement of bone thickness using second harmonic signal generation (blue fluorescence) during two-photon imaging (Drew et al. 2010). In acute uses of this window, application of artificial cerebral spinal fluid to the thinned skull surface will reduce light scattering by the bone (Yang et al. 2010). Acute thinned-skull windows are kept small in part to reduce motion artifacts, which can still occur with distention of the thinned bone. For long-term imaging, the scalp is sutured between imaging sessions and re-opened on subsequent imaging sessions. The skull may regrow during extended intervals and must then be re-thinned for imaging. In a chronic PoRTs version of the thinned-skull window, a layer of cyanoacrylate glue is placed between the bone and cover glass (Shih et al. 2012b). The glue, which has a similar optical index as bone and glass, helps to reduce light scattering for improved light penetration. A second benefit is that the hardened glue preserves optical clarity for longer term imaging, removing the need to re-thin the window. It further adds rigidity to the thinned skull allowing creation of larger windows. In applications seeking to understand microvascular changes across brain regions, entire hemispheres of the skull have been thinned for meso-scale two-photon imaging (Mateo et al. 2017). Thinned skull windows reduce disruption to the underlying brain tissue, though recent studies have shown that brain temperature decreases similarly in thinned-skull and craniotomies for several days post-op (Roche et al. 2019). The maximal depth of imaging achieved through thinned skull windows is approximately 300 μm from the pial surface. Under optimal conditions, thinned-skull windows can be imaged for over a month (Drew et al. 2010), though quality is typically best within the first 7 days after implantation. Loss of imaging quality is usually attributed to new skull growth, which occurs gradually and prevents light penetration. Thinned-skull windows are ideal for experiments where any surgically-induced inflammation or brain swelling would compromise the validity of the study. Thinned skull windows are more difficult to create in older animals (9+ months) due to their more brittle and porous skulls. However, with greater care during surgery, studies can still be performed in aged mice using this approach (Watson et al. 2020; Hyacinth et al. 2019).

1.5.4 Chronic Cranial Window

For long-term experiments over weeks to months requiring optimal imaging resolution, chronic cranial windows are the preparation of choice. Chronic windows are allowed to fully heal for 2–3 weeks between the time of surgery and start of imaging, at which point glial reactivity returns to near baseline levels (Holtmaat et al. 2009; Goldey et al. 2014). While earlier versions of this window used only a single layer of glass, recent studies have implemented a double glass coverslip "plug" that helps to reduce bone regrowth from the edges of the window (Berthiaume et al. 2018b; Goldey et al. 2014) (Fig. 1.3d). These are made by gluing a small circular coverglass (3 mm diameter) onto a larger circular coverglass (4 mm) with optical glue. This plug is then placed into a circular craniotomy that just fits the smaller coverslip, leaving the edges of the larger coverslip to sit atop the skull surface. This plug is then sealed into place with dental cement. For studies requiring a wider visual field, a 4 mm diameter window can also be created by combining a 4 mm and 5 mm round coverglass as the plug. The intracortical injection of a virus pairs well with the creation of a chronic cranial window. The virus can be micro-injected directly into the brain before the glass plug is inserted, and the 2–3 week surgical recovery period can simultaneously provide necessary viral incubation time. This also guarantees that the virally transduced region is within the cranial window. A chronic cranial window with an adjacent access port for the administration of desired compounds can also be created, thereby combining the major benefits of both a chronic and acute cranial window into one preparation (Roome and Kuhn 2014). This allows for longitudinal imaging sessions interleaved with cortical drug applications or intracortical injections to study the long-term effects of various pharmacological interventions. Chronic cranial windows can generally be imaged for many months, though this varies depending on initial quality of surgery (Holtmaat et al. 2009).

1.5.5 Anesthesia during Surgery and Imaging

The method of surgical anesthesia should be chosen based on cranial window type. For chronic cranial windows, fentanyl anesthesia is a good choice. Dexamethasone (4.8 mg/kg, s.c.) is typically administered 4 hours prior to surgery to minimize inflammation and edema (Goldey et al. 2014). Then a cocktail composed of fentanyl citrate (0.05 mg/kg), midazolam (5 mg/kg), dexmedetomidine (0.25 mg/kg) is delivered intraperitoneally. Both dexamethasone and the fentanyl cocktail help to reduce brain swelling upon removal of the skull, which can otherwise lead to compression of the cortex when the cover glass is overlaid. Following surgery, animals require the administration of a "wake-up cocktail" consisting of buprenorphine (0.1 mg/kg), flumazenil (0.5 mg/kg), antisedan (2.5 mg/kg), meloxicam (5 mg/kg), lactated Ringer's solution (1-2 ml/25 g mouse). After a 2–3 week period of recovery mice can be imaged in the anesthetized state, or awake state if properly habituated to head fixation. For acute cranial windows, imaging immediately follows surgery, and the choice of anesthetic depends upon the goal of the experiment. If stable structural features of the brain will be examined, such as pericyte or microvascular structure, it is be convenient to use isoflurane anesthesia throughout both surgery and subsequent imaging phase. Surgery would require a higher isoflurane concentration (2% MAC), while isoflurane should be minimized to 0.8-1.5% MAC during imaging. However, if vascular dynamics such as blood flow and neurovascular coupling will be measured, isoflurane is not ideal. Instead, surgery can be performed with first with isoflurane, and the animal can then be transitioned to alpha-chloralose anesthesia or chlorprothixene sedation, which better preserve neurovascular responses (Masamoto and Kanno 2012). While isoflurane is more convenient to use for achieving surgical-plane anesthesia, it is a vasodilator that can worsen brain swelling following a craniotomy. However, less invasive procedures such as thinned-skull or PORTS windows can be generated using isoflurane anesthesia without the concern of brain swelling as the skull is not completely breached.

For imaging of pericyte or capillary structure, as well as BBB leakage, the use of light isoflurane anesthesia (0.8–1.5% MAC) is necessary to obtain very stable, high-resolution imaging. Anesthesia may also be appropriate for studying slower aspects of blood flow change that might take place over days, such as those seen after optical ablation of pericytes (Berthiaume et al. 2018b). However, dynamic changes in blood flow or pericyte activity evoked by neurovascular coupling are ideally performed with an awake, habituated or lightly sedated subject. This is because anesthesia significantly dampens the neuronal activity and the neurovascular response (Gao et al. 2017). Awake-imaging involves the habituation of the mouse to head restraint in the days prior to an imaging experiment (Shih et al. 2014). Enclosures, such as hollow tubes, can be utilized to dampen the natural fear response of mice to open spaces. Other strategies include spherical treadmills for the animal to walk on at will during the course of imaging (Dombeck et al. 2007). Once habituated, mice do well under these awake-imaging conditions, and high quality two-photon data can be obtained. Generally speaking, the field is moving towards using awake-imaging as the standard for all two-photon imaging experiments, as this state of consciousness likely gives the most authentic results (Grutzendler and Nedergaard 2019; Gao et al. 2017). However, light sedation using drugs such as chlorprothixene remain a reasonable compromise if study results would be significantly altered by any movement artifact, as is often the case when imaging the characteristics of structures as small as capillaries (Grutzendler and Nedergaard 2019).

1.5.6 Head-Mounting/Stabilization

Stabilization of the mouse's head is necessary for both the cranial window surgery and for optical imaging. Head fixation can be achieved with a custom-made head-mount setup (Fig. 1.3a). This is usually an adjustable structure that allows the angle of the mouse head to be leveled to the imaging plane, and then fixed in position to prevent further movement. Proper head fixation is critical to minimize motion artifacts caused by breathing or heartbeat during acquisition of high-resolution optical images. The head fixation apparatus is also useful to maintain a similar head positioning for the same mouse on each subsequent imaging day. This facilitates longitudinal imaging studies where exact locations need to be repeatedly imaged and compared over time.

1.5.7 In vivo *Two-Photon Microscopy*

Once a cranial window has been generated, and the appropriate post-operative recovery period has passed, optical imaging experiments can begin. Two-photon microscopy has been revolutionary for the advancement of neuroscience by allowing visualization of cellular and subcellular processes *in vivo,* while minimizing photodamage to the delicate brain tissue (Denk et al. 1990). Two-photon imaging has significantly advanced cerebrovascular research, with the ability to visualize and quantify aspects of the dense vascular networks of the brain, their cellular components, and their physiological dynamics (Shih et al. 2012a).

1.5.8 Physiological Variables

It is important to maintain an animal's health during *in vivo* imaging. A pulse oximeter can be used to obtain a constant read out of an anesthetized animal's heart rate and breathing rate while enclosed within the light-tight enclosure of the microscope. Some studies require full control of physiological variables including implantation of femoral artery catheters for continuous blood pressure measurement, and occasional blood gas sampling. Tracheal tubes are also inserted for mechanical ventilation, which helps to ensure a normal breathing rate and proper balance of blood gases (Grubb et al. 2020). However, these implantation procedures are invasive and therefore reserved for acute imaging studies. A feed-back regulated heat pad is necessary to ensure body temperature at normal levels of ~37 °C. Brain temperature is also important to consider, as it can be drastically altered immediately beneath a craniotomy (Kalmbach and Waters 2012). The use of water-immersion microscope objectives poses some additional problems, as room-temperature water has been shown to decrease the temperature of the cortex directly beneath the window, which can affect both blood flow and tissue oxygenation (Roche et al. 2019). The use of objective heaters helps to address this issue.

1.5.9 I.v. Dyes as Vascular Labels

The intricate cerebrovascular networks of the mouse brain can be labeled by a single intravenous (i.v.) injection of fluorescent-dextran dye. Since the i.v. dye labels the blood plasma, any perfused vessel will fluoresce during two-photon imaging. The choice of i.v. dye depends upon which imaging channels are already occupied by genetically expressed fluorophores. The conjugated dextran must also be of sufficient molecular weight so that it is not filtered from the blood stream too quickly. In practice, dextran conjugates at or above 70 kDa are sufficient for hours of imaging. Some common commercially available dye choices are fluorescein isothiocyanate (FITC)-dextran (70 kDa-2 MDa), tetramethylrhodamine (TMR)-dextran (500 kDa), and Texas Red-dextran (70 kDa). Recent studies have also provided protocols to custom conjugate fluorophores to dextran, as was done for Alexa fluor 680 in a deep two-photon imaging study (Li et al. 2020). Ideally, the i.v. dye and fluorophore used to label pericytes in each experiment would have similar two-photon excitation cross-sections, enabling simultaneous imaging of pericytes and vascular structure and flow. For example, an experiment to visualize tdTomato in pericytes and FITC-dextran simultaneously could use a 975 nm excitation wavelength, and capture emissions in the red and green channels, respectively. Alternatively, if pericytes were to express YFP, a Texas Red-dextran could be used and both fluorophores imaged at 900 nm for green and red emission. However, sequential imaging with different excitation wavelengths can also done if the two-photon cross sections of the labels are too dissimilar.

1.5.10 I.v. Dyes as Vascular Permeability Probes

Vascular permeability metrics are often included in pericyte studies seeking to investigate their involvement in blood-brain barrier (BBB) function. For these studies, the molecular weight of the chosen dextran is an important consideration. A dextran weighing ~70,000 Da will be peripherally extravasated from the blood stream within a few hours, but is of an ideal size to test BBB integrity (Armulik et al. 2010). Several blood plasma proteins established as appropriate surrogates for BBB leakage, such as albumin, are almost identical in molecular weight. A dextran of this size may accumulate in the brain parenchyma outside of vessels for minutes to hours after extravasation, and can be readily visualized with two-photon microscopy (Fig. 1.4d). The smallest fluorophores, such as Alexa fluor 488-Cadaverine, are ≤1000 Da and are filtered from the blood stream within minutes. These can only be visualized around the time of bolus injection *in vivo (*Berthiaume et al. 2018b;

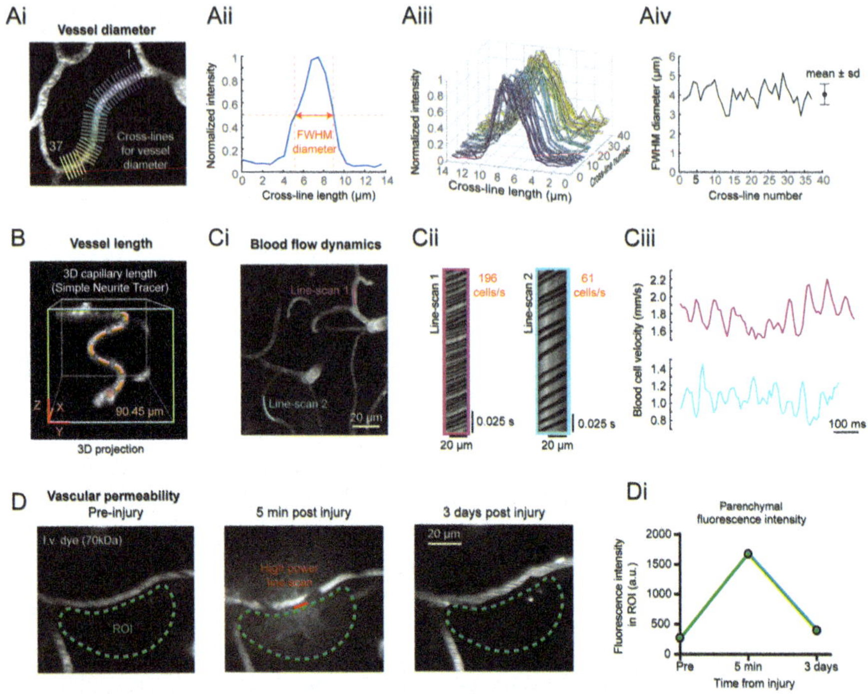

Fig. 1.4 Vascular metrics acquired with in vivo two-photon imaging. (**ai**) Capillary in maximally-projected image stack with cross-lines used for acquisition of average diameter measurement. (**aii**) Normalized fluorescence intensity plotted for a single capillary cross-line, used for calculation of full-width at half maximum capillary diameter. (**aiii**) Intensity plots of full dataset for cross-lines 1–37 from (**ai**). (**aiv**) Plot of diameter deviations along the length of a single capillary segment with average and standard deviation of diameter reported. (**b**) Demonstration of 3D capillary length measured using the ImageJ plugin Simple Neurite Tracer (Longair et al. 2011). (**ci**) Blood cell velocity is measured using high-speed line-scans during two-photon microscopy. The location of line-scanning is shown for collection of blood cell velocity and flux from two capillaries within the field of view. (**cii**) Raw image output from the line-scans marked in (**ci**). Dark bands are generated by passing blood cells that exclude the fluorescent dye, while light bands are i.v. dye within the blood plasma. The number of dark bands over time is used to determine blood cell flux in cells per second. The angle of the streaks is used to calculate the speed of blood cell flow, with streaks closer to the horizontal corresponding to faster flow speed. (**ciii**) Blood cell velocity in millimeters per second plotted as a function of time for each of the scanned capillaries. (**d**) Leakage of a 70 kDa i.v. dye immediately following capillary injury. (**di**) Quantification of fluorescence intensity within a region of interest (ROI) in (**d**) at pre-injury, 5 minutes post-injury, and 3 days post-injury timepoints

Lou et al. 2016) or in tissue slices following fixation (Armulik et al. 2010). This method is useful for detecting small, local breaches in the BBB that involve rapid dye extravasation. When using the smallest molecular weight dyes, it is often necessary to introduce an additional high molecular weight dye to the bloodstream to assess both vessel structure and permeability within one imaging session.

1.5.11 Vascular Characteristics and Dynamics

Pericyte control of cerebral blood flow is a very active area of research. Two-photon imaging is uniquely suited to study this process because it enables the real-time detection of blood flow changes at the level of arterioles, capillaries and venules. This type of imaging also allows for the quantification of vascular structure characteristics such as vessel length and diameter. Vascular length should be measured in three dimensions whenever possible, using a tool such as the ImageJ segmentation analysis tool Simple Neurite Tracer (Longair et al. 2011) (Fig. 1.4b). This allows the full length of the vessel in the z axis to be taken into consideration, rather than only the x and y planes. The diameter of a vessel can be obtained by measuring the full-width at half maximum of a fluorescence intensity profile taken across the width of the capillary (Driscoll et al. 2013) (Fig. 1.4ii). Since brain capillaries are often heterogeneous in diameter even within the same capillary segment, recent studies have also developed strategies to collect diameter at multiple regions along the vessel (Watson et al. 2020; Ivanova et al. 2017) (Fig. 1.4ai). The average of these measurements can then be obtained and reported as an unbiased measure of capillary diameter, while also providing a metric of the variability of diameter along the length of a vessel segment (Watson et al. 2020) (Fig. 1.4aiii, aiv). Blood flow measurements, including blood cell flux and velocity, can be measured at the level of brain capillaries. Since i.v. dyes are excluded by blood cells, they appear as dark shadows within the fluorescent blood plasma. A two-photon line scan through the central axis of the vessel length will reveal moving blood cell shadows as streaks within a resultant line-scan image (Fig 1.4 ci, cii). The number of streaks over time is the flux of blood cells, typically reported as cells per second. The angle of the streaks is used to calculate the velocity of blood cell flow, typically reported as millimeter per second (Shih et al. 2012a; Kleinfeld et al. 1998). Convenient algorithms have been developed to aid extraction of blood cell velocity from line-scan images (Kim et al. 2012; Chhatbar and Kara 2013) (Fig 1.4ciii).

1.6 *In Vivo* Observation of Pericyte Structure and Function

There exists a spectrum of mural cell morphologies in cortical vasculature, with pericytes being subcategorized into groups based on morphology and α-SMA expression. As capillary pericytes (>4th order from a penetrating vessel) are the most abundant in the mouse cortex, these populations will be the focus of the following section. When targeting the capillary pericyte network for *in vivo* optical imaging studies, a combination of cell morphology and branching distance from a penetrating arteriole can be used to confirm mural cell identity (Grant et al. 2017).

The unique structure of capillary pericytes, and how this structure relates to their *in vivo* function, has been a question of great interest in past decades. With constitutive genetic labeling, pericytes of the capillary bed form a contiguous network. The discrete domains of individual pericytes may be difficult to discern, but careful inspection reveals occasional small gaps between neighboring pericytes processes, ranging from only a few micrometers in width. The basal growth of individual pericyte processes can be tracked over time, if the boundaries between neighboring cells are sufficiently visible (Berthiaume et al. 2018a). Long-term tracking of these pericyte gaps has shown that pericytes slightly adjust their territories along the capillary bed over time (Fig. 1.5c). This repulsion between neighboring pericytes maintains a non-overlapping chain of pericytes along the capillary bed. The specific mechanisms driving this pericyte-to-pericyte repulsion have yet to be described, but may involve EphrinB2 signaling (Berthiaume et al. 2018a; Foo et al. 2006). In instances where capillary pericytes are lost or damaged, the modulation of this cellular repulsion could play a role in triggering a growth response of the remaining capillary pericytes.

Mice with inducible control of Cre activity provide sparse labeling of individual capillary pericytes, allowing their individual structures to be examined. Many structural characteristics of individual pericytes can be quantified from *in vivo* two-photon imaging data, including: (i) the location of their somata, (ii) the size of the territory (the summated length of capillary contacted by a single pericyte), and (iii) various aspects of pericyte processes, such as their number, length, complexity and branching pattern. Pericyte somata can be found both at capillary junctions and on capillary segments between junctions. Interestingly, pericyte somata are more common at junctions (~50% of all capillary pericytes in cortex) (Hartmann et al. 2015), suggesting that their occurrence may support physiological roles, such as blood flow control at capillary bifurcations. Abundance of capillary pericytes can be quantified over time or under different experimental conditions by counting the number of pericyte somata within a cortical volume (Fig. 1.5a). When tracked over days to months under basal conditions, pericyte somata are found to be structurally fixed and immobile (Berthiaume et al. 2018b; Cudmore et al. 2017).

Fig. 1.5 (continued) resolution imaging of histological sections, is the ratio of total pericyte area over total vessel area. (**b**) Image of a single capillary pericyte from a Myh11-CreERT2::Ai14 mouse, with measurements for individual and total process lengths, measured in 3D with ImageJ plugin Simple Neurite Tracer (Longair et al. 2011), and soma diameter. (**c**) Tracking neighboring capillary pericyte territories over time. Inset shows a shift in original boundaries of processes (yellow lines) over 77 days. Image from a Myh11-CreERT2::Ai14 mouse. Adapted from Berthiaume et al. 2018 (Berthiaume et al. 2018a). (**d**) Super-resolution confocal image of fixed tissue from Pdgfrb-Cre::Ai14 animal. Thin-strand pericyte processes, pseudocolored green, shows serrated-liked edges. Immunostaining for collagen IV, pseudocolored red, labels the basement membrane associated with the underlying capillary. (**ei**) Mesh pericyte imaged with *in vivo* two-photon microscopy in a Pdgfrb-Cre::Ai14 mouse. (**eii**) The pericyte imaged *in vivo* was re-located *post-mortem* in fixed tissue and imaged with super-resolution confocal microscopy, revealing the intricate substructure of pericyte processes that is unresolvable with *in vivo* two-photon imaging. (**f**) Image of a thin-strand pericyte from a NG2-GCaMP6f mouse. Numbers indicate regions of interest within a single pericyte process analyzed along the line-scan (yellow). (**g**) Spontaneous spikes in calcium transients plotted over time, separated by ROIs outlined in (**f**). (Reproduced with permission from Rungta et al. 2018 (Rungta et al. 2018))

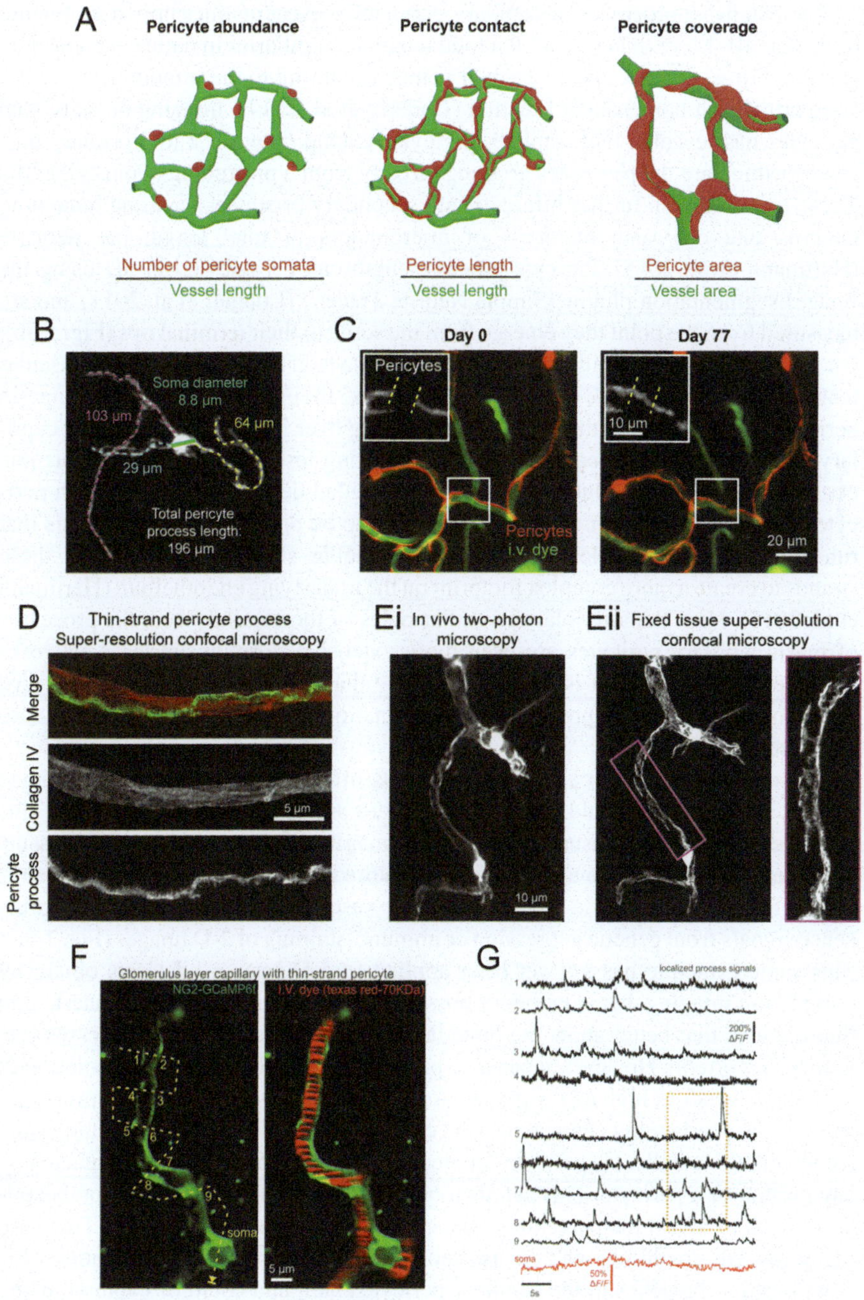

Fig. 1.5 Pericyte structural and functional characteristics. (**a**) Schematic of methods for quantifying pericytes relative to the capillary endothelium. (Left) Pericyte abundance is quantified as number of pericyte somata over vessel length. (Center) Pericyte contact is the ratio of total pericyte length in a volume over total vessel length. (Right) Pericyte coverage, typically assessed with high

The extensive processes of capillary pericytes present another important feature of these cells. More than 90% of the total length of capillaries in cortex is contact by pericyte processes, as opposed to their somata, pointing to the importance this cell compartment in homeostatic function (Underly et al. 2017). Imaging of individual pericytes has revealed that capillary pericytes extend between 2 to 3 primary processes, which are the processes growing directly from a protruding ovoid cell body. These processes can further bifurcate into secondary processes, creating branching patterns that can span hundreds of micrometers of total length per pericyte (Hartmann et al. 2015). Pericyte process length can be assessed in 3D using the ImageJ segmentation plugin "Simple Neurite Tracer," (Longair et al. 2011) and are measured from the point they emerge from the soma to their terminal tips (Fig. 1.5b). On average, a single capillary pericyte will contact 250 ± 41 μm of total vascular length in the adult mouse somatosensory cortex, though there is a large range in pericyte territory size, with some cells contacting over 300 μm of contiguous capillary length while other nearby pericytes contacting less than 200 μm (Berthiaume et al. 2018b). Super-resolution imaging has revealed detailed sub-structure in pericyte processes (Fig. 1.5d). While there appears to be one or two major strands that run along the capillary length, occasional lamellar sheaths form between these strands to create a more complex footprint on the underlying endothelium (Hartmann et al. 2015). Near the arteriolar or venular poles of the capillary bed, the processes of mesh capillary pericytes are even more extensive, forming intricate networks over the capillary wall. It is important to note that the resolution limits of *in vivo* two-photon microscopy preclude the examination of this intricate cellular substructure (Fig 1.5ei, eii).

Vessel "coverage" by capillary pericytes is often a reported metric in pericyte studies. However, it should be noted that there are discrepancies in the way this characteristic can be measured with optical imaging *in vivo* versus in fixed tissue. Pericyte "coverage" of the underlying vasculature in fixed tissue is typically quantified as the ratio of pericyte-labeled area over vascular-labeled area, using fluorescence signals from genetic expression or immuno-staining of 2-D images (Fig. 1.5a). This same principle has not yet been applied to 3-D imaging datasets obtain by two-photon imaging, but is in theory possible, especially if using membrane-bound fluorophores that better show the lamellar outcroppings of pericyte processes (i.e. mT/mG reporters). However, it can also present some challenge, as the resolution in the z plane is not as high as x-y planes. As discussed above, the extensive processes form a non-overlapping chain across the capillary network. Thus, an alternative and simple metric to collect is pericyte "contact", *i.e.*, total capillary length contacted by any portion of a pericyte. This can then be expressed as a proportion of total capillary length to reveal the percentage of capillaries contacted by pericytes. There are many disease conditions and mouse models in which pericyte loss occurs. This would lead to breaches in the capillary pericyte chain and could be captured using the pericyte "contact" metric (Fig. 1.5a). Quantifying length rather than area has the added advantage of supporting measurements in three dimensions, which leverages the 3-D data obtained by two-photon microscopy.

Longitudinal two-photon imaging enables researchers to track features of pericytes in order to gauge structural stability or reactivity to different experimental conditions. Pericytes are not at their maximum size under basal conditions, as they are able to extend their processes to cover a larger territory when challenged with loss of a neighbor (Berthiaume et al. 2018a). This structural plasticity of capillary pericytes can be measured with *in vivo* two-photon imaging by tracking the same cortical volume over time. Changes to pericyte structure occur on the timescale of days to weeks, requiring cell volume changes (Berthiaume et al. 2018b). This is in contrast to other cell types such as microglial cells that can change shape and migrate on the order of minutes, by re-distributing their existing cell volume. Capillary pericyte remodeling appears to be important for homeostatic functions of brain capillaries, such as blood flow regulation, though the full scope of its physiologic relevance needs to be better delineated. The loss of pericyte coverage or number, as seen in humans and animal models with small vessel disease (Halliday et al. 2016; Park et al. 2014), likely involves a complex interplay between pathological influences leading to decreased pericyte coverage, and reparative pericyte growth to attempt recovery of pericyte coverage. Small vessel disease models suggest that pericytes can also become mal-distributed along the capillary bed. In a mouse model of cerebral amyloid angiopathy, there are larger gaps between adjacent pericytes due to shorter processes (Park et al. 2014). Conversely, in Col4a1 mutant mice that develop spontaneous hemorrhage, mural cells in the transition zone between arterioles and capillaries are packed more tightly leading to hypermuscularization and increased upstream pressure (Ratelade et al. 2020). Both conditions may lead to abnormalities in capillary homeostasis.

Along with pericyte structure, other aspects of pericyte function can also be studied *in vivo* using two-photon imaging. Pericyte function can be inferred from metrics collected from their adjacent microvessels, such as vessel diameter, blood flow, and BBB integrity. Pericytes control these aspects of vascular function and correlative studies can be performed at the level of single pericytes and capillaries. Recent studies have expanded to the examination of pericyte calcium activity *in vivo* in relation to neurovascular coupling. With the NG2-CreERT2 mouse line crossed with GCaMP6f reporter mice, it is possible to study moment-to-moment changes in intracellular calcium in pericytes using two-photon imaging (Rungta et al. 2018) (Fig. 1.5f, g). Pericyte neurovascular coupling studies using the NG2-GCaMP6f mouse also benefit from the off-target labeling of OPCs. Because OPC activity has been shown to correlate with neural activity, increased calcium in OPCs can serve as a surrogate for neural activation (Rungta et al. 2018). Using this method, Rungta and colleagues demonstrated that calcium spikes occurred in small, seemingly disjointed compartments of pericyte processes, with no evidence of synchrony or wave-like signal propagation throughout the cell during functional hyperemia (Rungta et al. 2018). In another approach, an adeno-associated virus (AAV2/5) has also been used to transduce pericytes with GCaMP6 transgene for *in vivo* calcium imaging, with GCaMP6s expression driven under activity of the PDGFRβ promoter (Khennouf et al. 2018). Khennouf and colleagues reported a decrease in cytosolic calcium in pericytes located on dilated vessels during whisker stimulation.

Conversely, cortical spreading depression-mediated capillary constriction led to increased intracellular calcium in associated pericytes. As demonstrated by these two studies, cutting-edge calcium imaging approaches can be used to investigate both basal activity of pericytes as well as their response to pathology. For now, the importance of pericyte calcium activity *in vivo* remains obscure and requires further investigation.

1.7 *In Vivo* Manipulation of Pericytes

Given that pericytes live within a complex vascular network and continuum of related cells, understanding their functional roles *in vivo* requires cause-and-effect approaches to experimentally manipulated pericytes in adulthood. While studies with pericyte-specific manipulation are still rare, a few strategies have emerged to either ablate pericytes in the adult brain or to stimulate them. By ablating these cells, we gain a better understanding of the consequence of pericyte loss on the remaining neurovascular unit. Pericyte loss can be achieved on a large scale with partial or global pericyte ablation, or on the much smaller scale of individual pericytes. These two scales of pericyte manipulation can serve complementary roles in investigations of how pericyte loss contributes to cerebrovascular disease. The strategic stimulation of pericytes *in vivo*, on the other hand, can help our understanding of how capillary pericytes perform their functions in the context of normal brain health, for example in the regulation of capillary blood flow.

1.7.1 Genetic Pericyte Ablation

Ablation of brain pericytes can be accomplished with one of several approaches, depending on the degree and timing of pericyte loss required. There exists an abundance of work using congenital knockouts targeting important signaling components for pericyte recruitment in the developing vasculature. When created as heterozygous mutants, many of these pericyte-deficient mice are able to survive into adulthood. Some of these lines, such as the $Pdgfr\beta^{F7/F7}$ mice feature an age-related decrease in pericyte numbers, which may mirror some aspects of pericyte loss in age-related human diseases (Nikolakopoulou et al. 2017). One limitation of these congenital ablations is the loss of pericytes from the earliest developmental stages. Pericytes are involved in angiogenesis and formation of a stable vascular network, and vascular abnormalities resulting from reduced pericyte number would create differences in vasculature during adulthood. This is especially true because PDGFRβ mutations do not have effects exclusive to capillary pericytes, but may also affect the function of other cells including smooth muscle cells and neurons (Ishii et al. 2006). With this caveat in mind, pericyte-deficient mice are easily imaged with two-photon microscopy and have been used to show increased BBB disruption, cerebral hypoperfusion, white matter degeneration, and impairment of neurovascular coupling with pericyte loss (Kisler et al. 2017; Montagne et al. 2018; Bell et al. 2010).

A recent study developed an approach to selectively ablate capillary pericytes *en masse* in the normally developed adult brain. By using the previously described pericyte-CreER mouse line (Nikolakopoulou et al. 2019), non-specific targeting of other cell types is eliminated, including closely related smooth muscle cells of arterioles. When pericyte-CreER::Ai14 are combined with a line encoding an inducible diphtheria toxin receptor (iDTR), DTR expression is exclusive to capillary pericytes (Nikolakopoulou et al. 2019). This method allows for titratable levels of pericyte loss through adjusting the dose of either tamoxifen (to induce DTR expression in Cre-expressing cells) or diphtheria toxin (DT; to induce apoptosis in these cells). At the highest doses administered in this study, a 60% reduction in capillary pericytes was achieved from cortical microvasculature over the following days to weeks. The result of this rapid capillary pericyte loss was brain circulatory failure due to endothelial leakage and edema, which leads to loss of neurovascular coupling and eventually neuronal viability (Nikolakopoulou et al. 2019; Kisler et al. 2020). The administration of lower doses of DT intermittently over weeks to months could reveal a distinct phenotype to model more gradual pericyte drop-out.

1.7.2 Two-Photon Pericyte Ablation

Two-photon optical ablation of individual pericytes is a second approach to ablate pericytes *in vivo*. The method is tightly coupled with two-photon imaging of the vasculature, and relies on the precise targeting of pericyte somata with two-photon laser light to created localized thermal damage at the pericyte soma, inducing necrotic cell death (Fig. 1.6a). Thermal ablation of single or multiple pericytes is distinct from genetic approaches because it creates localized pericyte loss in an otherwise unperturbed vascular network. This allows researchers to study the effect of pericyte loss on local microvascular responses without disrupting upstream or downstream perfusion. Pericyte loss also occurs with no detectable disruption to the blood-brain barrier (Berthiaume et al. 2018b), which helps to isolate the effects of pericytes on blood flow from complex pathological influences. Further, optical pericyte ablation circumvents the complicating factors of pericyte loss in other tissues and organs that might indirectly influence brain function when using genetic ablation. Two-photon pericyte ablation has been used to study pericyte structural plasticity in reparative responses to maintain coverage of the endothelium in the event of pericyte loss (Berthiaume et al. 2018b) (Fig. 1.6c, d).

One drawback of two-photon thermal ablation is that pericytes can only be eliminated in a restricted volume brain volume, typically less than ~200 μm in depth below the pial surface in the cranial window. Additionally, the ablation process can cause off-target damage to the surrounding tissues and extensive controls must be added to understand these effects. Some pericyte cannot be accurately targeted without damaging the blood vessel because they are not oriented in a way that

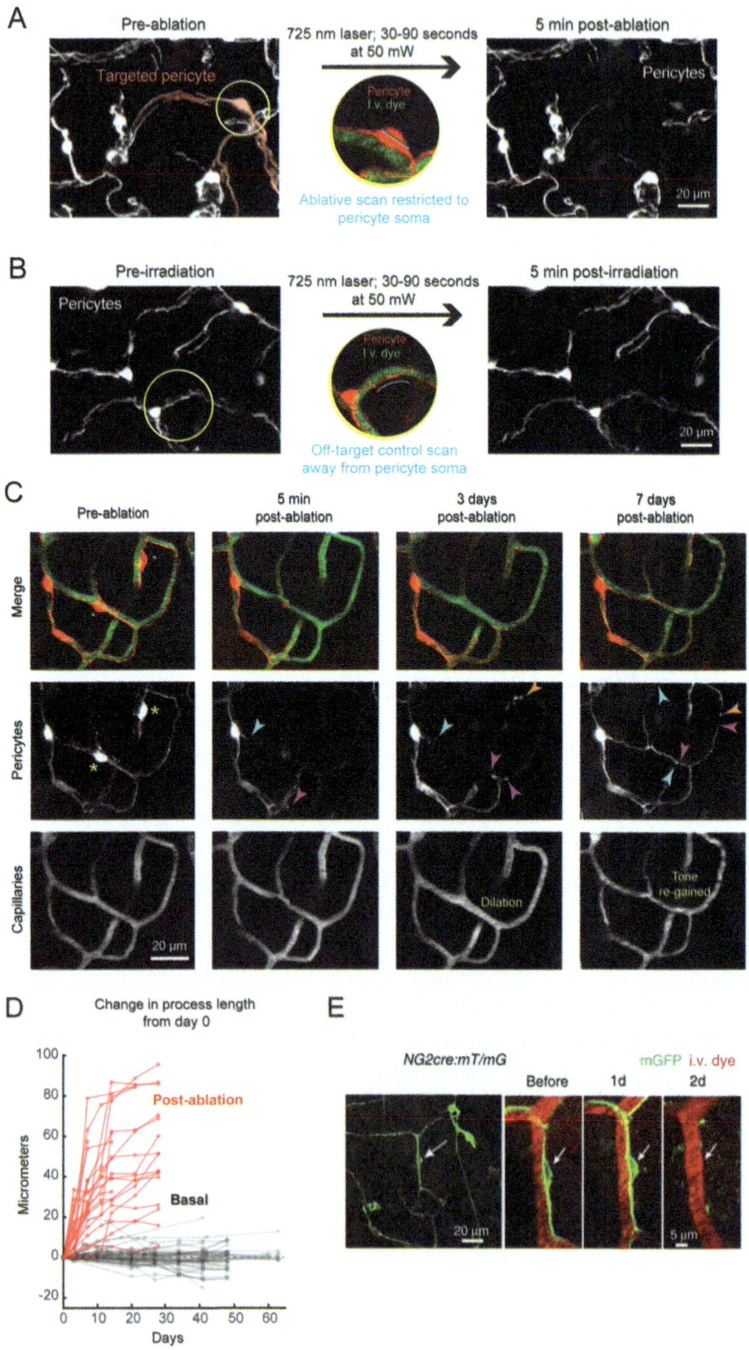

Fig. 1.6 Ablating pericyte *in vivo* with two-photon microscopy. (**a**) Image of capillary pericytes from NG2-CreER::Ai14 mouse, with pericyte targeted for two-photon thermal ablation encircled

allows for selective ablation. For pericytes genetically labeled with tdTomato, the ideal thermal ablation settings is between 30 and 50 mW of laser power at a wavelength of 725 nm (Berthiaume et al. 2018b). A line-scan is restricted to the pericyte soma and held for 30–90 seconds. In a successful ablation, the tdTomato signal specifically from the targeted pericyte is lost and there is no breach of the underlying capillary, which would indicate endothelial injury. Overt indications of off-target damage, such as photobleaching of mural cells in z-planes above or below the area or the creation of microbleeds from damage to vessel walls, indicate a need to reduce laser power. Ablation settings should be evaluated on a case-by-case basis. Factors such as increased depth of the soma below the pial surface and reduced quality of the cranial window may necessitate higher laser powers. Thermal ablation experiments should always be accompanied with laser-irradiation controls to understand the effect of laser exposure alone. This can be done by administering laser light with settings identical to those used for thermal ablation but to a perivascular area not containing a pericyte cell body (Fig. 1.6b).

To avoid some of the limitations of two-photon thermal ablation (unintended thermal injury to other cells, necrotic cell death), an alternative method of two-photon laser ablation of individual pericytes was developed, named 2Phatal (two-photon chemical apoptotic targeted ablation) (Hill et al. 2017). This technique relies on the administration of the nuclear-binding fluorescent dye Hoechst 33342 (H33342) to living animals, which generates deadly levels of reactive oxygen species (ROS) when selectively activated with 775 nm two-photon laser light. The apoptotic cascade triggered by this reaction causes progressive cell death over the course of several hours to days (Fig. 1.6e). The administration of H33342 via systemic injection or direct cortical application results in the entry of the dye indiscriminately into nuclei of several cell types. Cell specificity in ablation is achieved by controlling the area of two-photon laser irradiation. Capillary pericytes can be identified by pairing any mouse line with labeled pericytes with this dye. This

Fig. 1.6 (continued) in yellow. Thermal ablation line-scan is limited to pericyte soma only. Post-ablation image reveals the immediate disappearance of the targeted pericyte within minutes of ablation. Adapted from Berthiaume et al. 2018 (Berthiaume et al. 2018b). (**b**) Off-target control for two-photon thermal ablation involves creating a line-scan path in a perivascular location of similar distance to the capillary wall, but devoid of a pericyte cell body. Similar laser powers and time of irradiation must also be used for controls. (**c**) Longitudinal time-course of response to thermal ablation of two capillary pericytes. Asterisks indicated targeted cells. Both cells are successfully ablated, as determined by their complete disappearance 5 minutes post-ablation. Arrowheads mark the original boundaries of neighboring pericyte processes. Over the course of 7 days, neighboring pericytes extend their processes (arrowheads) into the uncovered capillary area to re-form a contiguous pericyte chain. Arrowheads of the same color denote the process tips that split from an original single process tip of a neighboring pericyte. (**d**) Change in pericyte process length plotted over time under basal conditions, observed at gaps between neighboring pericytes processes (gray), or following thermal ablation of a pericyte neighbor (red). Adapted from Berthiaume et al. 2018 (Berthiaume et al. 2018b). (**e**) Image from NG2-CreER::mT/mG mouse demonstrating the 2Phatal method of inducing apoptotic pericyte death over the course of days following laser exposure. (Reproduced with permission from Hill et al. 2017 (Hill et al. 2017))

method of induced cell apoptosis has the advantage of requiring shorter laser exposure time to trigger cell death than traditional thermal ablation (775 nm at ~21 to 45 mW of power for 5–20s). Reduced local microglial activation is also reported with 2Phatal (Hill et al. 2017). Further studies will be required to better understand how the presence of the DNA binding agent affects cells during routine two-photon imaging. Since all nuclei contain the DNA dye, any off-target laser light in or out of the plane of focus could create low levels of ROS. Though observational imaging would not be strong enough to directly kill cells, any incidental diffuse damage to the imaging area may affect normal brain and cerebrovascular function.

1.7.3 Two-Photon Optogenetic Manipulation of Pericytes

Optogenetic stimulation provides a means to reversibly depolarize mural cells *in vivo*. The approach was first applied to pericytes when Hill et al. used channelrhodopsin (ChR2) to depolarize brain pericytes in NG2-CreER::Ai32 mice *in vivo (*Hill et al. 2015*)*. They found that ChR2 stimulation caused contraction of SMCs, but not of capillary pericytes, supporting a conclusion that capillary pericytes do not regulate cerebral blood flow.

The study by Hill and colleagues used one relatively low 'dose' of light stimulation with short stimulation times to test the contractile ability of capillary pericytes. As recent studies have shown that stimulating laser intensity and duration determine the extent of ChR2 activation and mural cell contraction (Wu et al. 2015; Rorsman et al. 2018), we revisited this experiment using different stimulation parameters. In PDGFRβ::Ai32 mice, we carefully counted branch orders (e.g. >4th) to ensure that we examined capillary pericytes (Fig. 1.7a). Though PDGFRβ-Cre mice also target arteriolar SMCs, the activation of ChR2 occurs preferentially at the focal plane, which minimizes optogenetic perturbation of arterioles (Fig. 1.7b). Specialized line-scans can be used to simultaneously stimulate ChR2 and collect vasodynamic data, including diameter, blood cell velocity and flux (Fig. 1.7c). Using higher laser powers and longer illumination times (60 s) than used by Hill *et al*, we found that optogenetic activation of capillary pericytes can indeed reduce capillary diameter and blood flow in PDGFRβ-Cre:ChR2-YFP mice (Hartmann et al. 2020) (Fig. 1.7d,e). Further, we found that similar excitation protocols caused a much greater contraction of SMCs, suggesting that different mural cell types have a different threshold of ChR2 activation necessary for contraction. Capillary pericytes are contractile but exhibit slower kinetics, thereby requiring longer periods of stimulation than in SMCs.

A number of important practical considerations arise when using optogenetics to study mural cell contractility *in vivo*. First, there is a wavelength specificity for activation of ChR2, in this case the H134R variant. While 920–940 nm excitation is the most efficient to drive neuronal firing in cultured neurons (Prakash et al. 2012; Rickgauer and Tank 2009), 800 nm excitation appears most effective for driving contraction of SMCs and pericytes. We and others have found that imaging at

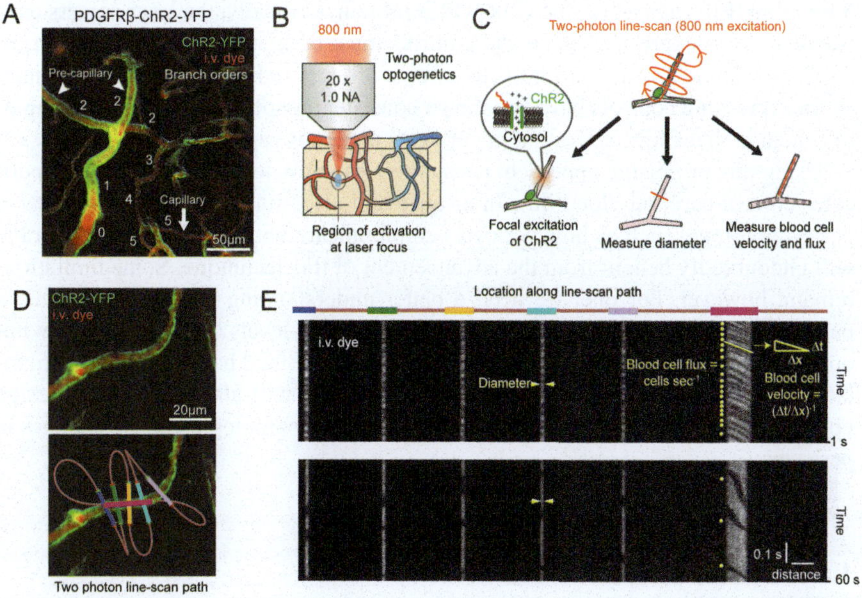

Fig. 1.7 Optogenetic stimulation of pericytes concurrent with vasodynamic measurements in vivo. (**a**) Representative image of the cerebrovascular network in a Pdgfrb-Cre::ChR2-YFP mouse. Precapillary and capillary zones are specified, as are vessel branch orders from a penetrating arteriole (0 order). (**b**) The ideal ChR2 excitation wavelength for mural cells using two-photon microscopy is set at 800 nm. Two-photon activation of ChR2 near the focal plane reduces possibility of stimulating upstream SMCs. (**c**) A schematic of the multifaceted purposed of the two-photon line-scan pattern utilized for optogenetic stimulation studies. The scan path allows for the two-photon activation of ChR2, the repeated measurement of capillary diameter, and the assessment of blood cell dynamics over time. (**d**) An example of pericyte stimulation using a multi-segmented two-photon line-scan. (**e**) (Top) Colored blocks in line correspond to locations along the scan-path in (**d**). (Middle, and bottom) Raw data acquired from a line-scan demonstrates a reduction on capillary diameter and blood cell flux over a 60 second period of stimulation. FWHM vessel diameter, blood cell flux, and blood cell velocity, can be extracted from the line-scan data. (Adapted from Hartmann et al., 2020 (Hartmann et al. 2020))

900 nm leads to a considerably lower level of contraction than when imaging at 800 nm, even while using the same laser powers. Although the basis for this discrepancy in optimal wavelengths is unclear, it provides the opportunity to observe and measure hemodynamics at 900 nm, then switch to 800 nm to simultaneously measure hemodynamics and stimulate mural cells. It is also important that laser power be graded based on depth below the pial surface, as deeper pericytes will require more power to stimulate. Second, it is important to recognize that light can alter vasodynamics even in the absence of exogenous opsins like ChR2. In naive animals, blue light (Sikka et al. 2014; Rungta et al. 2017), and two-photon light at 800 nm (Choi et al. 2010; Kimbrough et al. 2015), have been shown to cause vasodilation and vasoconstriction, respectively. To control for these effects, optogenetic stimulation studies should include control animals which express a fluorophore alone (i.e.

YFP or mGFP) rather than the ChR2-YFP protein. Critically, the laser powers used between opsin expressing mice and controls must be matched at different cortical depths for an adequate control of the effects of laser light. In our hands, the non-opsin expressing controls have not shown consistent responses to light, allowing us to conclude that ChR2 activation in pericytes mediates capillary constriction.

Using this promising approach, researchers can now test whether capillary pericytes control cerebral blood flow in a cause-and-effect manner. Other questions in pericyte biology, such as how pericytes communicate with other cells in the brain, will undoubtedly benefit from the advancement of this technique. Some limitations remain, however. For one, we need a better understanding of the fluctuations in pericyte membrane potential and intracellular calcium levels *in vivo*, with and without optogenetic stimulation. This would help us understand how optogenetic stimulation compares to physiologic pericyte function. It will also allow us to design optogenetic stimuli that mimic certain physiologic or pathologic voltage changes in pericytes.

1.8 Concluding Remarks

We have highlighted a variety of new tools and approaches to visualize the structure and function of brain pericytes in their native environment. Imaging provides not only a means to observe fine pericyte structure over time, but also an opportunity for causal studies to delineate pericyte roles in brain physiology. While much has been learned from *in vitro* and *ex vivo* studies on pericytes, the field has shifted to discerning which of the diverse functions ascribed to pericytes are fact or fiction *in vivo*. We have discussed how seemingly basic topics, such as how to name and identify a capillary pericyte, are crucial to the field. This allows for pericytes to be studied consistently among research labs. Careful selection of a pericyte targeting mouse line is similarly crucial, each with their own advantages and disadvantages. Even mice that appear problematic with non-specific targeting can be leveraged for elegant experiments in neurovascular coupling, or comparative analysis across vascular zones. Quality is always preferred over quantity with two-photon imaging, and therefore the utmost care with cranial window construction and attention to animal physiology is important. Optical imaging is not only a means to observe physiological processes, but also a means to delete or stimulate pericytes with single cell precision *in vivo*. These approaches have already revealed a phenomenon of structural plasticity of pericytes that may be targeted to improve pericyte coverage in ailing cerebrovascular networks. They have also revealed how different pericyte types can have distinct effects on microvascular tone and therefore blood flow regulation. With these tools in hand, the field is now poised for a new era of *in vivo* studies on pericyte function.

Acknowledgements Our work is supported by grants to A.Y.S. from the NINDS (NS106138, AG063031, NS097775), the Dana Foundation, the American Heart Association (14GRNT20480366), Alzheimer's Association NIRG award, and an Institutional Development Award (IDeA) from the NIGMS under grant number P20GM12345. A.A.B. is supported by a scholarship from the American Federation of Aging Research. V.C.S is supported by an American Heart Association Post-doctoral Fellowship (20POST35160001). D.A.H. is supported by awards NIH T32 GM08716, NIH - NCATS (UL1 TR001450 and TL1 TR001451), and NIH-NINDS F30NS096868. We thank Patrick J. Mulholland for use of a Zeiss LSM 880 with Airyscan for super-resolution imaging of pericytes (NIH S10 OD021532). We also thank Tiago Figueiredo for creating artwork used in Figs. 1.3 and 1.5a (www.behance.net/TiagoFigueiredoGD).

References

Arango-Lievano M et al (2018) Topographic reorganization of cerebrovascular mural cells under seizure conditions. Cell Rep 23(4):1045–1059

Armulik A, Abramsson A, Betsholtz C (2005) Endothelial/pericyte interactions. Circ Res 97(6):512–523

Armulik A et al (2010) Pericytes regulate the blood-brain barrier. Nature 468(7323):557–561

Bell RD et al (2010) Pericytes control key neurovascular functions and neuronal phenotype in the adult brain and during brain aging. Neuron 68(3):409–427

Berthiaume AA et al (2018a) Pericyte structural remodeling in cerebrovascular health and homeostasis. Front Aging Neurosci 10:210

Berthiaume AA et al (2018b) Dynamic remodeling of Pericytes in vivo maintains capillary coverage in the adult mouse brain. Cell Rep 22(1):8–16

Birbrair A et al (2014a) Type-1 pericytes accumulate after tissue injury and produce collagen in an organ-dependent manner. Stem Cell Res Ther 5(6):122

Birbrair A et al (2014b) Type-2 pericytes participate in normal and tumoral angiogenesis. Am J Physiol Cell Physiol 307(1):C25–C38

Blanchette M, Daneman R (2015) Formation and maintenance of the BBB. Mech Dev 138(Pt 1):8–16

Cai C et al (2018) Stimulation-induced increases in cerebral blood flow and local capillary vasoconstriction depend on conducted vascular responses. Proc Natl Acad Sci U S A 115(25):E5796–E5804

Chhatbar PY, Kara P (2013) Improved blood velocity measurements with a hybrid image filtering and iterative radon transform algorithm. Front Neurosci 7:106

Choi M, Yoon J, Choi C (2010) Label-free optical control of arterial contraction. J Biomed Opt 15(1):015006

Coucha M et al (2019) Inhibition of Ephrin-B2 in brain pericytes decreases cerebral pathological neovascularization in diabetic rats. PLoS One 14(1):e0210523

Cudmore RH, Dougherty SE, Linden DJ (2017) Cerebral vascular structure in the motor cortex of adult mice is stable and is not altered by voluntary exercise. J Cereb Blood Flow Metab 37(12):3725–3743

Cuttler AS et al (2011) Characterization of Pdgfrb-Cre transgenic mice reveals reduction of ROSA26 reporter activity in remodeling arteries. Genesis 49(8):673–680

Daigle TL et al (2018) A suite of transgenic driver and reporter mouse lines with enhanced brain-cell-type targeting and functionality. Cell 174(2):465–480

Damisah EC et al (2017) A fluoro-Nissl dye identifies pericytes as distinct vascular mural cells during in vivo brain imaging. Nat Neurosci 20(7):1023–1032

Daneman R et al (2010) Pericytes are required for blood-brain barrier integrity during embryogenesis. Nature 468(7323):562–566

Davidson BL, Breakefield XO (2003) Viral vectors for gene delivery to the nervous system. Nat Rev Neurosci 4(5):353–364
Denk W, Strickler JH, Webb WW (1990) Two-photon laser scanning fluorescence microscopy. Science 248(4951):73–76
Dombeck DA et al (2007) Imaging large-scale neural activity with cellular resolution in awake, mobile mice. Neuron 56(1):43–57
Dore-Duffy P et al (2000) Pericyte migration from the vascular wall in response to traumatic brain injury. Microvasc Res 60(1):55–69
Drew PJ et al (2010) Chronic optical access through a polished and reinforced thinned skull. Nat Methods 7(12):981–984
Driscoll JD et al (2013) Two-photon imaging of blood flow in the rat cortex. Cold Spring Harb Protoc 2013(8):759–767
Foo SS et al (2006) Ephrin-B2 controls cell motility and adhesion during blood-vessel-wall assembly. Cell 124(1):161–173
Gao YR et al (2017) Time to wake up: studying neurovascular coupling and brain-wide circuit function in the un-anesthetized animal. NeuroImage 153:382–398
Goldey GJ et al (2014) Removable cranial windows for long-term imaging in awake mice. Nat Protoc 9(11):2515–2538
Grant RI et al (2017) Organizational hierarchy and structural diversity of microvascular pericytes in adult mouse cortex. J Cereb Blood Flow Metab:271678X17732229
Grubb S et al (2020) Precapillary sphincters maintain perfusion in the cerebral cortex. Nat Commun 11(1):395
Grutzendler J, Nedergaard M (2019) Cellular control of brain capillary blood flow: in vivo imaging Veritas. Trends Neurosci 42(8):528–536
Guimaraes-Camboa N et al (2017) Pericytes of multiple organs do not behave as mesenchymal stem cells in vivo. Cell Stem Cell 20(3):345–359
Hall CN et al (2014) Capillary pericytes regulate cerebral blood flow in health and disease. Nature 508(7494):55–60
Halliday MR et al (2016) Accelerated pericyte degeneration and blood-brain barrier breakdown in apolipoprotein E4 carriers with Alzheimer's disease. J Cereb Blood Flow Metab 36(1):216–227
Hartmann DA et al (2015) Pericyte structure and distribution in the cerebral cortex revealed by high-resolution imaging of transgenic mice. Neurophotonics 2(4):041402
Hartmann DA et al (2020) Brain capillary pericytes exert a substantial but slow influence on blood flow. In: bioRxiv
Hellstrom M et al (2001) Lack of pericytes leads to endothelial hyperplasia and abnormal vascular morphogenesis. J Cell Biol 153(3):543–553
Hill RA et al (2015) Regional blood flow in the Normal and ischemic brain is controlled by arteriolar smooth muscle cell contractility and not by capillary Pericytes. Neuron 87(1):95–110
Hill RA et al (2017) Targeted two-photon chemical apoptotic ablation of defined cell types in vivo. Nat Commun 8:15837
Holtmaat A et al (2009) Long-term, high-resolution imaging in the mouse neocortex through a chronic cranial window. Nat Protoc 4(8):1128–1144
Huang W et al (2014) Novel NG2-CreERT2 knock-in mice demonstrate heterogeneous differentiation potential of NG2 glia during development. Glia 62(6):896–913
Hyacinth HI et al (2019) Higher prevalence of spontaneous cerebral vasculopathy and cerebral infarcts in a mouse model of sickle cell disease. J Cereb Blood Flow Metab 39(2):342–351
Ishii Y et al (2006) Mouse brains deficient in neuronal PDGF receptor-beta develop normally but are vulnerable to injury. J Neurochem 98(2):588–600
Ivanova E, Kovacs-Oller T, Sagdullaev BT (2017) Vascular Pericyte impairment and Connexin43 gap junction deficit contribute to vasomotor decline in diabetic retinopathy. J Neurosci 37(32):7580–7594
Kalmbach AS, Waters J (2012) Brain surface temperature under a craniotomy. J Neurophysiol 108(11):3138–3146

Kamouchi M et al (2004) Calcium influx pathways in rat CNS pericytes. Brain Res Mol Brain Res 126(2):114–120

Khennouf L et al (2018) Active role of capillary pericytes during stimulation-induced activity and spreading depolarization. Brain 141(7):2032–2046

Kim TN et al (2012) Line-scanning particle image velocimetry: an optical approach for quantifying a wide range of blood flow speeds in live animals. PLoS One 7(6):e38590

Kimbrough IF et al (2015) Vascular amyloidosis impairs the gliovascular unit in a mouse model of Alzheimer's disease. Brain 138(Pt 12):3716–3733

Kisler K et al (2017) Pericyte degeneration leads to neurovascular uncoupling and limits oxygen supply to brain. Nat Neurosci 20(3):406–416

Kisler K et al (2020) Acute ablation of cortical pericytes leads to rapid neurovascular uncoupling. Front Cell Neurosci 14:27

Kleinfeld D et al (1998) Fluctuations and stimulus-induced changes in blood flow observed in individual capillaries in layers 2 through 4 of rat neocortex. Proc Natl Acad Sci U S A 95(26):15741–15746

Lai AY et al (2015) Venular degeneration leads to vascular dysfunction in a transgenic model of Alzheimer's disease. Brain 138(Pt 4):1046–1058

Li B et al (2020) Two-photon microscopic imaging of capillary red blood cell flux in mouse brain reveals vulnerability of cerebral white matter to hypoperfusion. J Cereb Blood Flow Metab 40(3):501–512

Liao M et al (2017) An X-linked Myh11-CreER(T2) mouse line resulting from Y to X chromosome-translocation of the Cre allele. Genesis 55(9)

Lindahl P et al (1997) Pericyte loss and microaneurysm formation in PDGF-B-deficient mice. Science 277(5323):242–245

Lindblom P et al (2003) Endothelial PDGF-B retention is required for proper investment of pericytes in the microvessel wall. Genes Dev 17(15):1835–1840

Longair MH, Baker DA, Armstrong JD (2011) Simple neurite tracer: open source software for reconstruction, visualization and analysis of neuronal processes. Bioinformatics 27(17):2453–2454

Lou N et al (2016) Purinergic receptor P2RY12-dependent microglial closure of the injured blood-brain barrier. Proc Natl Acad Sci U S A 113(4):1074–1079

Lowery RL, Majewska AK (2010) Intracranial injection of adeno-associated viral vectors. J Vis Exp (45)

Madisen L et al (2010) A robust and high-throughput Cre reporting and characterization system for the whole mouse brain. Nat Neurosci 13(1):133–140

Masamoto K, Kanno I (2012) Anesthesia and the quantitative evaluation of neurovascular coupling. J Cereb Blood Flow Metab 32(7):1233–1247

Mateo C et al (2017) Entrainment of arteriole vasomotor fluctuations by neural activity is a basis of blood-oxygenation-level-dependent "Resting-State" connectivity. Neuron 96(4):936–948

Miners JS et al (2019) CSF evidence of pericyte damage in Alzheimer's disease is associated with markers of blood-brain barrier dysfunction and disease pathology. Alzheimers Res Ther 11(1):81

Mishra A et al (2014) Imaging pericytes and capillary diameter in brain slices and isolated retinae. Nat Protoc 9(2):323–336

Montagne A et al (2015) Blood-brain barrier breakdown in the aging human hippocampus. Neuron 85(2):296–302

Montagne A et al (2018) Pericyte degeneration causes white matter dysfunction in the mouse central nervous system. Nat Med 24(3):326–337

Muzumdar MD et al (2007) A global double-fluorescent Cre reporter mouse. Genesis 45(9):593–605

Nakagawa S et al (2007) Pericytes from brain microvessels strengthen the barrier integrity in primary cultures of rat brain endothelial cells. Cell Mol Neurobiol 27(6):687–694

Nikolakopoulou AM et al (2017) Regional early and progressive loss of brain pericytes but not vascular smooth muscle cells in adult mice with disrupted platelet-derived growth factor receptor-beta signaling. PLoS One 12(4):e0176225

Nikolakopoulou AM et al (2019) Pericyte loss leads to circulatory failure and pleiotrophin depletion causing neuron loss. Nat Neurosci 22(7):1089–1098
Nishimura N et al (2010) Limitations of collateral flow after occlusion of a single cortical penetrating arteriole. J Cereb Blood Flow Metab 30(12):1914–1927
Orlidge A, D'Amore PA (1987) Inhibition of capillary endothelial cell growth by pericytes and smooth muscle cells. J Cell Biol 105(3):1455–1462
Packer AM, Roska B, Hausser M (2013) Targeting neurons and photons for optogenetics. Nat Neurosci 16(7):805–815
Park L et al (2014) Age-dependent neurovascular dysfunction and damage in a mouse model of cerebral amyloid angiopathy. Stroke 45(6):1815–1821
Peppiatt CM et al (2006) Bidirectional control of CNS capillary diameter by pericytes. Nature 443(7112):700–704
Prakash R et al (2012) Two-photon optogenetic toolbox for fast inhibition, excitation and bistable modulation. Nat Methods 9(12):1171–1179
Ratelade J et al (2020) Reducing Hypermuscularization of the transitional segment between arterioles and capillaries protects against spontaneous intracerebral hemorrhage. Circulation
Reeves C et al (2019) Spatiotemporal dynamics of PDGFRbeta expression in pericytes and glial scar formation in penetrating brain injuries in adults. Neuropathol Appl Neurobiol 45(6):609–627
Rickgauer JP, Tank DW (2009) Two-photon excitation of channelrhodopsin-2 at saturation. Proc Natl Acad Sci U S A 106(35):15025–15030
Roche M et al (2019) In vivo imaging with a water immersion objective affects brain temperature, blood flow and oxygenation. Elife 8
Roome CJ, Kuhn B (2014) Chronic cranial window with access port for repeated cellular manipulations, drug application, and electrophysiology. Front Cell Neurosci 8:379
Rorsman NJG et al (2018) Defining the ionic mechanisms of optogenetic control of vascular tone by channelrhodopsin-2. Br J Pharmacol 175(11):2028–2045
Rungta RL et al (2017) Light controls cerebral blood flow in naive animals. Nat Commun 8:14191
Rungta RL et al (2018) Vascular compartmentalization of functional hyperemia from the synapse to the pia. Neuron 99(2):362–375
Rustenhoven J, Jansson D, Smyth LC, Drahunow M (2017) Brain Pericytes as mediators of Neuroinflammation. Trends Pharmacol Sci
Sato Y et al (1990) Characterization of the activation of latent TGF-beta by co-cultures of endothelial cells and pericytes or smooth muscle cells: a self-regulating system. J Cell Biol 111(2):757–763
Sengillo JD et al (2013) Deficiency in mural vascular cells coincides with blood-brain barrier disruption in Alzheimer's disease. Brain Pathol 23(3):303–310
Shih AY et al (2012a) Two-photon microscopy as a tool to study blood flow and neurovascular coupling in the rodent brain. J Cereb Blood Flow Metab 32(7):1277–1309
Shih AY et al (2012b) A polished and reinforced thinned-skull window for long-term imaging of the mouse brain. J Vis Exp (61)
Shih AY, Drew PJ, Kleinfeld D (2014) Imaging Vasodynamics in the awake mouse brain with two-photon microscopy. Neurovas Coupling Methods
Sikka G et al (2014) Melanopsin mediates light-dependent relaxation in blood vessels. Proc Natl Acad Sci U S A 111(50):17977–17982
Stark K, Pekayvaz K, Massberg S (2018) Role of pericytes in vascular immunosurveillance. Front Biosci (Landmark Ed) 23:767–781
Stoica L, et al. (2013) Gene transfer to the CNS using recombinant adeno-associated virus. Curr Protoc Microbiol. Chapter 14: p. Unit14D 5
Underly RG et al (2017) Pericytes as inducers of rapid, matrix Metalloproteinase-9-dependent capillary damage during ischemia. J Neurosci 37(1):129–140
Vanlandewijck M, Betsholtz C (2018) Single-cell mRNA sequencing of the mouse brain vasculature. Methods Mol Biol 1846:309–324

Vanlandewijck M et al (2018) A molecular atlas of cell types and zonation in the brain vasculature. Nature 554(7693):475–480

von Tell D, Armulik A, Betsholtz C (2006) Pericytes and vascular stability. Exp Cell Res 312(5):623–629

Wang Y et al (2018) Viral vectors as a novel tool for clinical and neuropsychiatric research applications. Gen Psychiatr 31(2):e000015

Watson AN et al (2020) Mild pericyte deficiency is associated with aberrant brain microvascular flow in aged PDGFRbeta(+/−) mice. J Cereb Blood Flow Metab:271678X19900543

Wei HS et al (2016) Erythrocytes are oxygen-sensing regulators of the cerebral microcirculation. Neuron 91(4):851–862

Wirth A et al (2008) G12-G13-LARG-mediated signaling in vascular smooth muscle is required for salt-induced hypertension. Nat Med 14(1):64–68

Wu Y et al (2015) Optogenetic approach for functional assays of the cardiovascular system by light activation of the vascular smooth muscle. Vasc Pharmacol 71:192–200

Xu HT et al (2007) Choice of cranial window type for in vivo imaging affects dendritic spine turnover in the cortex. Nat Neurosci 10(5):549–551

Yang G et al (2010) Thinned-skull cranial window technique for long-term imaging of the cortex in live mice. Nat Protoc 5(2):201–208

Yemisci M et al (2009) Pericyte contraction induced by oxidative-nitrative stress impairs capillary reflow despite successful opening of an occluded cerebral artery. Nat Med 15(9):1031–1037

Zehendner CM et al (2015) Traumatic brain injury results in rapid pericyte loss followed by reactive pericytosis in the cerebral cortex. Sci Rep 5:13497

Zhu X, Bergles DE, Nishiyama A (2008) NG2 cells generate both oligodendrocytes and gray matter astrocytes. Development 135(1):145–157

Chapter 2
The Complex and Integral Roles of Pericytes Within the Neurovascular Unit in Health and Disease

Brad A. Sutherland

Abstract
Introduction
Pericytes are multifunctional and heterogeneous cells that are embedded within the basement membrane of capillaries. Alongside other mural cells, pericytes help develop, stabilize and maintain the cerebral vasculature. Pericytes also form one part of a group of cells called the neurovascular unit (NVU). Pericytes interact closely with neighboring NVU cells including endothelial cells that line capillaries and astrocyte endfeet that ensheathe cerebral blood vessels. These interactions control many of the functions of pericytes. In this chapter, I detail the complex and novel roles of pericytes at the NVU in the healthy and diseased brain. In particular, I discuss the functions and signaling mechanisms of pericytes at the NVU and how these become dysfunctional in neurological disease.

Methods
Journal databases (Pubmed, Google Scholar and Scopus) were searched with the keywords "pericyte(s)", "brain", "neurovascular" and/or "disease" or with the relevant disease titles such as "stroke/ischemia", "Alzheimer's disease/dementia", and "amyotrophic lateral sclerosis/motor neuron disease". Relevant published works were selected for discussion.

Result
Multiple studies have revealed that there are a number of paracrine signaling mechanisms between pericytes and other cells of the NVU. These mechanisms help regulate many pericyte functions including cerebral blood flow, angiogenesis, blood-brain barrier maintenance and inflammatory responses in the brain. Many neurological diseases, including stroke and Alzheimer's disease, exhibit neurovascular dysfunction at the capillary level in which numerous studies have indicated that pericytes

B. A. Sutherland (✉)
Medical Sciences, Tasmanian School of Medicine, College of Health and Medicine, University of Tasmania, Hobart, Tasmania, Australia
e-mail: brad.sutherland@utas.edu.au

A. Birbrair (ed.), *Biology of Pericytes – Recent Advances*, Stem Cell Biology and Regenerative Medicine 68, https://doi.org/10.1007/978-3-030-62129-2_2

may drive this. In addition, the stem cell potential of pericytes provides an exciting opportunity to use these cells to replace or repair injured territories.

Conclusions
Pericytes regulate many critical processes involved in the maintenance of neurovascular function and are now become an attractive therapeutic target for neurological diseases as well as a biomarker for neurovascular damage. Future strategies will uncover methods to use or target pericytes therapeutically.

Keywords Pericyte · Neurovascular unit · Capillary · Blood vessel · Cerebral blood flow · Blood-brain barrier · Neurological disease · Angiogenesis · Neuroinflammation · Stem cells · Stroke · Alzheimer's disease · Amyotrophic lateral sclerosis · Multiple sclerosis

Abbreviations

20-HETE	20-hydroxyeicosatetraenoic acid
Aβ	amyloid beta
AD	Alzheimer's disease
ALS	amyotrophic lateral sclerosis
ApoE	apolipoprotein E
α-SMA	alpha-smooth muscle actin
ATP	adenosine triphosphate
BBB	blood-brain barrier
BOLD	blood oxygen level dependent
CADASIL	cerebral autosomal dominant arteriopathy with subcortical infarcts and leukoencephalopathy
CNS	central nervous system
COX	cyclooxygenase
CSF	cerebrospinal fluid
DTA	diptheria toxin A
fMRI	functional magnetic resonance imaging
ICAM	intracellular adhesion molecule
iDTR	Cre-dependent human diptheria toxin receptor
IgG	immunoglobulin G
IL	interleukin
LRP-1	low density lipoprotein receptor-related protein 1
MHC	major histocompatibility complex
MMP9	matrix metalloprotease 9
MS	multiple sclerosis
NADPH	nicotinamide adenine dinucleotide phosphate
NF-κB	nuclear factor-kappa B
NO	nitric oxide
NOX4	nicotinamide adenine dinucleotide phosphate oxidase 4
NVU	neurovascular unit

OPC	oligodendrocyte precursor cell
PDGF-BB	platelet-derived growth factor-BB
PDGFRβ	platelet-derived growth factor receptor beta
PGE_2	prostaglandin E_2
sPDGFRβ	soluble platelet-derived growth factor receptor beta
TGFβ	transforming growth factor beta
TGFβR2	transforming growth factor beta receptor 2
TNFα	tumor necrosis factor alpha
VCAM	vascular cell adhesion molecule
VEGF	vascular endothelial growth factor
VEGFR2	vascular endothelial growth factor receptor 2
VSMC	vascular smooth muscle cell

2.1 Pericytes Reside in the Brain's Microvasculature

The brain constitutes 2% of the mass of the human body but consumes 20% of its resting energy, making it arguably the body's most energy-dependent organ (Xing et al. 2017). To supply this energy the brain has an extremely complex vascular architecture. The surface of the brain contains pial arteries that branch off into penetrating arterioles which dive deep into the brain parenchyma. Branching off the penetrating arterioles are the first order capillaries, also known as pre-capillary arterioles. A complex network of capillaries is derived from the first order capillary and these are the primary locations for gas and nutrient exchange from the blood to the brain. Deoxygenated blood is then transported from the capillaries up ascending venules and out to the pial veins. See Fig. 2.1a for a depiction of the brain's internal vasculature.

Given the brain's high energy needs, blood flow to specific brain regions needs to be tightly controlled. This primarily occurs through two mechanisms – autoregulation and neurovascular coupling. Autoregulation is where the cerebral arteries feeding the brain's microvasculature maintains a constant flow, even with changing blood pressure. This is an important mechanism of vascular control to limit the catastrophic consequences to the brain of acute hypotensive or hypertensive episodes. Neurovascular coupling, on the other hand, is the coordinated signaling from active neurons to capillaries and arterioles to promote the blood flow that provides the energy substrates needed for those neurons to fire. This local control of blood flow is critically important to help maintain primary brain function and coordinated synchrony across multiple networks within the brain.

Supporting cerebral vessels in their primary role, that being delivering blood, are mural cells which lie embedded within the basement membrane of these blood vessels. There is a spectrum of mural cells which possess different morphologies based on the type/size of vessel they are associated with (see Fig. 2.1) including both vascular smooth muscle cells (VSMCs) and pericytes. The larger arteries possess rings of VSMCs, which encircle the entire vessel and have the contractile machinery capable of actively constricting and dilating these vessels. The penetrating arterioles have VSMCs that encircle the vessel but do contain distinct protruding cell bodies,

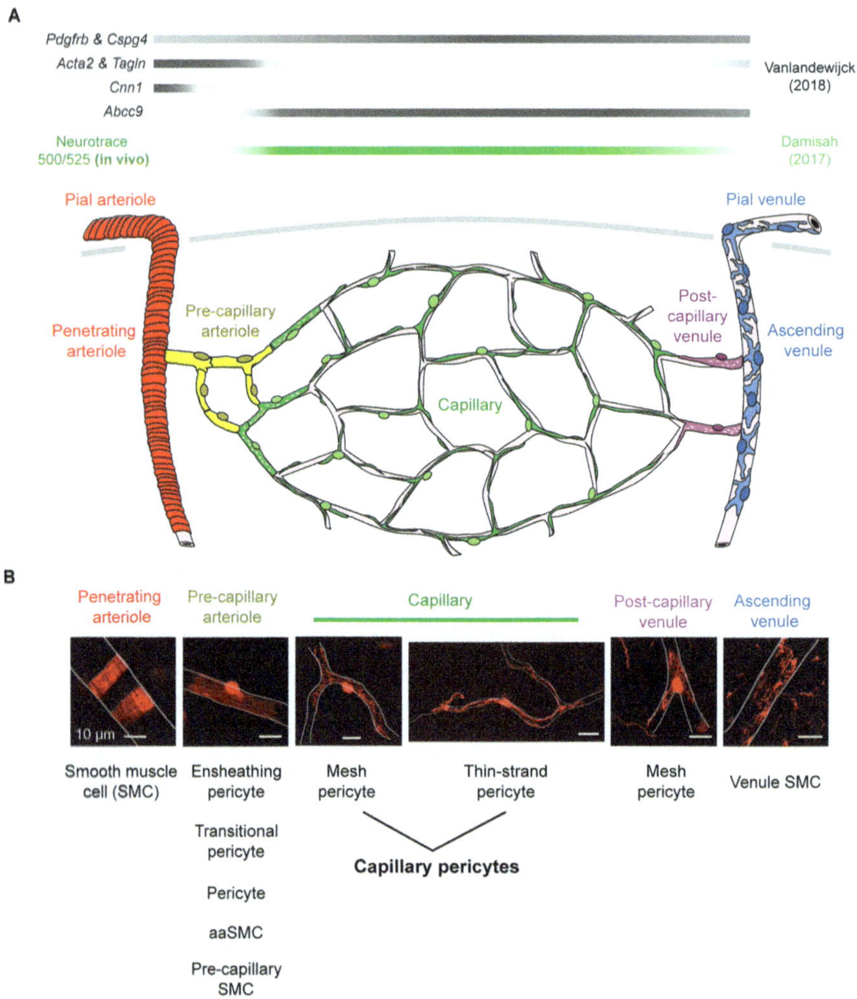

Fig. 2.1 The heterogeneity of the brain's vasculature with the varied morphologies and locations of pericytes. (**a**) Data from Vanlandewijck et al. (2018) and Damisah et al. (2017) showing the variability of gene expression and mural cell labelling across the microvasculature of the cortex (top). A schematic of the cortical microvasculature showing the transition of mural cells and their morphologies along the vascular tree (bottom). (**b**) In different zones of the microvasculature, mural cells have different morphologies and terminologies which are represented in the images. This figure has been reproduced from Berthiaume et al. 2018b)

suggesting these cells may be VSMC-pericyte hybrids. The branching first-order capillaries off the arteriole (the pre-capillary arteriole) contain a specialized VSMC-pericyte hybrid cell named the pre-capillary sphincter, recently identified to have an important role in maintaining cerebral perfusion (Grubb et al. 2020). Pericytes reside in the capillary bed, are evenly spaced and possess differing morphologies depending on which end of the capillary bed they are located. At the arteriolar end of the capillary bed, pericytes have processes that wrap around the capillary and have therefore been termed 'ensheathing pericytes'. In the centre of the capillary bed (third-sixth order capillaries), pericytes have a distinct cell body that protrudes from the vessel with longitudinal processes that run parallel with the vessel. These have been coined 'thin-strand' or 'string' pericytes. At the venule end of the capillary bed, pericyte cell bodies are present but do not have as well-defined processes. These are known as 'terminal capillary' or 'mesh' pericytes. On the ascending venules and pial veins, a mesh-like cell termed 'venous smooth muscle cells' are present, but these can also be referred to as pericytes due to their distinct cell body. There are a number of recently published reviews and research articles on this subject eloquently describing the heterogeneity of pericytes (Attwell et al. 2016; Berthiaume et al. 2018b; Brown et al. 2019; Grant et al. 2019; Hartmann et al. 2015).

Pericytes have several properties that make them critical for the maintenance of blood flow, energy delivery, molecular transport and brain function. Pericytes are particularly numerous within the capillary bed of the brain, with an estimated pericyte:endothelial cell ratio of 1:1–1:3, significantly higher than other tissues (Armulik et al. 2011). Within the capillary bed, the cell bodies of pericytes are evenly spaced approximately 50 μm apart (Hall et al. 2014). This enables the entire capillary bed to have functional capability while having limited number of pericytes. This is only made possible because pericytes are highly branched and wrap around as well as along capillaries with processes of a single pericyte estimated to cover approximately 250–300 μm of capillary (Berthiaume et al. 2018a; Hartmann et al. 2015). As a result, 70–80% of the capillary tube is estimated to be covered by pericytes and their branches (Armulik et al. 2010; Bell et al. 2010; Daneman et al. 2010; Sagare et al. 2013; Sengillo et al. 2013). Approximately 56% of pericyte cell bodies within the cortical capillary network are located at capillary junctions (Hartmann et al. 2015), enabling pericytes to control the distribution of flow. Pericytes can form an interconnected network within the microvasculature that is capable of transmitting signals upstream and downstream to activate adjacent pericytes (Peppiatt et al. 2006). A recent study calculated the speed of these conducted responses to be 5–20 μm/s, and illustrated that they can occur both up and downstream of the initiating pericyte (Cai et al. 2018). This action suggests that pericytes may meet the brain's metabolic demand through regulated and co-ordinated capillary network action, rather than by acting independently on single capillaries. This implies that signals originated in the capillary bed and transmitted upstream in the vascular network may even modulate arteriolar blood flow.

Ever since Zimmerman first coined the term pericytes including their various morphologies, there has been a reasonable distinction as to what constitutes a pericyte as opposed to a VSMC (Attwell et al. 2016). However, controversy has arisen as to how to precisely define mural subsets, whether it be based on function, morphology, location on the vasculature, or gene/protein expression. Pericyte diversity has in fact been recognized for at least 20 years, particularly in regard to how their location on the vascular tree governs their function (Sims 2000), and so this controversy is not new. In the brain, Hill and colleagues reported that VSMCs regulated cerebral blood flow in the brain, rather than pericytes (Hill et al. 2015). This claim was in stark contrast to one made in a landmark paper that it was pericytes that primarily regulate cerebral blood flow, not VSMCs (Hall et al. 2014). Subsequent studies by other groups have variably supported the findings of both Hall et al. (Biesecker et al. 2016; Cai et al. 2018; Fernandez-Klett et al. 2010; Grubb et al. 2020; Khennouf et al. 2018; Kisler et al. 2017, 2020; Kornfield and Newman 2014; Mishra et al. 2016; Neuhaus et al. 2017a; Peppiatt et al. 2006; Rungta et al. 2018) and Hill et al. (Wei et al. 2016; Zhou et al. 2019) and consequently there remains disagreement on the exact definition of a pericyte. A universal classification system of pericytes based on their function and morphology has been attempted, but is yet to be agreed upon (Berthiaume et al. 2018b; Grant et al. 2019).

One published classification system that further highlights the heterogeneity of pericytes is based on the expression of the neural progenitor gene nestin. In 2013, Birbrair and colleagues used Nestin-GFP/NG2-DsRed mice to reveal two types of pericytes: type 1 pericytes express NG2 but are nestin negative (Nestin-GFP-/NG2-DsRed+); type 2 pericytes express both NG2 and nestin (Nestin-GFP+/NG2-DsRed+) (Birbrair et al. 2013). While these subtypes of pericytes were first described in the skeletal muscle, they have since been demonstrated to exist in the brain's microvasculature, with 33% of pericytes being type 1 (nestin negative) and 67% of pericytes being type 2 (nestin positive) (Birbrair et al. 2014b). It is also apparent that both of these pericyte subtypes function differently, particularly in response to injury or disease. Type 2 pericytes can proliferate and promote angiogenesis (Birbrair et al. 2014b), which may be reflective of their stem cell and regenerative potential (Courtney and Sutherland 2020). On the other hand, type 1 pericytes have been shown to be adipogenic in the periphery (e.g. skeletal muscle) (Birbrair et al. 2013) but in the brain, the exact role of these pericytes is less clear except that they are reactive to injury (Birbrair et al. 2014a). In addition, there have been other reports of molecular heterogeneity in pericytes, such as expression of connexin-30 (a gap junction protein highly expressed on astrocyte endfeet) in a subset of pericytes (Mazare et al. 2018), but these studies require further confirmation.

Alongside the aforementioned functional, morphological and molecular studies, a single cell RNA-seq analysis of vascular mural cells revealed relatively homogeneous gene expression within the pericyte population but substantial differences in the profile of RNA expression between pericytes on capillaries and VSMCs on arterioles (Vanlandewijck et al. 2018). α-SMA was not expressed in pericytes within this dataset (He et al. 2018; Vanlandewijck et al. 2018) in a similar way to Hill and colleagues whom also classified pericytes as being negative for α-SMA (Hill et al. 2015). However, there are α-SMA positive isolated mural cells on the first order capillary which have been termed ensheathing or contractile pericytes based on morphology (Berthiaume et al. 2018b; Hall et al. 2014). Additionally, one group suggests that all pericytes express α-SMA and that depolymerization of α-SMA during tissue fixation leads to a lack of immunolabelling in higher order capillaries where thin-strand pericytes are present (Alarcon-Martinez et al. 2018). Overall, the contractile phenotype seen by all groups likely belongs to the effects of the same mural cell on the first order capillary/pre-capillary arteriole, with controversy over the nomenclature (Attwell et al. 2016). Despite the controversy regarding their definition, pericytes are clearly an important cell type for physiological cerebrovascular function and form an essential component of the neurovascular unit (NVU).

2.2 Pericytes Are a Central Part of the Neurovascular Unit

In order to maintain its energy need as well as protect its environment, the mammalian brain has developed a complex network of multiple cell types called the NVU. The NVU connects the cells residing within the brain parenchyma to the cells of the cerebral vasculature. Together they regulate blood flow to specific regions of the brain, whenever energy is required. In the brain parenchyma, as neurons fire, there is a greater energy requirement to maintain action potentials and synaptic transmission (Attwell and Laughlin 2001), and so neurons send neurotransmitters that can be relayed through other cell types to promote vasodilation and increase energy supply. Cell types in the brain parenchyma that can receive signals from the excitatory neurons and communicate them to the vasculature include inhibitory GABAergic interneurons, astrocytes, microglia, and oligodendrocyte precursor cells (OPCs). Astrocytes are an important supporting cell that forms contacts with neuronal synapses to create the tripartite synapse (Araque et al. 1999). They also extend endfeet that wrap around cerebral blood vessels (Macvicar and Newman 2015). On the vasculature itself, there are endothelial cells within the vessel that closely interact with mural cells but can also react to signals from within the blood. Mural cells such as VSMCs (on arteries, arterioles and veins) and pericytes (on capillaries and venules) reside on the abluminal side of the vessel and therefore do not have direct access to the blood stream. However, considering capillaries make up the majority of blood vessels in the brain (Blinder et al. 2013) and neurons lie closer to capillaries (~8.4 μm in hippocampus) than arterioles (~70 μm) (Lovick et al. 1999), pericytes on capillaries are central to NVU function due to their ability to sense and communicate to both the neuro- and vascular-sides of the NVU. The complex cellular interplay of the NVU is depicted in Fig. 2.2.

Some of the cells of the NVU (i.e. astrocytes and pericytes) also function as part of the blood-brain barrier (BBB) to prevent toxic species from entering the brain (Attwell et al. 2010; Mcconnell et al. 2017). The endothelial cells lining the brain's capillaries are non-fenestrated (i.e. lack openings), in contrast to endothelial cells in capillaries of certain peripheral tissues. The likely purpose of this barrier is to protect the brain from circulatory proteins that are neurotoxic, and to maintain brain function.

Their location at the nexus of the NVU make pericytes a key signaling hub that can control brain function. Pericytes have strong paracrine interactions with two main cell types: endothelial cells and the endfeet of astrocytes. It is known that pericytes physically interact with endothelial cells, can release factors to influence endothelial cells and astrocytes, and receive signals released from endothelial cells and astrocytes to alter pericyte function. These topics have been the subject of

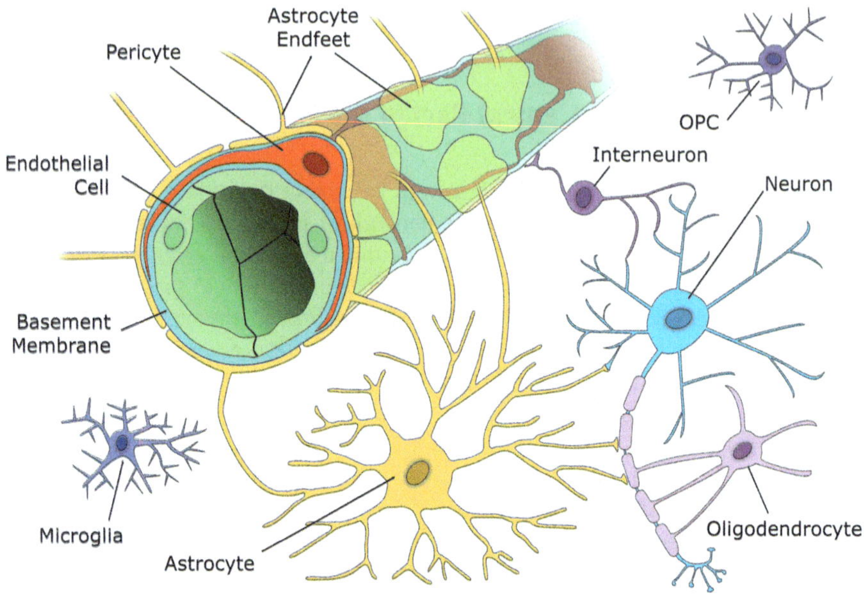

Fig. 2.2 The neurovascular unit of the mammalian brain showing the complex interplay of cells to mediate both cerebral and vascular function. Oligoedendrocyte precursor cell (OPC). (This figure is adapted from Brown et al. 2019))

several recent reviews (Brown et al. 2019; Sweeney et al. 2016). The major signaling pathways involved in these interactions will be described below (see Fig. 2.3).

2.2.1 Physical Pericyte-Endothelial Cell Interactions

Endothelial cells are connected by both tight junction proteins (including claudins and occludins), and adherens junction proteins (such as β-catenin and VE-cadherin). These connections are regulated by multiple signaling pathways including transforming growth factor beta (TGFβ) and angiopoietin/Tie2 (Luissint et al. 2012). To help stabilize the BBB, pericytes and endothelial cells are also physically connected through peg and socket contacts. Pericyte fingers are inserted into endothelial invaginations and junctional proteins such as connexins and N-cadherin help stabilize these connections. These pericyte-endothelial cell connections are regulated by the TGFβ and Notch pathways (Armulik et al. 2011; Geevarghese and Herman 2014; Sweeney et al. 2016).

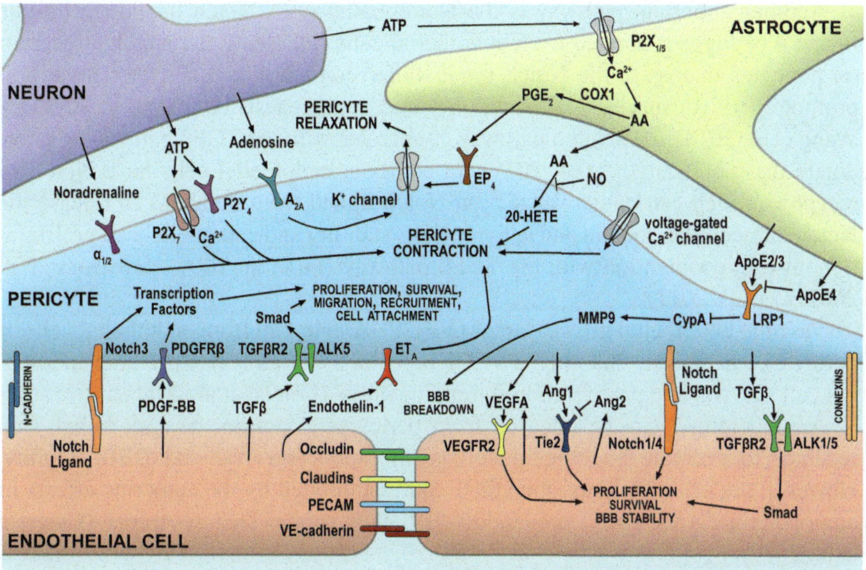

Fig. 2.3 Paracrine signaling between cells of the neurovascular unit. Pericytes interact with endothelial cells lining the capillary walls, astrocyte endfeet ensheathing the capillary, and neuronal processes that reside close to capillaries. Pericytes receive endothelial signals that modulate pericyte function including TGFβ, PDGF-BB, endothelin-1 and the Notch ligand. Endothelial cells can receive pericyte signals that modulate endothelial cell function including VEGF-A, Angiopoietin-1, TGFβ, and the Notch ligand. Pericytes and endothelial cells are connected through junctional proteins such as N-cadherin and connexins. Pericytes can also receive signals from astrocytes that modulate contractility and BBB function including PGE_2, arachidonic acid and ApoE isoforms. These astrocyte signals are regulated by neuronal ATP release. Neurons can also directly modulate pericyte contractility through adenosine, ATP and noradrenaline. 20-hydroxyeicosatetraenoic acid (20-HETE); arachidonic acid (AA); angiopoietin (Ang); apolipoprotein (ApoE); adenosine triphosphate (ATP); blood-brain barrier (BBB); cyclooxygenase-1 (COX1); cyclophilin A (CypA); endothelin A receptor (ET_A); low density lipoprotein receptor related protein-1 (LRP1); matrix metalloprotease 9 (MMP9); nitric oxide (NO); platelet derived growth factor BB (PDGF-BB); platelet derived growth factor receptor beta (PDGFRβ); prostaglandin E_2 (PGE_2); transforming growth factor beta (TGFβ); transforming growth factor beta receptor 2 (TGFβR2); vascular endothelial growth factor A (VEGFA); vascular endothelial growth factor receptor 2 (VEGFR2)

2.2.2 Pericyte-Endothelial Cell Communication

Endothelial cells have a close affinity with pericytes due to their adjacent location within the capillary wall. Endothelial cells can detect signals from within the blood and can signal to pericytes through paracrine signaling. One of the major mediators of pericyte function is platelet-derived growth factor-BB (PDGF-BB), which is released from endothelial cells to act on PDGF receptor beta (PDGFRβ) on pericytes. This activates several intracellular signaling pathways to promote pericyte proliferation, migration, survival, and recruitment as well as their attachment to endothelial cells (Tallquist et al. 2003).

Another important pathway is the Notch pathway. A Notch ligand (Delta-like ligand 4 or Jagged) derived from endothelial cells can cleave the Notch 3 receptor on pericytes to release the Notch intracellular domain which activates Smad4 to promote pericyte survival, BBB stability and angiogenesis (Henshall et al. 2015; Wang et al. 2014). The Notch pathway cooperates with the TGFβ pathway to coordinate these effects (Li et al. 2011). In addition, endothelial cells have Notch1/4 receptors which can be activated by a Notch ligand from pericytes to release the Notch intracellular domain and activate BBB stability and endothelial cell proliferation, also in cooperation with the TGFβ pathway (Fu et al. 2009; Hellstrom et al. 2007; Li et al. 2011).

Vascular endothelial growth factor (VEGF)-A, released from endothelial cells or pericytes, can activate endothelial VEGF receptor 2 (VEGFR2) to promote endothelial cell survival and proliferation for angiogenesis (Franco et al. 2011).

Angiopoietin-1 can be released from pericytes to act on the endothelial Tie2 receptor to promote endothelial cell proliferation, survival and BBB stability. However, the actions of angiopoietin-1 can be inhibited by the autocrine effects of angiopoietin-2, released from endothelial cells, upon Tie2 receptors (Gaengel et al. 2009).

Another bidirectional mediator is TGFβ which can be released from endothelial cells to act on pericyte TGFβ receptor 2 (TGFβR2), which phosphorylates Alk5 to activate the Smad signaling cascade and promote pericyte proliferation, attachment and survival (Van Geest et al. 2010). TGFβ can also be secreted from pericytes to bind to endothelial TGFβR2 which phosphorylates Alk1 and Alk5 to activate the Smad signaling cascade, and promote endothelial proliferation, survival and BBB stability (Li et al. 2011; Van Geest et al. 2010).

Another paracrine mediator acting between these two cell types is endothelin-1, which can be released by endothelial cells and act on endothelin-A receptors on pericytes to cause vasoconstriction (Dehouck et al. 1997; Neuhaus et al. 2017a).

The integrated signaling between pericytes and endothelial cells is complex and critical for both the maintenance of the BBB and for neurovascular function. The multiple pathways controlling these interactions allow for several regulatory steps in the signaling cascade that requires the cooperation of both cell types.

2.2.3 Pericyte-Astrocyte Communication

Astrocytes have endfeet that wrap around cerebral blood vessels (Fig. 2.2) and are therefore in close proximity to pericytes. It is well established that astrocytes can relay signals from neurons to pericytes and the vasculature in order to modulate blood flow (Attwell et al. 2010; Macvicar and Newman 2015; Mishra et al. 2016; Takano et al. 2006). One important molecule derived from astrocytes that can modulate pericyte function is prostaglandin E_2 (PGE_2). Upon stimulation, neurons release adenosine triphosphate (ATP) which acts on $P2X_{1/5}$ channels on astrocytes leading to an influx of Ca^{2+} (Mishra et al. 2016). This leads to the metabolism of arachidonic acid to PGE_2 by cyclooxygenase (COX)-1 within astrocytes (Howarth et al. 2017; Mishra et al. 2016). Once PGE_2 is formed, astrocytes release it to act on neighboring pericytes by binding the EP_4 receptor (Mishra et al. 2016). Activation of the EP_4 receptor by PGE_2 then leads to a relaxation of pericytes through promotion of K^+ efflux allowing vasodilation and, subsequently, increased blood flow (Hall et al. 2014; Hamilton et al. 2010). This dilatory action by PGE_2 can however only be performed when the vasoconstrictor 20-hydroxyeicosatetraenoic acid (20-HETE), which is also a metabolite of arachidonic acid, is inhibited in pericytes by nitric oxide (NO) produced from neurons and endothelial cells (Attwell et al. 2010; Hall et al. 2014). Astrocytes can also secrete apolipoprotein E (ApoE)2 and ApoE3, which bind to the low density lipoprotein receptor-related protein 1 (LRP-1) receptor on pericytes. LRP-1 then blocks cyclophilin-A activation of matrix metalloprotease 9 (MMP9), preventing BBB breakdown (Bell et al. 2012). In contrast to ApoE2 and ApoE3, ApoE4 inhibits LRP-1, thereby allowing cyclophilin-A to activate MMP9 in a nuclear factor kappa B (NF-κB) dependent manner, promoting BBB breakdown (Bell et al. 2012). Through these mechanisms, astrocytes can modulate pericyte activity and, as a result, can alter blood flow and BBB permeability.

2.2.4 *Pericyte-Neuron Communication*

Neurons can also communicate directly with pericytes. Pericyte-covered capillaries are located approximately 8.4 μm away from neurons (in the hippocampus) and are therefore in close enough proximity to readily receive signals directly from neurons (Lovick et al. 1999). 65% of noradrenergic innervations occur at the capillary compared to arterioles in the brain (Peppiatt et al. 2006), suggesting neuron-capillary contacts are important for the local control of cerebral blood flow. While some neuron-derived molecules modulate pericyte function through astrocytes (e.g. ATP (Mishra et al. 2016)), pericytes possess receptors capable of receiving direct signals from neurons, such as adenosine, ATP and noradrenaline (He et al. 2018). Adenosine released from neurons can promote pericyte relaxation and capillary dilation (Li and Puro 2001; Neuhaus et al. 2017a). Alternatively, neuronal release of noradrenaline and ATP can cause pericyte contraction and subsequent capillary constriction (Cai et al. 2018; Hall et al. 2014; Kawamura et al. 2003).

2.3 Functions of Pericytes in the Neurovascular Unit

As described above, pericytes act as a central signaling hub within the NVU and are capable of driving several functions of the NVU including the regulation of cerebral blood flow, vascular development, angiogenesis and stability, BBB maintenance, and neuroinflammation (Fig. 2.4). Many of these functions have been elucidated through the use of transgenic technologies that enable knockdown or depletion of

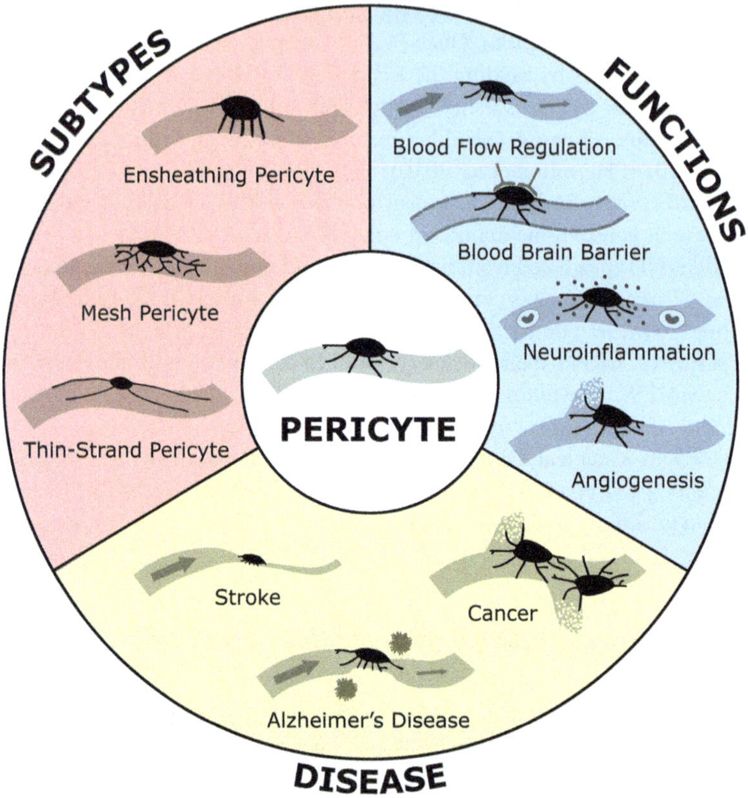

Fig. 2.4 The subtypes and functions of pericytes in the brain, as well as neurological disease states that implicate pericytes. (This figure has been adapted from Brown et al. 2019)

pericytes, allowing researchers to determine how this affects the various functions of the NVU.

2.3.1 *Experimental Methods to Reduce Pericyte Numbers* in vivo

The principal method employed to reduce pericyte numbers *in vivo* has been to genetically disrupt PDGF-BB/PDGFRβ signaling in mice (e.g. PDGFRβ$^{+/-}$ mice (Bell et al. 2010, Tallquist et al. 2003), PDGFRβ F7 mice (Bell et al. 2010; Daneman et al. 2010; Tallquist et al. 2003), PDGF-B$^{Ret/Ret}$ mice (Armulik et al. 2010), R26P$^{+/o}$ mice (Armulik et al. 2010), PDGF-B$^{lox/-}$ mice (Enge et al. 2002)). These disruptions can reduce the number of pericytes in the brain by 25–70%. However, because these mutations are constitutive, the phenotypes of these mutant mice are influenced by developmental deficits, limiting the interpretation of results derived from these lines. In addition, homozygote PDGFRβ$^{-/-}$ mice are embryonically lethal, likely due to an inability to form blood vessels, alongside other physiological problems (Leveen et al. 1994; Soriano 1994), but reveals the importance of pericytes and PDGF-BB signaling to vascular development.

Recently, new approaches have been developed to temporally induce pericyte knockdown, allowing researchers to study pericyte functions in adulthood. One such method is to cross a *Pdgfrb*-Cre ER^{T2} mouse with a *Rosa26*-diptheria toxin A (DTA) mouse, enabling the specific ablation of PDGFRβ-positive pericytes (Park et al. 2017). To date, this approach has been used to investigate the impact of reduced pericyte numbers in the retina. However, PDGFRβ is also expressed in other cells, for instance VSMCs, therefore this approach is not truly pericyte-specific. To overcome this limitation, one research group recently created a double cre mouse using both *Pdgfrb* and *Cspg4* as promoters (Nikolakopoulou et al. 2019). When crossed with a mouse expressing Cre-dependent human diptheria toxin receptor (iDTR) (i.e. pericyte-Cre;Ai14 x iDTR), a mouse was generated in which pericytes can be specifically ablated without off-target effects on other cells (Nikolakopoulou et al. 2019).

Non-genetic strategies for disrupting pericytes have also been trialled. In particular, receptor tyrosine kinase inhibitors, such as imatinib, can specifically block the PDGFRβ receptor and has been shown to reduce pericyte viability *in vitro* (Ruan et al. 2013). Imatinib administration *in vivo* has also been shown to induce pericyte loss in the retina (Wilkinson-Berka et al. 2004), but there is limited evidence of this in the brain. In addition, receptor tyrosine kinase inhibitors are known to have substantial off-target effects, therefore any results derived when using these compounds must be interpreted with caution.

As more research continues in this field and as technology for modulating pericytes improves, new roles for pericytes in NVU function will undoubtedly arise. Below we discuss the functional roles for pericytes that have been discovered using some of these pericyte-targeting strategies.

2.3.2 The Role of Pericytes in the Regulation of Cerebral Blood Flow

In order to maintain brain function, synaptic plasticity and neuronal firing, a constant energy supply to the brain is required. As described above, while autoregulation can account for blood flow changes in large vessels that feed into the smaller vessels, it is neurovascular coupling, the process by which arterioles and capillaries are regulated by the energy needs of local neurons, that allows for increased capacity and energy supply. Increasing neuronal activity leads to increasing energy needs which drives a vasodilatory response in the local microvasculature to direct cerebral blood flow to the area (Attwell et al. 2010). Pericytes play an important role in this response as they can actively dilate or constrict capillaries, altering cerebral blood flow in response to increased neuronal activity (Biesecker et al. 2016; Cai et al.

2018; Fernandez-Klett et al. 2010; Grubb et al. 2020; Hall et al. 2014; Khennouf et al. 2018; Kisler et al. 2017, 2020; Kornfield and Newman 2014; Mishra et al. 2016; Peppiatt et al. 2006; Rungta et al. 2018).

One useful marker of intact neurovascular coupling is the blood oxygen level dependent (BOLD) signal in functional magnetic resonance imaging (fMRI) (Attwell and Iadecola 2002). Deoxygenated hemoglobin is paramagnetic, therefore the level of deoxygenated hemoglobin is inversely proportional to the BOLD signal arising from T2*-weighted fMRI (Attwell et al. 2010). Although the BOLD signal can spatially correlate with neuronal activity following sensory stimuli tasks in fMRI studies (Kim et al. 2004), it is vascular changes in response to neuronal activity that the BOLD signal is actually measuring (Attwell et al. 2010; Attwell and Iadecola 2002). Therefore, the BOLD response is a direct measure of intact neurovascular coupling, and an impairment in the BOLD response could mean either a neuronal or vascular problem, and further studies would need to be performed with other techniques to specifically identify which component is impaired.

It follows that pericyte function is related to the BOLD response, considering their unique role in dilating and constricting capillaries. In fact, it has been calculated that pericyte relaxation is responsible for 84% of the increase in cerebral blood flow in response to increased neuronal activity (Hall et al. 2014). Pericytes are also the first responders to neuronal stimuli as pericyte-covered capillaries dilated before VSMC-covered arterioles (Hall et al. 2014). Recent studies have also illustrated that an acute ablation of pericytes causes neurovascular uncoupling (Kisler et al. 2020) and deficits in blood flow (Nikolakopoulou et al. 2019). This demonstrates that pericyte relaxation of capillaries following neuronal stimulation is critical for the cerebral blood flow increase and the appearance of the BOLD signal.

The signaling pathways that control pericyte contraction and relaxation to alter cerebral blood flow are still being characterized. As highlighted in Fig. 2.3, pericyte contraction can occur via several mechanisms including Ca^{2+} influx into the pericyte through voltage-gated calcium channels, activation of endothelin-1A receptors, production of 20-HETE from astrocyte-derived arachidonic acid, and neuronally-derived noradrenaline and ATP (Cai et al. 2018; Hall et al. 2014; Hamilton et al. 2010; Sweeney et al. 2016). Likewise, pericyte relaxation can be mediated via several mechanisms including neuronally-derived adenosine, activation of EP_4 receptors by astrocyte-derived PGE_2, and NO-mediated inhibition of 20-HETE, which leads to the activation of a number of potassium channels (Hall et al. 2014; Hamilton et al. 2010; Mishra et al. 2016; Sweeney et al. 2016).

It is known that pericytes possess the cellular machinery to contract by expressing contractile proteins such as alpha-smooth muscle actin (α-SMA), tropomyosin, and myosin (Alarcon-Martinez et al. 2018; Rucker et al. 2000). However, with the understanding that there are subtypes of pericytes, only some express these contractile markers (see Fig. 2.1), particularly the ensheathing pericytes at the arteriolar end of capillaries (Alarcon-Martinez et al. 2018), which may be a transitional cell between pericytes and VSMCs (Grant et al. 2019). There is now substantial evidence that it is these α-SMA positive pericytes at the arteriolar end of the capillary bed (that some groups label as VSMCs (Hill et al. 2015)) that modulate blood flow in response to neuronal activity from a variety of stimuli including odor (Rungta et al. 2018), light flicker (Biesecker et al. 2016; Kornfield and Newman 2014), whisker stimulation (Cai et al. 2018; Grubb et al. 2020; Hall et al. 2014; Khennouf et al. 2018), and hindlimb stimulation (Kisler et al. 2017; Kisler et al. 2020). However, further into the capillary bed, red blood cells still need to deform to be able to pass through capillaries (Jeong et al. 2006), and so any increased rigidity of the capillary by pericytes, without affecting diameter, may still modulate cerebral blood flow through alterations to the transit time of red blood cells across the capillary bed (Attwell et al. 2016). Overall, regardless of where on the vascular tree pericytes are located, they still have the ability to regulate cerebral blood flow either through direct constriction or relaxation of capillaries or through altering the rigidity of the capillary wall.

2.3.3 The Role of Pericytes in Vascular Development and Angiogenesis

For brain growth and development, as well as recovery after injury, it is crucial that the cerebral vasculature develops correctly. Angiogenesis, the process of new vessel formation, requires the sprouting of endothelial cells into a novel tube network. While pericyte recruitment to endothelial sprouts helps to stabilize the new vessels as they become functional (Bergers and Song 2005), pericytes can also be present at the growing front of endothelial sprouts guiding newly formed vessels (Nehls et al. 1992). Pericytes are directed to these new vessels through the secretion of growth factors by endothelial cells including PDGF-BB, TGF-β and angiopoietin 1 (Ribatti et al. 2011). A loss of endothelial PDGF-BB can lead to pericyte loss and microvascular abnormalities, illustrating the critical importance of endothelial-pericyte interactions through PDGF-BB signaling (Bjarnegard et al. 2004).

Pericytes can also influence endothelial cell activity through angiogenic factors such as VEGF as well as Notch signaling to promote angiogenesis (Ribatti et al. 2011). Additionally, pericytes appear to help form the basement membrane of new capillaries as well as regulating the cell cycle of endothelial cells, thereby controlling angiogenesis (Bergers and Song 2005). Supporting these roles for pericytes are studies in pericyte deficient (PDGFR$\beta^{+/-}$) mice which display vascular dysfunction leading to reduced microcirculation and perfusion as well as BBB deficits (Bell et al. 2010; Kisler et al. 2017). More recent evidence has shown even a mild pericyte deficiency can lead to aberrant cerebral microvascular flow irrespective of the maintenance of pericyte coverage (though pericyte cell somata was decreased by 27% (Watson et al. 2020)). Collectively, these studies show the importance of pericytes to the development and maintenance of the cerebral microvasculature, with the knowledge that pericytes are critical for vascular development across multiple vascular beds (Payne et al. 2020).

2.3.4 The Role of Pericytes in Vascular Stability

In addition to vascular development and angiogenesis, pericytes have an important role in stabilizing capillaries. As noted above, the processes of pericytes are extensive and wrap around and along all capillaries, covering approximately 70–80% of their surface (Armulik et al. 2010; Bell et al. 2010; Daneman et al. 2010; Sagare et al. 2013; Sengillo et al. 2013). Without this stabilization, capillaries cannot constrict or dilate effectively, BBB integrity is compromised, and vessels become more fragile (Bell et al. 2010). The interaction between endothelial cells and pericytes, which in part is regulated by the angiopoietin-Tie2 receptor pathway, is critical for this stabilization (Sundberg et al. 2002). A recent study investigated the hemodynamics of capillaries after a single pericyte was ablated using an *in vivo* two-photon microscope (Berthiaume et al. 2018a). It was shown that while pericyte processes did not physically touch each other, when a single pericyte was laser ablated to create a portion of uncovered capillary, the neighboring pericyte's processes extended to cover the ablated site. This process took time and while the uncovered capillary did not have an impaired BBB, there was uncontrolled dilation of the capillary, suggesting that pericyte processes maintain the basal tone of the capillary.

2.3.5 The Role of Pericytes in Blood-Brain Barrier Maintenance

The BBB is extremely important for the maintenance of brain function as it prevents material from within the circulation passively entering the brain. Of course, there are a number of molecules/peptides/proteins/cells that require entry into the brain from the blood and there exist mechanisms to enable them to cross the BBB including diffusion, active transport, paracellular transport and efflux pumps (Wong et al. 2013). The barrier itself consists of the endothelial cells that make up the capillary wall connected by tight junctions, which is supported by astrocytes and pericytes (Figs. 2.2 and 2.3). In 2010, three papers were published all demonstrating the importance of pericytes to the regulation of the BBB using genetic knockdown of pericytes (Armulik et al. 2010; Bell et al. 2010; Daneman et al. 2010). Pericyte deficient-mice not only have issues with vascular development, but have a cerebral capillary bed characterized by a lack of tight junctions between endothelial cells and increased permeability of the BBB (Sengillo et al. 2013). The inducible knockdown of pericytes can also lead to acute changes in the permeability of the BBB as evidenced by increased immunoglobulin G (IgG) labelling in the brain (Nikolakopoulou et al. 2019). The main role for pericytes at the BBB is to release signaling factors that control astrocyte endfeet polarization and govern the number of endothelial cell tight junctions (Armulik et al. 2010), and so are clearly important cells for the formation and maintenance of the BBB.

2.3.6 The Role of Pericytes in Neuroinflammation

The regulation of inflammatory processes in the central nervous system (CNS) (i.e. neuroinflammation, a distinct concept from inflammation in the periphery) is often discussed in the context of microglia, the brain's resident immune cells (Streit et al. 2004). While microglia undoubtedly have strong immune functions in the brain including immune surveillance and the production of cytokines and chemokines in response to insults, neuroinflammation is more complex than microglial responses alone. All cells of the CNS can contribute to neuroinflammation and it has recently become apparent that so too can pericytes (Jansson et al. 2014).

There are several ways in which pericytes may influence neuroinflammation and their location at the interface between the peripheral circulation and the brain parenchyma puts them in a prime position to do so. Pericytes have recently been described as the first-responders in the brain to a systemic inflammatory insult, through the release of cytokines and chemokines which alter brain function (Duan et al. 2018). Pericytes can secrete a range of chemokines such as CC chemokines (e.g. CCL2), CX3 chemokines (e.g. CXCL1) and CX3C chemokines (e.g. CX3CL1) as well as cytokines (interleukin (IL)-1, IL-4, IL-6, IL-10 and tumor necrosis factor alpha (TNFα) (Rustenhoven et al. 2017)) that enables them to perform both anti-inflammatory and pro-inflammatory functions. When pericytes are exposed to cytokines like IL-1β and TNFα, they not only secrete other cytokines and chemokines, but also MMP9 which causes the breakdown of the BBB (Herland et al. 2016). The activation of the pro-inflammatory CypA-NF-κB-MMP9 pathway in pericytes, which alters both BBB and vascular function, can be induced through a number stimuli such as a lack of murine ApoE (Bell et al. 2012). The secretion of cytokines such as CCL2, CXCL8 and IL-1β by pericytes in combination with expression of adhesion molecules can lead to the recruitment of immune cells or promote inflammatory pathways in other cell types such as microglia and astrocytes (Rustenhoven et al. 2017). Adhesion molecule expression in pericytes such as vascular cell adhesion molecule (VCAM) and intracellular adhesion molecule (ICAM)-1 (Proebstl et al. 2012; Verbeek et al. 1995) can regulate the passage of leukocytes from the circulation into tissue (Stark et al. 2013). Pericytes can also directly alter the BBB during neuroinflammation which can assist with this leukocyte trafficking into the brain (Jansson et al. 2014).

In addition to their first-responder properties, pericytes may assist in the uptake and presentation of antigens through the presence of major histocompatability complex (MHC)-II complexes (Pieper et al. 2014). Although pericytes could reactivate T cells previously primed by antigen presenting cells, pericytes do not possess the necessary co-stimulatory molecules to stimulate naïve T cells (Maier and Pober 2011).

Another role for pericytes in the context of neuroinflammation is the ability for pericytes to phagocytose other cells, material and debris allowing the clearance of these potentially toxic species, including Aβ, out of the brain (Kristensson and Olsson 1973; Sagare et al. 2013; Schultz et al. 2017; Thomas 1999). It appears pericytes work in concert with microglia to clean up damaged areas of the brain (Ozen et al. 2014). This function of pericytes will promote the resolution of inflammation but also limit the extent of neuroinflammation in the healthy brain. Overall, pericytes play an important role in neuroinflammation have the potential to become a key target for therapeutically regulating these processes in neurological diseases.

2.3.7 Pericytes and Their Stem Cell Potential

A statement was recently made that "*pericytes...are the only true pluripotent adult stem cells*" (Davidoff 2019). While there is considerable debate as to exactly whether pericytes are a type of pluripotent stem cell, there is burgeoning evidence that pericytes can differentiate into other cell types and may have reparative and regenerative properties in the brain (Courtney and Sutherland 2020). Pericytes can express mesenchymal stem cell markers that include CD44, CD73, CD90 and CD105 (Crisan et al. 2008) but it remains difficult to confirm whether pericytes are mesenchymal stem cells due to the heterogeneous nature of these cell types. In the brain specifically, evidence suggests pericytes have the potential to differentiate into neural, glial and vascular cells, including microglia (Nakagomi et al. 2015; Ozen et al. 2014). However, this differentiation capacity may only occur under pathological circumstances such as cerebral ischemia. It remains to be seen whether pericytes differentiate into other cell types in the healthy or healthy-aging brain. Curiously, recent research has shown that even peripherally-derived pericytes (from dental pulp) can be differentiated into both glial and neuronal cell types (Farahani et al. 2019). This and other novel findings regarding the potential of pericytes to act as stem cells will become an interesting therapeutic avenue to explore for neurological diseases in future studies.

2.4 Pericytes in Neurological Diseases

Given the improved understanding of the importance of pericytes for physiological brain function, it is perhaps not surprising that pericyte dysfunction is increasingly being implicated in a wide range of neurological diseases (Fig. 2.4). Although pericytes reside on the microvasculature, the brain diseases in which pericytes appear to be dysfunctional are not strictly vascular (i.e. stroke or vascular dementia). Pericyte dysfunction has been implicated in chronic neurodegenerative diseases such as Alzheimer's disease (AD) and amyotrophic lateral sclerosis (ALS), where the vascular contributions to these diseases are poorly understood. For an excellent review that highlights the large range of neurological diseases pericyte dysfunction or deficiency has been implicated in, see Cheng et al. (2018). In the following sections, we focus on the role of pericytes in stroke, AD, ALS and multiple sclerosis (MS).

2.4.1 *The Role of Pericytes in Stroke*

There are two types of stroke: (1) Ischemic stroke, which is an abrupt cessation of blood flow to the brain due to an occluding thrombus or embolus in a cerebral artery, and (2) Hemorrhagic stroke, wherein a cerebral artery ruptures leading to excessive bleeding in the brain. Both types of strokes have dramatic effects on the functions of the NVU (Balbi et al. 2017; Sutherland et al. 2017), including disruptions to pericyte function.

In 1968, Adelbert Ames III and colleagues published a study illustrating that in the rabbit brain, even though cerebral arteries were recanalized following ischemia, there was a profound deficit in capillary flow. They coined this observation the 'no-reflow phenomenon' (Ames 3rd et al. 1968). Subsequently, there have been many hypotheses as to what causes this no-reflow phenomenon, including cortical spreading depression and post-ischemic swelling of endothelial cells and astrocyte endfeet (Kloner et al. 2018). In 2009, Yemisci and colleagues showed that pericytes could constrict blood vessels during ischemia and illustrated that this could give rise to no-reflow (Yemisci et al. 2009). Remarkably, subsequent research has demonstrated that pericytes can contract during stroke and then die in rigor (i.e. in a contracted position), thereby limiting blood flow up to 24 h after reperfusion post-stroke (Hall et al. 2014). Other experiments have shown that pericyte contraction can be mediated by oxidative stress (Yemisci et al. 2009), and pericyte death is related to a large calcium influx into pericytes during stroke and reperfusion (Hall et al. 2014). These findings heavily implicate pericyte dysfunction as a major player in the no-reflow phenomenon.

There are several other ways in which pericytes could contribute to injury post-stroke. MMP9 is known to disrupt the binding of astrocyte endfeet to the vascular wall while affecting tight junction integrity between endothelial cells, and it has been shown that following ischemia, pericytes can rapidly release MMP9 to promote BBB breakdown (Underly et al. 2017). In addition, the enzyme nicotinamide adenine dinucleotide phosphate (NADPH) oxidase 4 (NOX4), which produces superoxide, is expressed by pericytes and is upregulated following middle cerebral artery occlusion, ultimately leading to MMP9 expression and BBB breakdown (Nishimura et al. 2016).

The influence of pericyte dysfunction to the pathophysiology of hemorrhagic stroke is less-well studied. In 2016, one study found that α-SMA-positive pericytes could constrict microvessels which was associated with neurological deficits in a rat model of subarachnoid hemorrhage (Li et al. 2016). In addition, hemoglobin can upregulate cyclophilin A which can activate pericyte release of MMP9 to disrupt the BBB, thereby worsening outcome following subarachnoid hemorrhage (Pan et al. 2020). In the context of intracerebral hemorrhage, while bleeding into the parenchyma will cause contraction of pericytes and closure of microvessels, significant remodeling of the NVU and subsequent neovascularization can occur to repair the brain. Thrombin, which is activated by intracerebral hemorrhage, can promote angiogenesis as well as pericyte coverage of vessels through a Tie2-dependent mechanism (Hu et al. 2019). However, more research needs to be conducted to determine the full extent of the role of pericytes in both the acute effects and the recovery of the brain following a subarachnoid or intracerebral hemorrhage. Further research will be particularly important in the context of vasospasms and delayed cerebral ischemia following subarachnoid hemorrhage where severe vasoconstriction occurs, which may have pericyte involvement (Ostergaard et al. 2013).

In summary, these findings suggest that pericytes could be therapeutically targeted post-stroke in an effort to prevent both the constriction of capillaries and the opening of the BBB. For more information, the reader is referred to a number of excellent recent reviews on this subject (Dalkara et al. 2019; Gautam and Yao 2018; Neuhaus et al. 2017b).

2.4.2 The Role of Pericytes in Alzheimer's Disease

The neuropathological hallmarks of AD, amyloid plaques (Aβ accumulation) and neurofibrillary tangles (tau hyperphosphorylation), are well described (Dubois et al. 2014; Montine et al. 2012). However, epidemiological studies have revealed that many vascular factors (e.g. hypertension, hypercholesterolemia, diabetes, obesity, heart disease and stroke) increase the risk of AD (de la Torre 2012; Norton et al. 2014). It has been shown that in AD patients, brain regions with hypoperfusion overlap with regions of AD pathology (Austin et al. 2011). A reduction in both oxygen and glucose not only prevent neurons from getting these important energy substrates, but hypoxia can also trigger Aβ accumulation (Sun et al. 2006) and a glucose deficit can trigger tau phosphorylation (Lauretti et al. 2017). In AD mouse models, there is evidence of stalled capillaries, i.e. capillaries where flow does not occur, which contributes to neurodegeneration (Cruz Hernandez et al. 2019), while strategies that increase cerebral blood flow can improve cognitive outcomes (Bracko et al. 2019). Vascular dysfunction, characterized by hypoperfusion and BBB disruption where pericytes have important roles, could therefore lead to an energy deficit within the brain, and drive AD pathogenesis and subsequent cognitive decline, independently or in concert with other AD-associated pathologies (Winkler et al. 2014). This vascular dysfunction, which possibly precedes the onset of other characteristic AD-associated pathologies (Iturria-Medina et al. 2016), has led to calls for vascular biomarkers to be included in all human AD studies (Sweeney et al. 2019a).

In regards to pericytes specifically, one study has shown that pericyte number is reduced in the hippocampus and cortex of post-mortem human AD brains (Sengillo et al. 2013). This corresponds with evidence from animal models showing that pericyte degeneration fosters an AD phenotype, including neurodegeneration (Sagare et al. 2013). Subsequent studies have observed increased shedding of the soluble PDGFRβ (sPDGFRβ) into the cerebrospinal fluid (CSF), considered reflective of pericyte injury, in human dementia patients (Montagne et al. 2015; Nation et al. 2019). These increased CSF sPDGFRβ levels were associated with hippocampal BBB permeability and were independent of Aβ or tau changes.

Multiple *in vitro* studies have established that Aβ exposure can cause pericyte death (Schultz et al. 2018; Wilhelmus et al. 2007). More recently, evidence derived from both human brain slices and AD mouse models has illustrated that Aβ can constrict capillaries, thereby reducing blood flow, through endothelin-1 acting on pericytes (Nortley et al. 2019). In addition, pericytes also express LRP-1 (Kovac et al. 2011), which is a receptor for ApoE, and can bind and phagocytose Aβ (Ma et al. 2018; Wilhelmus et al. 2007). When LRP-1 is genetically deleted from pericytes in mice, Aβ can no longer be taken up by pericytes, suggesting that this mechanism is critical for the clearance of Aβ from the brain (Ma et al. 2018). ApoE4, the strongest genetic risk factor for sporadic AD (Corder et al. 1993), derived from human astrocytes can not resolve impaired Aβ uptake in LRP-1 deficient mouse pericytes whereas ApoE3 could (Ma et al. 2018). In fact, ApoE4 carriers have accelerated pericyte degeneration and BBB breakdown (Halliday et al. 2016).

Overall, this evidence suggests issues associated with the uptake of Aβ by pericytes can lead to the constriction of capillaries. This would limit blood flow and impair the clearance of Aβ, causing pericyte toxicity and amyloid plaque buildup as well as BBB damage. These multiple lines of evidence suggest there are several mechanisms by which pericyte dysfunction could be involved in the development of the neuropathological and, subsequently, cognitive features of AD. For more detailed information, see these recent reviews on pericyte dysfunction in AD (Erdener and Dalkara 2019; Sweeney et al. 2019b; Winkler et al. 2014).

2.4.3 The Role of Pericytes in Amyotrophic Lateral Sclerosis

Another neurodegenerative disease that could possibly have an underlying vascular component is ALS, otherwise known as motor neuron disease or Lou Gehrig's disease. Genetic mutations in the *Sod1* (Cu/Zn superoxide dismutase), *C9orf72* (Chromosome 9 Open Reading Frame 72), *Tardbp* (TDP-43) and *Fus* (Fused in sarcoma) genes can give rise to this disease, but only account for a minority of cases (Taylor et al. 2016). Interestingly, neurovascular dysfunction has been shown to occur both in human and animal models of ALS, as observed by an upregulation of MMP9 prior to motor neuron degeneration (Miyazaki et al. 2011). Subsequently, it was shown that there is a 54% reduction in pericyte number in human ALS, possibly leading to the breakdown of the blood-spinal cord barrier and extravasation of blood proteins in the spinal cord (Winkler et al. 2013). Pericyte coverage of capillaries in the ventral horn is also reduced in ALS patients, which may be linked to the predominant neurodegeneration that occurs in this location in ALS (Yamadera et al. 2015). Although only a few studies have investigated the role of pericytes in the development of ALS, therapeutically targeting pericytes could be one possible strategy to prevent the blood-spinal cord barrier opening in ALS.

2.4.4 The Role of Pericytes in Multiple Sclerosis

Although MS is highly prevalent, the precise underlying causes are unknown. Some of the modifiable risk factors for MS, such as obesity, are broadly considered vascular risk factors, indicating there may be an association between MS and neurovascular dysfunction. There is significant support for the notion that the BBB is disrupted in MS, as evidenced by the deposition of the blood protein fibrinogen in the cortex preceding demyelination (Petersen et al. 2017; Vos et al. 2005; Yates et al. 2017). Fibrinogen can cause the death of oligodendrocytes (Montagne et al. 2018), as well as an autoimmune response and demyelination (Ryu et al. 2015).

As described earlier, pericytes form a critical component of the BBB, as clearly evidenced by transgenic mice with reduced pericyte numbers exhibiting BBB opening and fibrinogen entry into the CNS (Armulik et al. 2010; Bell et al. 2010; Daneman et al. 2010). One recent study demonstrated that when pericyte coverage in the corpus callosum was reduced, this caused the opening of the BBB that was associated with increased deposition of fibrinogen in the white matter tracts (Montagne et al. 2018). This led to fewer oligodendrocytes, reduced axon number and thinner myelin around axons. Fibrinogen was likely the cause of these axonal and oligodendrocytic changes as reducing plasma fibrinogen through pharmacological or genetic means in these transgenic mice rescued pericyte coverage, BBB integrity, oligodendrocyte number and axonal health (Montagne et al. 2018). Pericyte reductions in transgenic rats can also lead to an increased density of OPCs but reduced numbers of oligodendrocytes during development, as pericytes can secrete laminin α2, an important extracellular matrix protein involved in the myelination of the CNS (de la Fuente et al. 2017). Furthermore, when a demyelinating insult is instigated with lysolecithin, pericyte numbers are reduced leading to a reduction in oligodendrogenesis (de la Fuente et al. 2017).

In summary, pericytes have an important function in preventing fibrinogen-induced oligodendrocyte toxicity as well as roles in the process of myelination and oligodendrogenesis, suggesting dysfunction to these roles may contribute to MS.

2.4.5 The Role of Pericytes in Other Neurological Diseases

The dysfunction of pericytes may be important in the pathophysiology of multiple other neurological disorders including traumatic brain injury (Choi et al. 2016), cerebral autosomal dominant arteriopathy with subcortical infarcts and leukoencephalopathy (CADASIL) (Ghosh et al. 2015), glioma (Sun et al. 2014), Huntington's disease (Padel et al. 2018) and epilepsy (Leal-Campanario et al. 2017). A recent review highlighted that pericytes may play a role in the substantial microvascular pathology that occurs in neurometabolic diseases such as homocystinuria, mitochondrial diseases, Fabry disease, and organic acidurias/acidemias (Isasi and

Olivera-Bravo 2020). Pericyte dysfunction in the context of other neurological diseases has been extensively reviewed elsewhere (Cheng et al. 2018; Liu et al. 2020).

2.4.6 *The Role of Pericyte Subtypes in Neurological Disease*

As outlined above, pericytes have functional, morphological and molecular heterogeneity. This makes the investigation into pericyte dysfunction in the pathophysioly of disease even more difficult. For example, Hall et al. showed that pericytes constrict capillaries and die in rigor following ischemic stroke (Hall et al. 2014), but it is likely that pericyte contraction following ischemic stroke (Yemisci et al. 2009) or hemorrhagic stroke (Li et al. 2016) only occurs in αSMA-expressing pericytes at the arteriolar end of the capillary bed. In the context of tumor vascularization, it has been demonstrated that only type 2 pericytes (nestin-positive) can invade the tumor within the brain to proliferate and promote angiogenesis, thereby encouraging tumor growth (Birbrair et al. 2014b). On the other hand, type 1 pericytes (nestin-negative) accumulate in lesioned tissue of the brain following an acute traumatic injury as well as contributing to scar formation in spinal cord injury (Birbrair et al. 2014a). Overall, there is diversity in the contribution of pericytes to the pathophysiology of multiple neurological diseases, and this may impact upon the characterization of pericytes in disease as well as the development of treatments that can specifically target subtypes of pericytes.

2.5 Are Pericytes in the Brain Different to Pericytes in the Periphery?

The spectrum of mural cells across the vascular tree in the brain is comparable throughout the vascular beds of different tissues, with VSMCs residing on the larger arteries/arterioles and pericytes residing on the capillaries. The functions of these cells are broadly similar in regard to the regulation of blood flow and the modulation of inflammation, but there may be tissue specific functions that differentiate them from one another (Attrill et al. 2020). For example, pericytes in the brain regulate the BBB, while in the periphery there is no NVU meaning that pericytes do not need to maintain a tight barrier. In fact, many tissue beds have fenestrated capillaries allowing a greater exchange of nutrients between the blood and tissue parenchyma.

Pericytes are located at specific branch points of the vascular tree to direct perfusion to the capillary bed that requires energy supply, a feature of pericytes that appears to be consistent across tissues (including the brain, heart, skeletal muscle and pancreas (Attrill et al. 2020)). However, recent single cell RNA-seq data of brain pericytes and lung pericytes revealed substantial transcriptomic differences (Vanlandewijck et al. 2018). Both brain and lung pericytes express canonical pericyte genes such as *Pdgfrb*, *Cspg4* and *Des*, but *Anpep* (the gene encoding CD13, another canonical pericyte marker) was absent in lung pericytes (Vanlandewijck et al. 2018). Further gene ontology analysis showed that an overexpression of transmembrane transporter activity was exhibited in brain pericytes compared to lung pericytes, reflective of their BBB properties (Vanlandewijck et al. 2018). Pericytes have different developmental origins in various tissue beds, which may account for the differences in gene expression and function described above (Dias Moura Prazeres et al. 2017; Yamamoto et al. 2017; Yamazaki and Mukouyama 2018). Interestingly, nestin expression can be used to classify pericytes into type 1 (nestin negative) and type 2 (nestin positive), and both types are present in peripheral and cerebral circulatory beds (Birbrair et al. 2014a; Birbrair et al. 2014b). Overall, there are similarities but also striking differences between pericytes in different tissue beds.

2.6 Future Opportunities in Brain Pericyte Research

As more research into these fascinating cells is conducted, the more opportunities arise to influence their function and potentially treat disease states. Outlined below are potential avenues in which pericytes could be used to help prevent or treat various neurological diseases.

2.6.1 Therapeutically Targeting Pericytes

Although there is growing evidence of dysfunctional pericytes in several neurological diseases, brain pericytes are difficult cells to target therapeutically. This is partly due to their heterogeneity along the vascular tree (see Fig. 2.1), but also due to the overlap with other cell types in the transmembrane receptors and molecular pathways they use. Furthermore, pericytes reside on the abluminal side of capillaries, meaning that access to pericytes can only occur if the therapeutic agent can cross the endothelial wall forming the BBB. Although there is a significant research effort currently being devoted to packaging small molecules into vehicles such as nanoparticles, exosomes and viral vectors to cross the BBB (Dong 2018), very few studies have attempted to target pericytes in this way.

Intranasal administration is one possible method to circumvent the restrictive properties of the BBB. One study has described the delivery of human glial-derived neurotrophic factor plasmid DNA nanoparticles intranasally and reported transfection, primarily in pericytes, using this approach (Aly et al. 2019). Nanoparticles are being formulated specifically to target pericytes using peptide sequences that can bind pericytes (Kang and Shin 2016). Successful targeting of anticancer agents to pericytes using peptide-conjugated nanoparticles has already been reported in rodent models of neuroblastoma (Loi et al. 2010) and lung melanoma (Guan et al.

2014). Unfortunately, these methods have not yet been developed yet for disorders of the brain, but may form an effective strategy for targeting brain-resident pericytes in the future. Nanoparticle-based therapy that can deliver vasodilators such as adenosine have been developed and have demonstrated efficacy at reducing lesion volume following ischemic stroke (Gaudin et al. 2014), showing the potential promise of this therapeutic strategy.

2.6.2 Pericyte Biomarkers of Neurological Disease

Biomarkers are important for the diagnostic, prognostic and therapeutic monitoring of disease, but discovering biomarkers with enough sensitivity and specificity for neurological diseases has proven difficult. Given the strong evidence of pericyte dysfunction in multiple neurological diseases, as well as their importance for BBB integrity (see above), assessing the status of pericytes through biomarkers could aid in the understanding of several diseases, as well as help to monitor disease prognosis.

PDGFRβ is one of the major cell surface receptors expressed by pericytes, and importantly, is more highly expressed in brain pericytes than brain VSMCs (Sagare et al. 2015). In response to hypoxia or Aβ, pericytes can shed sPDGFRβ, which is thought to be reflective of pericyte injury (Sagare et al. 2015). Furthermore, elevated sPDGFRβ has been detected in the CSF of mild cognitively impaired people, compared with individuals that have no cognitive impairment (Montagne et al. 2015), independent of $A\beta_{42}$ or pTau levels in CSF (classical indicators of AD progression (Nation et al. 2019)). Thus, these biomarker studies suggest pericyte injury and BBB breakdown could be detectable through the measurement of sPDGFRβ in the CSF. If this is ultimately the case, measurement of CSF sPDGFRβ may be useful for detecting BBB changes prior to the onset of several neurological diseases.

2.6.3 Therapeutic Administration of Pericytes

As outlined above, pericytes may have stem cell potential, i.e. an ability to differentiate into other cell types that may assist in the repair processes of the brain. Their functions in regulating angiogenesis, inflammation and clearance of toxic substances (Ma et al. 2018; Ribatti et al. 2011; Rustenhoven et al. 2017), along with their potential pluripotency make them ideal candidates as a therapeutic strategy in brain repair.

Promise has already been shown in a mouse of model of AD, where administration of fibroblast-derived pericytes increased blood flow and the clearance of Aβ to alleviate plaque burden (Tachibana et al. 2018). In the field of stroke, one published abstract illustrated that perivascular $CD146^{+}$ $NG2^{+}$ $PDGFR\beta^{+}$ stem cells could increase vascular density and promote behavioral recovery of rats following middle cerebral artery occlusion (Ogay et al. 2017). Another recent study administered pericytes intraperitoneally into mice with a mutation in *Sod1*, a model of ALS. This led to an extended survival of mutant mice compared to untreated or mesenchymal cell-treated mice (Coatti et al. 2017).

Excitingly, pericytes can now be derived from induced human pluripotent stem cells, which allows for the generation of autologous pericytes, bypassing the complications of immune rejection in cellular therapy (Faal et al. 2019; Stebbins et al. 2019). Another current barrier to pericyte therapy is the BBB. However, the transient opening of the BBB with agents such as mannitol can increase the number of autologous pericytes accessing the brain after an intra-arterial injection (Youn et al. 2015). If the delivery of pericytes to the brain can be overcome and pericytes are able to initiate their repair processes, this therapeutic strategy may be a very promising avenue for the treatment of several neurological diseases.

2.7 Conclusion

In recent years, our knowledge about pericytes and their functions in the brain has rapidly expanded, as evidenced by ~200 publications with the keywords 'pericytes' AND 'brain' currently being published per year (Brown et al. 2019). Their location in the heart of the NVU allows pericytes to influence both brain parenchymal function and vascular function. As technology has advanced, the critical roles of pericytes in vascular development and stability, cerebral blood flow regulation, BBB function and neuroinflammation have been revealed. The flip side of these integral roles is that pericyte dysfunction may be a critical component of multiple neurological diseases, although this provides the opportunity to target pericytes therapeutically. I suspect that as we develop a greater understanding of pericyte biology and the vascular underpinnings of neurological diseases, we will soon see exponential advances in our abilities to detect, monitor, visualize and specifically target pericytes in the context of brain health and disease.

Acknowledgements I would like to acknowledge the assistance of Dr. Jo-Maree Courtney in figure generation and Drs Jo-Maree Courtney and Gary Morris for constructive feedback on the chapter.

Ethical Approval This article does not contain any studies with human participants or animals performed by the author.

Disclosure of Interests The author declares he has no conflict of interest.

Funding BAS receives funding from the National Health and Medical Research Council (APP**1137776**, APP**1163384**) and the Rebecca L. Cooper Foundation.

References

Alarcon-Martinez L, Yilmaz-Ozcan S, Yemisci M, Schallek J, Kilic K, Can A, DI Polo A, Dalkara T (2018) Capillary pericytes express alpha-smooth muscle actin, which requires prevention of filamentous-actin depolymerization for detection. elife 7

Aly AE, Harmon B, Padegimas L, Sesenoglu-Laird O, Cooper MJ, Yurek DM, Waszczak BL (2019) Intranasal delivery of hGDNF plasmid DNA nanoparticles results in long-term and widespread transfection of perivascular cells in rat brain. Nanomedicine 16:20–33

Ames A 3rd, Wright RL, Kowada M, Thurston JM, Majno G (1968) Cerebral ischemia. II. The no-reflow phenomenon. Am J Pathol 52:437–453

Araque A, Parpura V, Sanzgiri RP, Haydon PG (1999) Tripartite synapses: glia, the unacknowledged partner. Trends Neurosci 22:208–215

Armulik A, Genove G, Mae M, Nisancioglu MH, Wallgard E, Niaudet C, He L, Norlin J, Lindblom P, Strittmatter K, Johansson BR, Betsholtz C (2010) Pericytes regulate the blood-brain barrier. Nature 468:557–561

Armulik A, Genove G, Betsholtz C (2011) Pericytes: developmental, physiological, and pathological perspectives, problems, and promises. Dev Cell 21:193–215

Attrill E, Ramsay C, Ross R, Richards S, Sutherland BA, Keske MA, Eringa E, Premilovac D (2020) Metabolic-vascular coupling in skeletal muscle: A potential role for capillary pericytes? Clin Exp Pharmacol Physiol 47:520–528

Attwell D, Iadecola C (2002) The neural basis of functional brain imaging signals. Trends Neurosci 25:621–625

Attwell D, Laughlin SB (2001) An energy budget for signaling in the grey matter of the brain. J Cereb Blood Flow Metab 21:1133–1145

Attwell D, Buchan AM, Charpak S, Lauritzen M, Macvicar BA, Newman EA (2010) Glial and neuronal control of brain blood flow. Nature 468:232–243

Attwell D, Mishra A, Hall CN, O'farrell FM, Dalkara T (2016) What is a pericyte? J Cereb Blood Flow Metab 36:451–455

Austin BP, Nair VA, Meier TB, Xu G, Rowley HA, Carlsson CM, Johnson SC, Prabhakaran V (2011) Effects of hypoperfusion in Alzheimer's disease. J Alzheimers Dis 26(Suppl 3):123–133

Balbi M, Koide M, Wellman GC, Plesnila N (2017) Inversion of neurovascular coupling after subarachnoid hemorrhage in vivo. J Cereb Blood Flow Metab 37:3625–3634

Bell RD, Winkler EA, Sagare AP, Singh I, Larue B, Deane R, Zlokovic BV (2010) Pericytes control key neurovascular functions and neuronal phenotype in the adult brain and during brain aging. Neuron 68:409–427

Bell RD, Winkler EA, Singh I, Sagare AP, Deane R, Wu Z, Holtzman DM, Betsholtz C, Armulik A, Sallstrom J, Berk BC, Zlokovic BV (2012) Apolipoprotein E controls cerebrovascular integrity via cyclophilin A. Nature 485:512–516

Bergers G, Song S (2005) The role of pericytes in blood-vessel formation and maintenance. Neuro-Oncology 7:452–464

Berthiaume AA, Grant RI, Mcdowell KP, Underly RG, Hartmann DA, Levy M, Bhat NR, Shih AY (2018a) Dynamic remodeling of pericytes in vivo maintains capillary coverage in the adult mouse brain. Cell Rep 22:8–16

Berthiaume AA, Hartmann DA, Majesky MW, Bhat NR, Shih AY (2018b) Pericyte structural remodeling in cerebrovascular health and homeostasis. Front Aging Neurosci 10:210

Biesecker KR, Srienc AI, Shimoda AM, Agarwal A, Bergles DE, Kofuji P, Newman EA (2016) Glial cell calcium signaling mediates capillary regulation of blood flow in the retina. J Neurosci 36:9435–9445

Birbrair A, Zhang T, Wang ZM, Messi ML, Enikolopov GN, Mintz A, Delbono O (2013) Role of pericytes in skeletal muscle regeneration and fat accumulation. Stem Cells Dev 22:2298–2314

Birbrair A, Zhang T, Files DC, Mannava S, Smith T, Wang ZM, Messi ML, Mintz A, Delbono O (2014a) Type-1 pericytes accumulate after tissue injury and produce collagen in an organ-dependent manner. Stem Cell Res Ther 5:122

Birbrair A, Zhang T, Wang ZM, Messi ML, Olson JD, Mintz A, Delbono O (2014b) Type-2 pericytes participate in normal and tumoral angiogenesis. Am J Phys Cell Physiol 307:C25–C38

Bjarnegard M, Enge M, Norlin J, Gustafsdottir S, Fredriksson S, Abramsson A, Takemoto M, Gustafsson E, Fassler R, Betsholtz C (2004) Endothelium-specific ablation of PDGFB leads to pericyte loss and glomerular, cardiac and placental abnormalities. Development 131:1847–1857

Blinder P, Tsai PS, Kaufhold JP, Knutsen PM, Suhl H, Kleinfeld D (2013) The cortical angiome: an interconnected vascular network with noncolumnar patterns of blood flow. Nat Neurosci 16:889–897

Bracko O, Njiru BN, Swallow M, Ali M, Haft-Javaherian M, Schaffer CB (2019) Increasing cerebral blood flow improves cognition into late stages in Alzheimer's disease mice. J Cereb Blood Flow Metab:271678X19873658

Brown LS, Foster CG, Courtney JM, King NE, Howells DW, Sutherland BA (2019) Pericytes and neurovascular function in the healthy and diseased brain. Front Cell Neurosci 13:282

Cai C, Fordsmann JC, Jensen SH, Gesslein B, Lonstrup M, Hald BO, Zambach SA, Brodin B, Lauritzen MJ (2018) Stimulation-induced increases in cerebral blood flow and local capillary vasoconstriction depend on conducted vascular responses. Proc Natl Acad Sci U S A 115:E5796–E5804

Cheng J, Korte N, Nortley R, Sethi H, Tang Y, Attwell D (2018) Targeting pericytes for therapeutic approaches to neurological disorders. Acta Neuropathol 136:507–523

Choi YK, Maki T, Mandeville ET, Koh SH, Hayakawa K, Arai K, Kim YM, Whalen MJ, Xing C, Wang X, Kim KW, Lo, E. H. (2016) Dual effects of carbon monoxide on pericytes and neurogenesis in traumatic brain injury. Nat Med 22:1335–1341

Coatti GC, Frangini M, Valadares MC, Gomes JP, Lima NO, Cavacana N, Assoni AF, Pelatti MV, Birbrair A, de Lima ACP, Singer JM, Rocha FMM, da Silva GL, Mantovani MS, Macedo-Souza LI, Ferrari MFR, Zatz M (2017) Pericytes extend survival of ALS SOD1 mice and induce the expression of antioxidant enzymes in the murine model and in IPSCs derived neuronal cells from an ALS patient. Stem Cell Rev Rep 13:686–698

Corder EH, Saunders AM, Strittmatter WJ, Schmechel DE, Gaskell PC, Small GW, Roses AD, Haines JL, Pericak-Vance MA (1993) Gene dose of apolipoprotein E type 4 allele and the risk of Alzheimer's disease in late onset families. Science 261:921–923

Courtney JM, Sutherland BA (2020) Harnessing the stem cell properties of pericytes to repair the brain. Neural Regen Res 15:1021–1022

Crisan M, Yap S, Casteilla L, Chen CW, Corselli M, Park TS, Andriolo G, Sun B, Zheng B, Zhang L, Norotte C, Teng PN, Traas J, Schugar R, Deasy BM, Badylak S, Buhring HJ, Giacobino JP, Lazzari L, Huard J, Peault B (2008) A perivascular origin for mesenchymal stem cells in multiple human organs. Cell Stem Cell 3:301–313

Cruz Hernandez JC, Bracko O, Kersbergen CJ, Muse V, Haft-Javaherian M, Berg M, Park L, Vinarcsik LK, Ivasyk I, Rivera DA, Kang Y, Cortes-Canteli M, Peyrounette M, Doyeux V, Smith A, Zhou J, Otte G, Beverly JD, Davenport E, Davit Y, Lin CP, Strickland S, Iadecola C, Lorthois S, Nishimura N, Schaffer CB (2019) Neutrophil adhesion in brain capillaries reduces cortical blood flow and impairs memory function in Alzheimer's disease mouse models. Nat Neurosci 22:413–420

Dalkara T, Alarcon-Martinez L, Yemisci M (2019) Pericytes in Ischemic Stroke. Adv Exp Med Biol 1147:189–213

Damisah EC, Hill RA, Tong L, Murray KN, Grutzendler J (2017) A fluoro-Nissl dye identifies pericytes as distinct vascular mural cells during in vivo brain imaging. Nat Neurosci 20:1023–1032

Daneman R, Zhou L, Kebede AA, Barres BA (2010) Pericytes are required for blood-brain barrier integrity during embryogenesis. Nature 468:562–566

Davidoff MS (2019) The Pluripotent Microvascular Pericytes Are the Adult Stem Cells Even in the Testis. Adv Exp Med Biol 1122:235–267

de la Fuente AG, Lange S, Silva ME, Gonzalez GA, Tempfer H, van Wijngaarden P, Zhao C, di Canio L, Trost A, Bieler L, Zaunmair P, Rotheneichner P, O'sullivan A, Couillard-Despres S, Errea O, Mae MA, Andrae J, He L, Keller A, Batiz LF, Betsholtz C, Aigner L, Franklin RJM,

Rivera FJ (2017) Pericytes Stimulate Oligodendrocyte Progenitor Cell Differentiation during CNS Remyelination. Cell Rep 20:1755–1764

de la Torre JC (2012) Cardiovascular risk factors promote brain hypoperfusion leading to cognitive decline and dementia. Cardiovasc Psychiatry Neurol 2012:367516

Dehouck MP, Vigne P, Torpier G, Breittmayer JP, Cecchelli R, Frelin C (1997) Endothelin-1 as a mediator of endothelial cell-pericyte interactions in bovine brain capillaries. J Cereb Blood Flow Metab 17:464–469

Dias Moura Prazeres PH, Sena IFG, Borges IDT, De Azevedo PO, Andreotti JP, De Paiva AE, De Almeida VM, De Paula Guerra DA, Pinheiro Dos Santos GS, Mintz A, Delbono O, Birbrair A (2017) Pericytes are heterogeneous in their origin within the same tissue. Dev Biol 427:6–11

Dong X (2018) Current Strategies for Brain Drug Delivery. Theranostics 8:1481–1493

Duan L, Zhang XD, Miao WY, Sun YJ, Xiong G, Wu Q, Li G, Yang P, Yu H, Li H, Wang Y, Zhang M, Hu LY, Tong X, Zhou WH, Yu X (2018) PDGFRbeta Cells Rapidly Relay Inflammatory Signal from the Circulatory System to Neurons via Chemokine CCL2. Neuron 100:183-200 e8

Dubois B, Feldman HH, Jacova C, Hampel H, Molinuevo JL, Blennow K, Dekosky ST, Gauthier S, Selkoe D, Bateman R, Cappa S, Crutch S, Engelborghs S, Frisoni GB, Fox NC, Galasko D, Habert MO, Jicha GA, Nordberg A, Pasquier F, Rabinovici G, Robert P, Rowe C, Salloway S, Sarazin M, Epelbaum S, DE Souza LC, Vellas B, Visser PJ, Schneider L, Stern Y, Scheltens P, Cummings JL (2014) Advancing research diagnostic criteria for Alzheimer's disease: the IWG-2 criteria. Lancet Neurol 13:614–629

Enge M, Bjarnegard M, Gerhardt H, Gustafsson E, Kalen M, Asker N, Hammes HP, Shani M, Fassler R, Betsholtz C (2002) Endothelium-specific platelet-derived growth factor-B ablation mimics diabetic retinopathy. EMBO J 21:4307–4316

Erdener SE, Dalkara T (2019) Small vessels are a big problem in neurodegeneration and neuroprotection. Front Neurol 10:889

Faal T, Phan DTT, Davtyan H, Scarfone VM, Varady E, Blurton-Jones M, Hughes CCW, Inlay MA (2019) Induction of mesoderm and neural crest-derived pericytes from human pluripotent stem cells to study blood-brain barrier interactions. Stem Cell Reports 12:451–460

Farahani RM, Rezaei-Lotfi S, Simonian M, Xaymardan M, Hunter N (2019) Neural microvascular pericytes contribute to human adult neurogenesis. J Comp Neurol 527:780–796

Fernandez-Klett F, Offenhauser N, Dirnagl U, Priller J, Lindauer U (2010) Pericytes in capillaries are contractile in vivo, but arterioles mediate functional hyperemia in the mouse brain. Proc Natl Acad Sci U S A 107:22290–22295

Franco M, Roswall P, Cortez E, Hanahan D, Pietras K (2011) Pericytes promote endothelial cell survival through induction of autocrine VEGF-A signaling and Bcl-w expression. Blood 118:2906–2917

Fu Y, Chang A, Chang L, Niessen K, Eapen S, Setiadi A, Karsan A (2009) Differential regulation of transforming growth factor beta signaling pathways by Notch in human endothelial cells. J Biol Chem 284:19452–19462

Gaengel K, Genove G, Armulik A, Betsholtz C (2009) Endothelial-mural cell signaling in vascular development and angiogenesis. Arterioscler Thromb Vasc Biol 29:630–638

Gaudin A, Yemisci M, Eroglu H, Lepetre-Mouelhi S, Turkoglu OF, Donmez-Demir B, Caban S, Sargon MF, Garcia-Argote S, Pieters G, Loreau O, Rousseau B, Tagit O, Hildebrandt N, LE Dantec Y, Mougin J, Valetti S, Chacun H, Nicolas V, Desmaele D, Andrieux K, Capan Y, Dalkara T, Couvreur P (2014) Squalenoyl adenosine nanoparticles provide neuroprotection after stroke and spinal cord injury. Nat Nanotechnol 9:1054–1062

Gautam J, Yao Y (2018) Roles of pericytes in stroke pathogenesis. Cell Transplant 27:1798–1808

Geevarghese A, Herman IM (2014) Pericyte-endothelial crosstalk: implications and opportunities for advanced cellular therapies. Transl Res 163:296–306

Ghosh M, Balbi M, Hellal F, Dichgans M, Lindauer U, Plesnila N (2015) Pericytes are involved in the pathogenesis of cerebral autosomal dominant arteriopathy with subcortical infarcts and leukoencephalopathy. Ann Neurol 78:887–900

Grant RI, Hartmann DA, Underly RG, Berthiaume AA, Bhat NR, Shih AY (2019) Organizational hierarchy and structural diversity of microvascular pericytes in adult mouse cortex. J Cereb Blood Flow Metab 39:411–425
Grubb S, Cai C, Hald BO, Khennouf L, Murmu RP, Jensen AGK, Fordsmann J, Zambach S, Lauritzen M (2020) Precapillary sphincters maintain perfusion in the cerebral cortex. Nat Commun 11:395
Guan YY, Luan X, Xu JR, Liu YR, Lu Q, Wang C, Liu HJ, Gao YG, Chen HZ, Fang C (2014) Selective eradication of tumor vascular pericytes by peptide-conjugated nanoparticles for anti-angiogenic therapy of melanoma lung metastasis. Biomaterials 35:3060–3070
Hall CN, Reynell C, Gesslein B, Hamilton NB, Mishra A, Sutherland BA, O'farrell FM, Buchan AM, Lauritzen M, Attwell D (2014) Capillary pericytes regulate cerebral blood flow in health and disease. Nature 508:55–60
Halliday MR, Rege SV, Ma Q, Zhao Z, Miller CA, Winkler EA, Zlokovic BV (2016) Accelerated pericyte degeneration and blood-brain barrier breakdown in apolipoprotein E4 carriers with Alzheimer's disease. J Cereb Blood Flow Metab 36:216–227
Hamilton NB, Attwell D, Hall CN (2010) Pericyte-mediated regulation of capillary diameter: A component of neurovascular coupling in health and disease. Front Neuroenerg 2:5
Hartmann DA, Underly RG, Grant RI, Watson AN, Lindner V, Shih AY (2015) Pericyte structure and distribution in the cerebral cortex revealed by high-resolution imaging of transgenic mice. Neurophotonics 2:041402
He L, Vanlandewijck M, Mae MA, Andrae J, Ando K, del Gaudio F, Nahar K, Lebouvier T, Lavina B, Gouveia L, Sun Y, Raschperger E, Segerstolpe A, Liu J, Gustafsson S, Rasanen M, Zarb Y, Mochizuki N, Keller A, Lendahl U, Betsholtz C (2018) Single-cell RNA sequencing of mouse brain and lung vascular and vessel-associated cell types. Sci Data 5:180160
Hellstrom M, Phng LK, Hofmann JJ, Wallgard E, Coultas L, Lindblom P, Alva J, Nilsson AK, Karlsson L, Gaiano N, Yoon K, Rossant J, Iruela-Arispe ML, Kalen M, Gerhardt H, Betsholtz C (2007) Dll4 signalling through Notch1 regulates formation of tip cells during angiogenesis. Nature 445:776–780
Henshall TL, Keller A, He L, Johansson BR, Wallgard E, Raschperger E, Mae MA, Jin S, Betsholtz C, Lendahl U (2015) Notch3 is necessary for blood vessel integrity in the central nervous system. Arterioscler Thromb Vasc Biol 35:409–420
Herland A, VAN DER Meer AD, Fitzgerald EA, Park TE, Sleeboom JJ, Ingber DE (2016) Distinct Contributions of Astrocytes and Pericytes to Neuroinflammation Identified in a 3D Human Blood-Brain Barrier on a Chip. PLoS One 11:e0150360
Hill RA, Tong L, Yuan P, Murikinati S, Gupta S, Grutzendler J (2015) Regional blood flow in the normal and ischemic brain is controlled by arteriolar smooth muscle cell contractility and not by capillary pericytes. Neuron 87:95–110
Howarth C, Sutherland BA, Choi HB, Martin C, Lind BL, Khennouf L, Ledue JM, Pakan JM, Ko RW, Ellis-Davies GC, Lauritzen MJ, Sibson NR, Buchan AM, Macvicar BA (2017) A critical role for astrocytes in hypercapnic vasodilation in brain. J Neurosci 37:2403–2414
Hu E, Hu W, Yang A, Zhou H, Zhou J, Luo J, Wang Y, Tang T, Cui H (2019) Thrombin promotes pericyte coverage by Tie2 activation in a rat model of intracerebral hemorrhage. Brain Res 1708:58–68
Isasi E, Olivera-Bravo S (2020) Pericytes in neurometabolic diseases. Curr Tissue Microenviron Rep
Iturria-Medina Y, Sotero RC, Toussaint PJ, Mateos-Perez JM, Evans AC, Alzheimer's Disease Neuroimaging, I (2016) Early role of vascular dysregulation on late-onset Alzheimer's disease based on multifactorial data-driven analysis. Nat Commun 7:11934
Jansson D, Rustenhoven J, Feng S, Hurley D, Oldfield RL, Bergin PS, Mee EW, Faull RL, Dragunow M (2014) A role for human brain pericytes in neuroinflammation. J Neuroinflammation 11:104
Jeong JH, Sugii Y, Minamiyama M, Okamoto K (2006) Measurement of RBC deformation and velocity in capillaries in vivo. Microvasc Res 71:212–217
Kang E, Shin JW (2016) Pericyte-targeting drug delivery and tissue engineering. Int J Nanomedicine 11:2397–2406

Kawamura H, Sugiyama T, Wu DM, Kobayashi M, Yamanishi S, Katsumura K, Puro DG (2003) ATP: a vasoactive signal in the pericyte-containing microvasculature of the rat retina. J Physiol 551:787–799
Khennouf L, Gesslein B, Brazhe A, Octeau JC, Kutuzov N, Khakh BS, Lauritzen M (2018) Active role of capillary pericytes during stimulation-induced activity and spreading depolarization. Brain 141:2032–2046
Kim DS, Ronen I, Olman C, Kim SG, Ugurbil K, Toth LJ (2004) Spatial relationship between neuronal activity and BOLD functional MRI. NeuroImage 21:876–885
Kisler K, Nelson AR, Rege SV, Ramanathan A, Wang Y, Ahuja A, Lazic D, Tsai PS, Zhao Z, Zhou Y, Boas DA, Sakadzic S, Zlokovic BV (2017) Pericyte degeneration leads to neurovascular uncoupling and limits oxygen supply to brain. Nat Neurosci 20:406–416
Kisler K, Nikolakopoulou AM, Sweeney MD, Lazic D, Zhao Z, Zlokovic BV (2020) Acute Ablation of Cortical Pericytes Leads to Rapid Neurovascular Uncoupling. Front Cell Neurosci 14:27
Kloner RA, King KS, Harrington MG (2018) No-reflow phenomenon in the heart and brain. Am J Physiol Heart Circ Physiol 315:H550–H562
Kornfield TE, Newman EA (2014) Regulation of blood flow in the retinal trilaminar vascular network. J Neurosci 34:11504–11513
Kovac A, Erickson MA, Banks WA (2011) Brain microvascular pericytes are immunoactive in culture: cytokine, chemokine, nitric oxide, and LRP-1 expression in response to lipopolysaccharide. J Neuroinflammation 8:139
Kristensson K, Olsson Y (1973) Accumulation of protein tracers in pericytes of the central nervous system following systemic injection in immature mice. Acta Neurol Scand 49:189–194
Lauretti E, Li JG, di Meco A, Pratico D (2017) Glucose deficit triggers tau pathology and synaptic dysfunction in a tauopathy mouse model. Transl Psychiatry 7:e1020
Leal-Campanario R, Alarcon-Martinez L, Rieiro H, Martinez-Conde S, Alarcon-Martinez T, Zhao X, Lamee J, Popp PJ, Calhoun ME, Arribas JI, Schlegel AA, Stasi LL, Rho JM, Inge L, Otero-Millan J, Treiman DM, Macknik SL (2017) Abnormal Capillary Vasodynamics Contribute to Ictal Neurodegeneration in Epilepsy. Sci Rep 7:43276
Leveen P, Pekny M, Gebre-Medhin S, Swolin B, Larsson E, Betsholtz C (1994) Mice deficient for PDGF B show renal, cardiovascular, and hematological abnormalities. Genes Dev 8:1875–1887
Li Q, Puro DG (2001) Adenosine activates ATP-sensitive K(+) currents in pericytes of rat retinal microvessels: role of A1 and A2a receptors. Brain Res 907:93–99
Li F, Lan Y, Wang Y, Wang J, Yang G, Meng F, Han H, Meng A, Wang Y, Yang X (2011) Endothelial Smad4 maintains cerebrovascular integrity by activating N-cadherin through cooperation with Notch. Dev Cell 20:291–302
Li Q, Chen Y, Li B, Luo C, Zuo S, Liu X, Zhang JH, Ruan H, Feng H (2016) Hemoglobin induced NO/cGMP suppression Deteriorate Microcirculation via Pericyte Phenotype Transformation after Subarachnoid Hemorrhage in Rats. Sci Rep 6:22070
Liu Q, Yang Y, Fan X (2020) Microvascular pericytes in brain-associated vascular disease. Biomed Pharmacother 121:109633
Loi M, Marchio S, Becherini P, di Paolo D, Soster M, Curnis F, Brignole C, Pagnan G, Perri P, Caffa I, Longhi R, Nico B, Bussolino F, Gambini C, Ribatti D, Cilli M, Arap W, Pasqualini R, Allen TM, Corti A, Ponzoni M, Pastorino F (2010) Combined targeting of perivascular and endothelial tumor cells enhances anti-tumor efficacy of liposomal chemotherapy in neuroblastoma. J Control Release 145:66–73
Lovick TA, Brown LA, Key BJ (1999) Neurovascular relationships in hippocampal slices: physiological and anatomical studies of mechanisms underlying flow-metabolism coupling in intraparenchymal microvessels. Neuroscience 92:47–60
Luissint AC, Artus C, Glacial F, Ganeshamoorthy K, Couraud PO (2012) Tight junctions at the blood brain barrier: physiological architecture and disease-associated dysregulation. Fluids Barriers CNS 9:23
Ma Q, Zhao Z, Sagare AP, Wu Y, Wang M, Owens NC, Verghese PB, Herz J, Holtzman DM, Zlokovic BV (2018) Blood-brain barrier-associated pericytes internalize and clear aggregated

amyloid-beta42 by LRP1-dependent apolipoprotein E isoform-specific mechanism. Mol Neurodegener 13:57

Macvicar BA, Newman EA (2015) Astrocyte regulation of blood flow in the brain. Cold Spring Harb Perspect Biol 7

Maier CL, Pober JS (2011) Human placental pericytes poorly stimulate and actively regulate allogeneic CD4 T cell responses. Arterioscler Thromb Vasc Biol 31:183–189

Mazare N, Gilbert A, Boulay AC, Rouach N, Cohen-Salmon M (2018) Connexin 30 is expressed in a subtype of mouse brain pericytes. Brain Struct Funct 223:1017–1024

Mcconnell HL, Kersch CN, Woltjer RL, Neuwelt EA (2017) The translational significance of the neurovascular unit. J Biol Chem 292:762–770

Mishra A, Reynolds JP, Chen Y, Gourine AV, Rusakov DA, Attwell D (2016) Astrocytes mediate neurovascular signaling to capillary pericytes but not to arterioles. Nat Neurosci 19:1619–1627

Miyazaki K, Ohta Y, Nagai M, Morimoto N, Kurata T, Takehisa Y, Ikeda Y, Matsuura T, Abe K (2011) Disruption of neurovascular unit prior to motor neuron degeneration in amyotrophic lateral sclerosis. J Neurosci Res 89:718–728

Montagne A, Barnes SR, Sweeney MD, Halliday MR, Sagare AP, Zhao Z, Toga AW, Jacobs RE, Liu CY, Amezcua L, Harrington MG, Chui HC, Law M, Zlokovic BV (2015) Blood-brain barrier breakdown in the aging human hippocampus. Neuron 85:296–302

Montagne A, Nikolakopoulou AM, Zhao Z, Sagare AP, Si G, Lazic D, Barnes SR, Daianu M, Ramanathan A, Go A, Lawson EJ, Wang Y, Mack WJ, Thompson PM, Schneider JA, Varkey J, Langen R, Mullins E, Jacobs RE, Zlokovic BV (2018) Pericyte degeneration causes white matter dysfunction in the mouse central nervous system. Nat Med 24:326–337

Montine TJ, Phelps CH, Beach TG, Bigio EH, Cairns NJ, Dickson DW, Duyckaerts C, Frosch MP, Masliah E, Mirra SS, Nelson PT, Schneider JA, Thal DR, Trojanowski JQ, Vinters HV, Hyman BT, National Institute ON, A. & Alzheimer's, A (2012) National Institute on Aging-Alzheimer's Association guidelines for the neuropathologic assessment of Alzheimer's disease: a practical approach. Acta Neuropathol 123:1–11

Nakagomi T, Kubo S, Nakano-Doi A, Sakuma R, Lu S, Narita A, Kawahara M, Taguchi A, Matsuyama T (2015) Brain vascular pericytes following ischemia have multipotential stem cell activity to differentiate into neural and vascular lineage cells. Stem Cells 33:1962–1974

Nation DA, Sweeney MD, Montagne A, Sagare AP, D'orazio LM, Pachicano M, Sepehrband F, Nelson AR, Buennagel DP, Harrington MG, Benzinger TLS, Fagan AM, Ringman JM, Schneider LS, Morris JC, Chui HC, Law M, Toga AW, Zlokovic BV (2019) Blood-brain barrier breakdown is an early biomarker of human cognitive dysfunction. Nat Med 25:270–276

Nehls V, Denzer K, Drenckhahn D (1992) Pericyte involvement in capillary sprouting during angiogenesis in situ. Cell Tissue Res 270:469–474

Neuhaus AA, Couch Y, Sutherland BA, Buchan AM (2017a) Novel method to study pericyte contractility and responses to ischaemia in vitro using electrical impedance. J Cereb Blood Flow Metab 37:2013–2024

Neuhaus AA, Sutherland BA, Buchan AM (2017b) Targeting pericytes and the microcirculation for ischemic stroke therapy. In: Lapchak PA, Zhang J (eds) Springer series in translational stroke research: neuroprotective therapy for stroke and ischemic disease. Springer

Nikolakopoulou AM, Montagne A, Kisler K, Dai Z, Wang Y, Huuskonen MT, Sagare AP, Lazic D, Sweeney MD, Kong P, Wang M, Owens NC, Lawson EJ, Xie X, Zhao Z, Zlokovic BV (2019) Pericyte loss leads to circulatory failure and pleiotrophin depletion causing neuron loss. Nat Neurosci 22:1089–1098

Nishimura A, Ago T, Kuroda J, Arimura K, Tachibana M, Nakamura K, Wakisaka Y, Sadoshima J, Iihara K, Kitazono T (2016) Detrimental role of pericyte Nox4 in the acute phase of brain ischemia. J Cereb Blood Flow Metab 36:1143–1154

Nortley R, Korte N, Izquierdo P, Hirunpattarasilp C, Mishra A, Jaunmuktane Z, Kyrargyri V, Pfeiffer T, Khennouf L, Madry C, Gong H, Richard-Loendt A, Huang W, Saito T, Saido TC, Brandner S, Sethi H, Attwell D (2019) Amyloid beta oligomers constrict human capillaries in Alzheimer's disease via signaling to pericytes. Science 365

Norton S, Matthews FE, Barnes DE, Yaffe K, Brayne C (2014) Potential for primary prevention of Alzheimer's disease: an analysis of population-based data. Lancet Neurol 13:788–794
Ogay V, Kumasheva V, Baidosova S, Li Y, Shpekov A, Makhambetov Y, Kaliyev A, Zhetpisbayev B, Olzhayev F, Ramankulov Y (2017) Adipose-derived perivascular stem cells promote sensorimotor recovery after ischemic stroke in rats. J Biotechnol 256:S116
Ostergaard L, Aamand R, Karabegovic S, Tietze A, Blicher JU, Mikkelsen IK, Iversen NK, Secher N, Engedal TS, Anzabi M, Jimenez EG, Cai C, Koch KU, Naess-Schmidt ET, Obel A, Juul N, Rasmussen M, Sorensen JC (2013) The role of the microcirculation in delayed cerebral ischemia and chronic degenerative changes after subarachnoid hemorrhage. J Cereb Blood Flow Metab 33:1825–1837
Ozen I, Deierborg T, Miharada K, Padel T, Englund E, Genove G, Paul G (2014) Brain pericytes acquire a microglial phenotype after stroke. Acta Neuropathol
Padel T, Roth M, Gaceb A, Li JY, Bjorkqvist M, Paul G (2018) Brain pericyte activation occurs early in Huntington's disease. Exp Neurol 305:139–150
Pan P, Zhao H, Zhang X, Li Q, Qu J, Zuo S, Yang F, Liang G, Zhang JH, Liu X, He H, Feng H, Chen Y (2020) Cyclophilin a signaling induces pericyte-associated blood-brain barrier disruption after subarachnoid hemorrhage. J Neuroinflammation 17:16
Park DY, Lee J, Kim J, Kim K, Hong S, Han S, Kubota Y, Augustin HG, Ding L, Kim JW, Kim H, He Y, Adams RH, Koh GY (2017) Plastic roles of pericytes in the blood-retinal barrier. Nat Commun 8:15296
Payne LB, Hoque M, Houk C, Darden J, Chappell JC (2020) Pericytes in vascular development. Curr Tissue Microenviron Rep 1:143–154
Peppiatt CM, Howarth C, Mobbs P, Attwell D (2006) Bidirectional control of CNS capillary diameter by pericytes. Nature 443:700–704
Petersen MA, Ryu JK, Chang KJ, Etxeberria A, Bardehle S, Mendiola AS, Kamau-Devers W, Fancy SPJ, Thor A, Bushong EA, Baeza-Raja B, Syme CA, Wu MD, Rios Coronado PE, Meyer-Franke A, Yahn S, Pous L, Lee JK, Schachtrup C, Lassmann H, Huang EJ, Han MH, Absinta M, Reich DS, Ellisman MH, Rowitch DH, Chan JR, Akassoglou K (2017) Fibrinogen activates BMP signaling in oligodendrocyte progenitor cells and inhibits remyelination after vascular damage. Neuron 96(1003-1012):e7
Pieper C, Marek JJ, Unterberg M, Schwerdtle T, Galla HJ (2014) Brain capillary pericytes contribute to the immune defense in response to cytokines or LPS in vitro. Brain Res 1550:1–8
Proebstl D, Voisin MB, Woodfin A, Whiteford J, D'acquisto F, Jones GE, Rowe D, Nourshargh S (2012) Pericytes support neutrophil subendothelial cell crawling and breaching of venular walls in vivo. J Exp Med 209:1219–1234
Ribatti D, Nico B, Crivellato E (2011) The role of pericytes in angiogenesis. Int J Dev Biol 55:261–268
Ruan J, Luo M, Wang C, Fan L, Yang SN, Cardenas M, Geng H, Leonard JP, Melnick A, Cerchietti L, Hajjar KA (2013) Imatinib disrupts lymphoma angiogenesis by targeting vascular pericytes. Blood 121:5192–5202
Rucker HK, Wynder HJ, Thomas WE (2000) Cellular mechanisms of CNS pericytes. Brain Res Bull 51:363–369
Rungta RL, Chaigneau E, Osmanski BF, Charpak S (2018) Vascular compartmentalization of functional hyperemia from the synapse to the pia. Neuron 99(362-375):e4
Rustenhoven J, Jansson D, Smyth LC, Dragunow M (2017) Brain Pericytes As Mediators of Neuroinflammation. Trends Pharmacol Sci 38:291–304
Ryu JK, Petersen MA, Murray SG, Baeten KM, Meyer-Franke A, Chan JP, Vagena E, Bedard C, Machado MR, Rios Coronado PE, Prod'homme T, Charo IF, Lassmann H, Degen JL, Zamvil SS, Akassoglou K (2015) Blood coagulation protein fibrinogen promotes autoimmunity and demyelination via chemokine release and antigen presentation. Nat Commun 6:8164
Sagare AP, Bell RD, Zhao ZE (2013) Pericyte loss influences Alzheimer-like neurodegeneration in mice. Nat Commun 4:2932

Sagare AP, Sweeney MD, Makshanoff J, Zlokovic BV (2015) Shedding of soluble platelet-derived growth factor receptor-beta from human brain pericytes. Neurosci Lett 607:97–101

Schultz N, Byman E, Fex M, Wennstrom M (2017) Amylin alters human brain pericyte viability and NG2 expression. J Cereb Blood Flow Metab 37:1470–1482

Schultz N, Brannstrom K, Byman E, Moussaud S, Nielsen HM, Netherlands Brain B, Olofsson A, Wennstrom M (2018) Amyloid-beta 1-40 is associated with alterations in NG2+ pericyte population ex vivo and in vitro. Aging Cell 17:e12728

Sengillo JD, Winkler EA, Walker CT, Sullivan JS, Johnson M, Zlokovic BV (2013) Deficiency in mural vascular cells coincides with blood-brain barrier disruption in Alzheimer's disease. Brain Pathol 23:303–310

Sims DE (2000) Diversity within pericytes. Clin Exp Pharmacol Physiol 27:842–846

Soriano P (1994) Abnormal kidney development and hematological disorders in PDGF beta-receptor mutant mice. Genes Dev 8:1888–1896

Stark K, Eckart A, Haidari S, Tirniceriu A, Lorenz M, von Bruhl ML, Gartner F, Khandoga AG, Legate KR, Pless R, Hepper I, Lauber K, Walzog B, Massberg S (2013) Capillary and arteriolar pericytes attract innate leukocytes exiting through venules and 'instruct' them with pattern-recognition and motility programs. Nat Immunol 14:41–51

Stebbins MJ, Gastfriend BD, Canfield SG, Lee MS, Richards D, Faubion MG, Li WJ, Daneman R, Palecek SP, Shusta EV (2019) Human pluripotent stem cell-derived brain pericyte-like cells induce blood-brain barrier properties. Sci Adv 5:eaau7375

Streit WJ, Mrak RE, Griffin WS (2004) Microglia and neuroinflammation: a pathological perspective. J Neuroinflammation 1:14

Sun X, He G, Qing H, Zhou W, Dobie F, Cai F, Staufenbiel M, Huang LE, Song W (2006) Hypoxia facilitates Alzheimer's disease pathogenesis by up-regulating BACE1 gene expression. Proc Natl Acad Sci U S A 103:18727–18732

Sun H, Guo D, Su Y, Yu D, Wang Q, Wang T, Zhou Q, Ran X, Zou Z (2014) Hyperplasia of pericytes is one of the main characteristics of microvascular architecture in malignant glioma. PLoS One 9:e114246

Sundberg C, Kowanetz M, Brown LF, Detmar M, Dvorak HF (2002) Stable expression of angiopoietin-1 and other markers by cultured pericytes: phenotypic similarities to a subpopulation of cells in maturing vessels during later stages of angiogenesis in vivo. Lab Investig 82:387–401

Sutherland BA, Fordsmann JC, Martin C, Neuhaus AA, Witgen BM, Piilgaard H, Lonstrup M, Couch Y, Sibson NR, Lauritzen M, Buchan AM (2017) Multi-modal assessment of neurovascular coupling during cerebral ischaemia and reperfusion using remote middle cerebral artery occlusion. J Cereb Blood Flow Metab 37:2494–2508

Sweeney MD, Ayyadurai S, Zlokovic BV (2016) Pericytes of the neurovascular unit: key functions and signaling pathways. Nat Neurosci 19:771–783

Sweeney MD, Montagne A, Sagare AP, Nation DA, Schneider LS, Chui HC, Harrington MG, Pa J, Law M, Wang DJJ, Jacobs RE, Doubal FN, Ramirez J, Black SE, Nedergaard M, Benveniste H, Dichgans M, Iadecola C, Love S, Bath PM, Markus HS, Salman RA, Allan SM, Quinn TJ, Kalaria RN, Werring DJ, Carare RO, Touyz RM, Williams SCR, Moskowitz MA, Katusic ZS, Lutz SE, Lazarov O, Minshall RD, Rehman J, Davis TP, Wellington CL, Gonzalez HM, Yuan C, Lockhart SN, Hughes TM, Chen CLH, Sachdev P, O'brien JT, Skoog I, Pantoni L, Gustafson DR, Biessels GJ, Wallin A, Smith EE, Mok V, Wong A, Passmore P, Barkof F, Muller M, Breteler MMB, Roman GC, Hamel E, Seshadri S, Gottesman RF, Van Buchem MA, Arvanitakis Z, Schneider JA, Drewes LR, Hachinski V, Finch CE, Toga AW, Wardlaw JM, Zlokovic BV (2019a) Vascular dysfunction-The disregarded partner of Alzheimer's disease. Alzheimers Dement 15:158–167

Sweeney MD, Zhao Z, Montagne A, Nelson AR, Zlokovic BV (2019b) Blood-brain barrier: from physiology to disease and back. Physiol Rev 99:21–78

Tachibana M, Yamazaki Y, Liu CC, Bu, G. & Kanekiyo, T. (2018) Pericyte implantation in the brain enhances cerebral blood flow and reduces amyloid-beta pathology in amyloid model mice. Exp Neurol 300:13–21

Takano T, Tian GF, Peng W, Lou N, Libionka W, Han X, Nedergaard M (2006) Astrocyte-mediated control of cerebral blood flow. Nat Neurosci 9:260–267
Tallquist MD, French WJ, Soriano P (2003) Additive effects of PDGF receptor beta signaling pathways in vascular smooth muscle cell development. PLoS Biol 1:E52
Taylor JP, Brown RH, Jr. & Cleveland, D. W. (2016) Decoding ALS: from genes to mechanism. Nature 539:197–206
Thomas WE (1999) Brain macrophages: on the role of pericytes and perivascular cells. Brain Res Brain Res Rev 31:42–57
Underly RG, Levy M, Hartmann DA, Grant RI, Watson AN, Shih AY (2017) Pericytes as inducers of rapid, matrix metalloproteinase-9-dependent capillary damage during ischemia. J Neurosci 37:129–140
Van Geest RJ, Klaassen I, Vogels IM, Van Noorden CJ, Schlingemann RO (2010) Differential TGF-{beta} signaling in retinal vascular cells: a role in diabetic retinopathy? Invest Ophthalmol Vis Sci 51:1857–1865
Vanlandewijck M, He L, Mae MA, Andrae J, Ando K, DEL Gaudio F, Nahar K, Lebouvier T, Lavina B, Gouveia L, Sun Y, Raschperger E, Rasanen M, Zarb Y, Mochizuki N, Keller A, Lendahl U, Betsholtz C (2018) A molecular atlas of cell types and zonation in the brain vasculature. Nature 554:475–480
Verbeek MM, Westphal JR, Ruiter DJ, de Waal RM (1995) T lymphocyte adhesion to human brain pericytes is mediated via very late antigen-4/vascular cell adhesion molecule-1 interactions. J Immunol 154:5876–5884
Vos CM, Geurts JJ, Montagne L, van Haastert ES, Bo L, van der Valk P, Barkhof F, de Vries HE (2005) Blood-brain barrier alterations in both focal and diffuse abnormalities on postmortem MRI in multiple sclerosis. Neurobiol Dis 20:953–960
Wang Y, Pan L, Moens CB, Appel B (2014) Notch3 establishes brain vascular integrity by regulating pericyte number. Development 141:307–317
Watson AN, Berthiaume AA, Faino AV, Mcdowell KP, Bhat NR, Hartmann DA, Shih AY (2020) Mild pericyte deficiency is associated with aberrant brain microvascular flow in aged PDGFRbeta(+/-) mice. J Cereb Blood Flow Metab:271678X19900543
Wei HS, Kang H, Rasheed ID, Zhou S, Lou N, Gershteyn A, Mcconnell ED, Wang Y, Richardson KE, Palmer AF, Xu C, Wan J, Nedergaard M (2016) Erythrocytes are oxygen-sensing regulators of the cerebral microcirculation. Neuron 91:851–862
Wilhelmus MM, Otte-Holler I, van Triel JJ, Veerhuis R, Maat-Schieman ML, Bu G, de Waal RM, Verbeek MM (2007) Lipoprotein receptor-related protein-1 mediates amyloid-beta-mediated cell death of cerebrovascular cells. Am J Pathol 171:1989–1999
Wilkinson-Berka JL, Babic S, de Gooyer T, Stitt AW, Jaworski K, Ong LG, Kelly DJ, Gilbert RE (2004) Inhibition of platelet-derived growth factor promotes pericyte loss and angiogenesis in ischemic retinopathy. Am J Pathol 164:1263–1273
Winkler EA, Sengillo JD, Sullivan JS, Henkel JS, Appel SH, Zlokovic BV (2013) Blood-spinal cord barrier breakdown and pericyte reductions in amyotrophic lateral sclerosis. Acta Neuropathol 125:111–120
Winkler EA, Sagare AP, Zlokovic BV (2014) The pericyte: a forgotten cell type with important implications for Alzheimer's disease? Brain Pathol 24:371–386
Wong AD, Ye M, Levy AF, Rothstein JD, Bergles DE, Searson PC (2013) The blood-brain barrier: an engineering perspective. Front Neuroeng 6:7
Xing CY, Tarumi T, Liu J, Zhang Y, Turner M, Riley J, Tinajero CD, Yuan LJ, Zhang R (2017) Distribution of cardiac output to the brain across the adult lifespan. J Cereb Blood Flow Metab 37:2848–2856
Yamadera M, Fujimura H, Inoue K, Toyooka K, Mori C, Hirano H, Sakoda S (2015) Microvascular disturbance with decreased pericyte coverage is prominent in the ventral horn of patients with amyotrophic lateral sclerosis. Amyotroph Lateral Scler Frontotemporal Degener 16:393–401

Yamamoto S, Muramatsu M, Azuma E, Ikutani M, Nagai Y, Sagara H, Koo BN, Kita S, O'donnell E, Osawa T, Takahashi H, Takano KI, Dohmoto M, Sugimori M, Usui I, Watanabe Y, Hatakeyama N, Iwamoto T, Komuro I, Takatsu K, Tobe K, Niida S, Matsuda N, Shibuya M, Sasahara M (2017) A subset of cerebrovascular pericytes originates from mature macrophages in the very early phase of vascular development in CNS. Sci Rep 7:3855

Yamazaki T, Mukouyama YS (2018) Tissue specific origin, development, and pathological perspectives of pericytes. Front Cardiovasc Med 5:78

Yates RL, Esiri MM, Palace J, Jacobs B, Perera R, Deluca GC (2017) Fibrin(ogen) and neurodegeneration in the progressive multiple sclerosis cortex. Ann Neurol 82:259–270

Yemisci M, Gursoy-Ozdemir Y, Vural A, Can A, Topalkara K, Dalkara T (2009) Pericyte contraction induced by oxidative-nitrative stress impairs capillary reflow despite successful opening of an occluded cerebral artery. Nat Med 15:1031–1037

Youn SW, Jung KH, Chu K, Lee JY, Lee ST, Bahn JJ, Park DK, Yu JS, Kim SY, Kim M, Lee SK, Han MH, Roh JK (2015) Feasibility and safety of intra-arterial pericyte progenitor cell delivery following mannitol-induced transient blood-brain barrier opening in a canine model. Cell Transplant 24:1469–1479

Zhou S, Giannetto M, Decourcey J, Kang H, Kang N, Li Y, Zheng S, Zhao H, Simmons WR, Wei HS, Bodine DM, Low PS, Nedergaard M, Wan J (2019) Oxygen tension-mediated erythrocyte membrane interactions regulate cerebral capillary hyperemia. Sci Adv 5:eaaw4466

Chapter 3
Biology of Pericytes – Recent Advances: Role of Pericytes in Brain Metastasis

Yoichi Morofuji, Shinsuke Nakagawa, Takashi Fujimoto, Susumu Yamaguchi, Kenta Ujifuku, and Takayuki Matsuo

Abstract

Introduction

Brain metastases are the most common brain neoplasms observed clinically in adults, comprising more than half of all brain tumors. **Brain metastases** have been historically understudied, and they represent an emerging and urgent unmet medical need. The major impediment to effectively treating brain metastatic disease is the blood-brain barrier (BBB). This barrier excludes most chemotherapeutics from the brain and creates a sanctuary site for metastatic cancers. Recent studies have provided some evidence that astrocytes play a major role in brain metastasis by engaging different modes of interactions with incoming cancer cells. By contrast, little is known concerning the underlying mechanisms of brain metastasis formation, especially the role of pericytes. In this review, we discuss the recent advances in our understanding of the biology of pericytes, especially regarding its role in brain metastasis.

Methods

Literature review.

Results

The primary function of pericytes is stabilizing the function of the BBB. By contrast, cancer cells can activate pericytes to participate in pathological angiogenesis and thus promote brain metastasis progression. It is important to understand both the stimulatory and inhibitory effects of pericytes on cells to reveal the mechanism of brain metastasis.

Y. Morofuji (✉) · T. Fujimoto · S. Yamaguchi · K. Ujifuku · T. Matsuo
Department of Neurosurgery, Nagasaki University Graduate School of Biomedical Sciences, Nagasaki, Japan

S. Nakagawa
Department of Medical Pharmacology, Nagasaki University Graduate School of Biomedical Sciences, Nagasaki, Japan

A. Birbrair (ed.), *Biology of Pericytes – Recent Advances*, Stem Cell Biology and Regenerative Medicine 68, https://doi.org/10.1007/978-3-030-62129-2_3

Conclusions
Based on the findings of this review, pericytes could represent a key cellular target for the treatment and prevention of brain metastasis. These data provided new insights into the biology of pericytes and their potential roles in the pathogenesis of brain metastasis. These findings should spur additional studies to better clarify the mechanisms by which pericytes promote brain metastasis and stimulate the development of treatment strategies targeting these cells.

Keywords Anergy · Angiogenesis · Cell-cell interaction · Degenerated pericytes · Extracellular vehicles · Extravasation · Heterogeneity · In vitro blood-brain barrier model · Ischemic pericyte · Metastasis · Neurovascular unit · Pericytic mimicry · Regenerative therapy · Stem cell transplantation · Stemness · Tumor co-option

3.1 Introduction

Brain metastases are the most common brain neoplasms encountered clinically in adults, representing more than half of all brain tumors (Patel and Mehta 2007). **Brain metastases** have been historically understudied, and they therefore represent an emerging and urgent unmet medical need. Actual treatment options for brain metastases, including surgical resection, radiotherapy, and chemotherapy, are rarely curative. Brain metastases contribute significantly to the morbidity and mortality of cancer. As the incidence of brain metastases has increased because of improved therapy for primary lesions, it is imperative to also improve the treatment of brain metastases. Progress in treating brain metastases has been hampered by a lack of model systems, a lack of human tissue samples, and the exclusion of patients with brain metastases from many clinical trials (Gril et al. 2010). Whereas each of these issues is significant, the major impediment to effectively treating brain metastatic disease is the blood-brain barrier (BBB) (Gril et al. 2010). This barrier excludes most chemotherapeutics from the brain and creates a sanctuary site for metastatic cancers. The interaction between the BBB and cancer cells plays a key role in the implantation and growth of brain metastases in the central nervous system. A better understanding of the molecular features of the BBB, its interrelation with metastatic cancer cells, and the cellular mechanisms responsible for establishing brain metastasis is necessary to improve treatment modalities, especially chemotherapy. This would improve the survival and quality of life of patients with brain metastasis and potentially increase the remission rate. Recent studies have provided some evidence that astrocytes play a major role in brain metastasis by engaging in different interactions with incoming cancer cells (Wasilewski et al. 2017). Conversely, little is known about the underlying mechanisms of brain metastasis formation, especially those involving pericytes. In this review, we discussed the recent advances in our understanding of the biology of pericytes with a particular focus on their role in brain metastasis.

3.2 Basic Role of Pericytes in Brain Metastasis

Circulating cancer cells reach the brain vasculature system, allowing their attachment to endothelial cells. Cancer cells transmigrate through endothelial cells into the brain parenchyma by disrupting tight junctions (TJs) and degrading extracellular matrix proteins. In the brain parenchyma, cancer cells encounter a microenvironment that differs from the niche of the primary site. Intracranial metastases need to adapt to the situation and grow effectively through angiogenesis. Pericytes are recruited to enclose and stabilize endothelial cells to form the BBB (Bergers and Song 2005; Bautch 2011), and they act as a physical barrier between the circulation and the brain parenchyma. Endothelial cells and pericytes communicate through mediators (Sweeney et al. 2016; Brown et al. 2019). Pericytes are regulated in brain metastasis through the strengthening of these barrier functions and release of various mediators.

There is no doubt that the function of the BBB has a profound effect on the formation of brain metastasis because cancer cells need to enter the brain parenchyma. The function of the BBB is reliant on endothelial cells supported by both pericytes and astrocytes. Pericytes can modulate and maintain the BBB through the release of signaling factors that enhance TJs and direct the polarization of astrocyte endfeet (Armulik et al. 2010). Reductions in the numbers of pericytes can cause a loss of TJs between endothelial cells, leading to increased BBB permeability (Sengillo et al. 2013). In addition, defects or the absence of pericytes may lead to several central nervous system (CNS) diseases, including neurodegeneration, neurovascular diseases, and injury (Dalkara et al. 2011). TJs are intercellular components of endothelial cells that are essential in controlling the paracellular permeability of the intact BBB (Tietz and Engelhardt 2015). Claudin-5, occludin, and zona occludens-1 are major TJ proteins that have important roles in BBB permeability, and the loss of these proteins contributes to the progression of brain metastasis (Nitta et al. 2003; Uzunalli et al. 2019). In the presence of pericytes, brain endothelial cells express elevated levels of these proteins (Nakagawa et al. 2009). In a study of claudin-5, the ability of the BBB to act against small molecules, but not larger molecules, was selectively reduced by this protein (Nitta et al. 2003). This size-selective loosening of the BBB supports the hypothesis that claudin-5 is important in the selective regulation of BBB permeability, which is a determinant of brain metastasis (Wang et al. 2016).

Factors of pericytes involved in angiogenesis are important for brain metastasis formation. A variety of signaling cascades, including platelet-derived growth factor (PDGF), vascular endothelial growth factor (VEGF), and transforming growth factor-beta (TGF-β), are involved in angiogenesis via interactions between pericytes and endothelial cells (Shen et al. 2012; Sieczkiewicz and Herman 2003; Sun et al. 2003; Gautam and Yao 2018). Both loss of the stabilizing functions of pericytes and the acquisition of a pro-angiogenic phenotype facilitate the generation of a vasculature that is proliferative, leaky, and dysfunctional (Barlow et al. 2013). Normally, PDGF-β is secreted by sprouting endothelial cells, whereas pericytes express PDGF receptors

(PDGFRs) (Lindblom et al. 2003; Thijssen et al. 2018). Pericytes are critically involved in the normal development and formation of microvessels, and deficiency of pericytes leads to the formation of microvessels with many of the typical hallmarks of tumor vessels through the reactivity of PDGF-β (Abramsson et al. 2003). Conversely, degenerated pericytes are implicated in tumor angiogenesis and involved in brain metastasis progression. Metastatic lesions display decreased expression of PDGFR-β, a specific protein of pericytes. In metastatic tumors with higher permeability, vessels are surrounded by desmin-positive pericytes. Moreover, a tendency of reduced CD13 staining in pericytes was observed (Lyle et al. 2016; Wilhelm et al. 2018). VEGF and TGF-β are considered to have major roles in the angiogenesis of brain metastasis. Degenerated pericytes, which produce elevated levels of VEGF, strengthen the link between angiogenesis and tumor formation. Thus, these pericytes may be of central importance for both brain metastasis and cancer growth (Boire et al. 2020). The TGF-β pathway changes from a tumor-suppressive pathway in early tumorigenesis to a metastasis-promoting pathway in the late stage of cancer. Pericytes also secrete angiogenic-promoting factors such as VEGF and neurogenic locus notch homolog protein 3 (NOTCH3) to activate angiogenic processes in the adult central nervous system (Ribatti et al. 2011). Abrogation of NOTCH3 expression significantly reduces the self-renewal and invasive capacity of cancer cells. Thus, NOTCH3 plays a critical role in promoting epithelial-to-mesenchymal transition (EMT, tumor stemness, and metastasis (Leontovich et al. 2018).

Activation of matrix proteases and degradation of the basement membrane are necessary for angiogenesis to occur (van Hinsbergh and Koolwijk 2008), and there is evidence that tumors secrete factors that activate pericytes to degrade the basement membrane and liberate matrix-bound growth factors, thereby contributing to angiogenesis and possibly facilitating tumor growth and invasion. Matrix metalloproteinases (MMPs), a family of zinc-dependent endopeptidases, degrade substrates of the extracellular matrix (ECM) (Yong 2005; Takata et al. 2008). MMPs have been implicated in suppressing angiogenesis as well as the anomalous relationships of cancer with the related processes of angiogenesis and vasculogenesis. Pericytes are the most common MMP-9–releasing cells that respond to thrombin stimulation among BBB-related cells (Machida et al. 2015). MMP-9 enhances the metastasis of cancer cells by degrading collagen proteins of the ECM after being activated by extracellular proteases under different physiological and pathological conditions. In addition, MMP-9 also generates the angiogenic and tumor repressor tumstatin. Tumstatin inhibits EC proliferation and induces apoptosis through its anti-angiogenic properties (Hamano et al. 2003; Quintero-Fabian et al. 2019).

Pericytes release low levels of chemokines spontaneously. After stimulation by inflammatory molecules such as lipopolysaccharide, the production of chemokines and cytokines by pericytes is significantly increased (Kovac et al. 2011). Interleukin (IL)-12 is a pro-inflammatory cytokine and pericyte-related protein with anti-angiogenic properties (Colombo and Trinchieri 2002). IL-12 production induced anti-tumor immune responses and significantly enhanced survival in a brain metastasis status (Shah et al. 2006; Cody et al. 2012). Extracellular vehicles (EVs) have emerged as important means of intercellular communication that contribute to the interaction of cancer cells with the microenvironment and promote brain metastasis (Husemann et al. 2008). EVs include a broad range of vesicles including exosomes, microvesicles, and oncosomes. The metastasis-related interactions of cancer cell-derived exosomes with BBB-related cells involve the increase of vascular permeability, transfer of EVs to the BBB, and degradation of TJ proteins (Weidle et al. 2017). Contrarily, exosomes from pericytes improved the ability of endothelial cells to regulate blood flow, reduce edema, decrease MMP-2 expression, increase claudin-5 expression, and inhibit apoptosis in a study of spinal cord injury (Yuan et al. 2019).

The primary function of pericytes is stabilizing the function of the BBB. By contrast, cancer cells can activate pericytes to participate in pathological angiogenesis and thus promote brain metastasis progression. It is important to understand both the stimulatory and inhibitory effects of pericytes on cells to reveal the mechanism of brain metastasis.

3.3 *In vitro* BBB Model

To achieve hematogenous brain metastasis formation, tumor cells must adhere to the luminal surface of endothelial cells, migrate to the endothelial monolayer, and then grow in the brain parenchyma (Heyder et al. 2002; Kienast et al. 2010). The BBB plays a crucial role in hematogenous metastasis as the gateway to tumor cells. The BBB is a highly specialized system that maintains brain homeostasis. Brain capillary endothelial cells (BECs) are tightly sealed by TJs, and they express specific transporters and receptors that properly control and regulate the passage of substances and cells from peripheral fluids into the brain (Abbott et al. 2010; Zhao et al. 2015). BBB properties are induced and maintained through crosstalk between BECs and surrounding elements of the neurovascular unit (NVU), namely pericytes, astrocytes, microglia, and neurons (Zhao et al. 2015; Liebner et al. 2018). In the process of brain metastasis formation, the crosstalk between components of the NVU and tumor cells may act as both a progressive and suppressive factor of metastasis depending on its response to interacting factors. Hence, studies of the interaction between components of the NVU and tumor cells may provide new insights into the mechanisms of brain metastasis.

Among the NVU components, astrocytes were the first to be recognized as regulators of brain endothelial characteristics and function (Abbott et al. 2006; Haseloff et al. 2005). As astrocytes represent a fundamental BBB cell type that maintains BBB properties, their dysfunction disrupts BBB function and disturbs brain homeostasis (Liebner et al. 2018; Parkes et al. 2018). In that sense, astrocytes appear to have an important role in restricting the migration of tumor cells by maintaining BBB function under physiological conditions. Conversely, previous studies indicated that the interaction between astrocytes and metastatic cancer cells stimulates transmigration and subsequent tumor growth (Lorger and Felding-Habermann 2010). Metastatic tumor cells affect the functions and characteristics of astrocytes through secreted factors and/or stimulation through direct contact (Brandao et al.

2019; Henrik Heiland et al. 2019). Furthermore, tumor cell-stimulated astrocytes produce several soluble factors and cytokines that stimulate cancer cell proliferation in the brain parenchyma (Henrik Heiland et al. 2019; Shumakovich et al. 2017). Accordingly, it is considered that the bidirectional interaction between astrocytes and tumor cells is important to the success of metastasis formation.

Compared with the role of astrocytes, the role of brain pericytes in brain metastasis is not well understood. Pericytes are the cells situated closest to BECs morphologically, and they share a common basement membrane. Accumulated evidence indicates that pericytes have several important roles in NVUs. The roles of pericytes in NVUs are (Patel and Mehta 2007) stabilizing vascular formation, (Gril et al. 2010) maintaining BBB function, (Wasilewski et al. 2017) regulating blood flow through a synergistic relationship with neuronal activation, (Bergers and Song 2005) and acting as immune-like cells (Sweeney et al. 2016; Su et al. 2019). In addition, pericyte dysfunction may lead to numerous CNS diseases including neurodegeneration, cerebrovascular diseases, and injury (Sweeney et al. 2016). Although the mechanism by which pericytes regulate brain metastasis is poorly understood, considering the important role of pericytes in NVUs, pericytes may play a critical role in brain metastasis by interacting with metastatic cancer cells.

Among several methods to elucidate the role of pericytes in brain metastasis formation such as *in vitro* studies, animal models, and analyses of human specimens, cell culture-based BBB models are widely accepted as powerful tools for evaluating the interaction between BBB cells and tumor cells. *In vitro* BBB models are particularly effective for elucidating the mechanisms of cancer cell extravasation and the survival of cancer cells in the brain environment. To examine the mechanism of brain metastasis using *in vitro* BBB models, it is necessary to secure a niche for the development of metastasis in the cerebral parenchyma. Accordingly, we consider that the triple co-culture model consisting of endothelial cells, pericytes, and astrocytes is a suitable model for mimicking the process of hematogenous brain metastasis formation (Nakagawa et al. 2009; Nakagawa and Aruga 2019). We examined the role of pericytes in brain metastasis using a series of *in vitro* co-culture models. KNS62 cells, which were derived from the brain metastasis of a patient with squamous cell carcinoma of the lungs, significantly decreased the barrier function of triple co-cultured models compared with the effects of the non-metastatic cell line A549. Conversely, the reduction of barrier function induced by KNS62 cells was decreased in the co-culture model composed of endothelial cells and pericytes (Fujimoto et al. 2020). These data suggest that pericyte-derived factors and/or the interaction between endothelial cells-pericytes and cancer cells suppresses barrier disruption, thereby suppressing the transmigration of cancer cells. The TJ function of the BBB is a fundamental property. TJs both regulate paracellular permeability and play a role in the regulation of cancer cell motility. Other studies revealed that TJs influence brain metastasis by regulating the proliferation and migration of a brain microvascular endothelial cell line (Jia et al. 2014; Ma et al. 2017). Considering the role of pericytes in the maintenance of TJ function (Sweeney et al. 2016; Nakagawa et al. 2009), pericytes may regulate the migration of cancer cells into the brain parenchyma under physiological conditions. In fact, pericytes appear to regulate cancer cell survival in the endothelial lumen. To form brain metastases, tumor cells need to survive intravascularly for some time (Kienast et al. 2010; Lorger and Felding-Habermann 2010). Some surviving cells may start proliferating on endothelial cells, and some of these cells transmigrate into the brain parenchyma. Fujimoto et al. examined the effect of pericytes on the survival of cancer cells in the endothelial lumen using an endothelial cell-pericyte co-culture model. When endothelial cells and pericytes were co-cultured with tumor cells, pericytes suppressed the proliferation of cancer cells on the endothelial cells. Of interest, these findings were related to soluble factors derived from tumor cell-stimulated pericytes but not pericytes alone (Fujimoto et al. 2020). Therefore, under physiological conditions, pericytes may play a critical role in the elimination of tumor cells in the vasculature.

Contrarily, there is little evidence regarding the interaction of pericytes and tumor cells after extravasation. The survival of extravasated tumor cells is extremely restricted in the brain environment, and only a small population survives. Recent studies indicated that surviving tumor cells tend to attach to the extraluminal surface of blood vessels (Heyder et al. 2002; Kienast et al. 2010); in particular, the vascular basement membrane is a key component of tumor cell survival (Valiente et al. 2014; Yoshimasu et al. 2004). As pericytes are located on the extraluminal surface of endothelial cells and they produce several extracellular matrix proteins, they are undoubtedly involved in the crosstalk of metastatic cancer cells with the brain microenvironment. Several studies suggested the important role of pericytes in brain metastasis. The multipotent stemness of pericytes may be associated with the proliferation of tumor cells (Ribeiro and Okamoto 2015). In addition, pericytes enhance the proliferation and migration of triple-negative breast cancer cells *in vitro* (Bagley et al. 2005). Meanwhile, the role of pericytes in vascularization in the brain-tumor barrier (BTB) has been well reported (Wilhelm et al. 2018; Ribeiro and Okamoto 2015; Chen et al. 2016). These findings indicated that the properties of pericytes differ between the intact BBB and BTB. Several studies indicated that BTB pericytes exhibit decreased expression of PDGFR-β, a specific protein of pericytes (Uzunalli et al. 2019; Lyle et al. 2016). In addition, another specific marker of pericytes, regulator of G-protein signaling 5 (RGS5), has been reported as a master gene responsible for the abnormal tumor vascular morphology (Hamzah et al. 2008). Although RGS proteins act as inhibitors of G-protein-coupled receptor signaling by stimulating the GTPase activity of activated Gα proteins (Hollinger and Hepler 2002), little is known about the role of RGS5 *in vivo*. However, recent reports described the role of RGS5 in the vascular system. Specifically, RGS5 was found to regulate pericyte maturation and vascular normalization (Berger et al. 2005; Manzur et al. 2009). In fact, several reports indicated that the loss of RGS5 in pericytes reduced vascular leakage and preserved tight junctions in an animal model of stroke and tumors (Hamzah et al. 2008; Ozen et al. 2018; Roth et al. 2019). Therefore, elevated RGS5 expression in pericytes might be linked to the induction of BBB leakage, thereby contributing to the progression of metastasis formation. In addition, pericytes are known to be heterogeneous depending on their expression markers, morphology, and origin. On the basis of their subpopulation, pericytes may

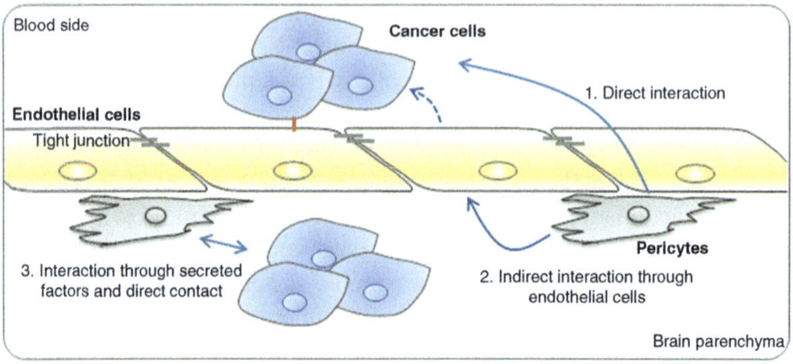

Fig. 3.1 Possible roles of pericytes in brain metastasis. **1**. Pericytes derived factors may directly affect cancer cells in the brain metastasis process. **2**. Interaction between endothelial cells and cancer cells is the first step of the brain metastasis process. As functions of endothelial cells are maintained by pericytes, pericytes-endothelium interaction has the potential to affect cancer cells in the brain metastasis process. **3**. Pericytes may influence on proliferation of extravasated cancer cells and vascularization in the brain-tumor barrier through secreted factors and direct contact

have a different function (Birbrair et al. 2013, 2014; Sims 2000; Dias Moura Prazeres et al. 2017). Birbrair et al. reported that type-2 pericytes (nestin and NG2 proteoglycan double-positive cells) participate in both normal and tumor angiogenesis, but type-1 pericytes (nestin negative and NG2 proteoglycan positive cells) did not participate (Birbrair et al. 2014). These findings indicate that revealing of pericytes subpopulation leads to a deeper understanding of the role of pericytes on pre- and post-extravasation of cancer cells.

Taken together, although our current knowledge about the role of pericytes in brain metastasis formation is limited, further understanding of the role of pericytes in brain metastasis may provide new insights into the development of targeted inhibitors of brain metastasis. A schema of possible role of pericytes in brain metastasis in presented in Fig. 3.1.

3.4 Differentially Expressed Genes of Pericytes in Brain Metastasis

Pericytes are heterogeneous cell populations. Organotypic differentiation, functional plasticity, and heterogenic origins have been reported for pericytes (Holm et al. 2018; Yamazaki and Mukouyama 2018). The phenotypic heterogeneity of pericytes and the lack of specific molecular markers represent challenges in pericyte research. Metastasis of tumors is a complex multistep process. Traditionally, brain metastasis was exclusively recognized as the hematogenous spread/intravascular dissemination of cancer cells. In this process, multiple processes (invasion, intravasation, transport of circulating cancer cells, extravasation, metastatic colonization, and angiogenesis), serially occur (Pecorino 2016). Recently, however, the concept of distant metastasis occurring along the vessel abluminal surface (angiotropism/pericytic mimicry [PM]/extravascular migratory metastasis [EVMM]) has been described (Lugassy et al. 2020). In these multistep processes, multiple biological mechanisms, such as EMT, cell adhesion, ECM formation, angiogenesis, microenvironment formation, maintenance of cancer stem cells, inflammation, immune surveillance, factor secretion (secretome), and their crosstalk, have been implicated in brain metastasis.

In this section, we summarized pericyte-related molecular factors, especially in the brain.

3.4.1 Pericyte Origins and Molecular Markers

Adult pericytes are considered to be derived from the mesoderm through preexisting mural microvascular cells, blood pluripotent stem cells, tissue progenitor stem cells, endothelial cells, and tissue myeloid stem cells (Holm et al. 2018). No general pan-pericytic marker has been established because of the heterogeneity and organ specificity of pericytes. PDGFR-β is one of the most accepted molecular markers of pericytes. PDGF-β/PDGFR-β signaling is essential for the proliferation and recruitment of pericytes to blood vessels (Betsholtz 1995; Lindahl et al. 1997). Neural glial antigen 2 (NG2) and membrane chondroitin sulfate proteoglycan 4 are expressed on the pericyte surface, and they are important for vascular pericyte recruitment. Alpha-smooth muscle actin and desmin are contractile filaments that have been described as pericyte (or mural cell) markers. RGS5 is a GTPase-activating protein that has been implicated in vascular remodeling and tumor development. Immunohistochemical pericyte markers should be selected depending on the tissue and microvessel type. The detection of more than two markers and morphological confirmation are recommended for the definition of pericytes, especially in histological analyses *in vivo* (Yamazaki and Mukouyama 2018; Harrell et al. 2018).

Single-cell genome-wide quantitative transcriptomes described the pericyte-specific 1088 transcriptome of the mouse brain, but pericytes remain poorly defined (Vanlandewijck et al. 2018).

3.4.2 Cancer-Associated Fibroblasts (CAFs), Angiogenesis, and Pericytes

CAFs are known to facilitate tumor invasion and metastasis (Kalluri and Zeisberg 2006). The normal brain parenchyma generally lacks fibroblasts, whereas metastatic lesions have connective tissue. In the specimens from 40 surgically resected brain metastases, pericytes were considered a source of connective tissue cells including CAFs (Teglasi et al. 2019).

PDGF-β/PDGFR-β signaling induced pericyte-to-fibroblast transition (PFT) and promoted tumor invasion and metastasis. EMT genes were revealed to be activated in genome-wide array analysis during PFT. PFT cells obtained stromal fibroblast and myofibroblast markers. Co-implantation of PFT and tumor cells increased tumor dissemination and invasion (Hosaka et al. 2016).

Perivascular cells (vascular smooth muscle cells and pericytes) lose the traditional pericyte markers in response to tumor-secreted factors and exhibit increased proliferation, migration, and ECM production. Kruppel-like factor 4 (Klf4), a stem cell factor, promoted the less differentiated state and fibronectin-rich pro-metastatic niche formation (Murgai et al. 2017).

In breast cancer metastasis, CAFs significantly increased the lumen size of blood microvessels. Inactivation of TGF-β signaling decreased microvessel density, lumen size, and cancer proliferation. CAFs are less numerous in TGF-β–inactivated tumors. CAFs enhance vascular coverage by pericytes. TGF-β receptor 1-SMAD signaling upregulates fibronectin, a prominent regulator of the endothelial cell-pericyte interaction (Zonneville et al. 2018).

Thus, pericytes are almost certainly an important source of fibroblasts in metastatic brain tumors. CAFs provide a metastasis-promoting microenvironment for cancer cells. EMT is implicated in ECM-producing processes. Cancer-fibroblast-vascular crosstalk activates tumor angiogenesis (De Palma et al. 2017).

A molecular atlas revealed the existence of fibroblast-like cells in the mouse brain, and thus, further research is required (Vanlandewijck et al. 2018).

3.4.3 Angiogenesis Patterns and Pericytes

Six patterns of angiogenesis-related processes (sprouting angiogenesis, vasculogenesis from endothelial precursor cells, tumor cell endothelial cell differentiation, intussusceptions, vascular co-option, and vascular mimicry) have been described in tumors. The latter three processes are independent of VEGF, and they can promote resistance to VEGF inhibitors (Carmeliet and Jain 2011). Pericyte-covered tumor vessels were implicated in the resistance to anti-angiogenic therapy at least in the field of glioma research (Lu-Emerson et al. 2015). Terms and definitions of angiogenesis are listed in Table 3.1.

Pericytes regulate both angiogenesis and vascular permeability in tumor vessels (Harrell et al. 2018). The paucity of stable pericyte-endothelial cell interactions enables sprouting angiogenesis, but it also generates a dysfunctional vascular network (De Palma et al. 2017). In an *in vivo* lung cancer brain metastasis model, the desmin-positive pericyte subpopulation was larger in the BTB than in the BBB. The desmin-positive pericyte subpopulation was related to high vascular permeability (Uzunalli et al. 2019; Lyle et al. 2016). However, the BTB remained a significant impediment to standard chemotherapeutic delivery in breast cancer brain metastasis (Lockman et al. 2010). Muscle derived nestin positive/NG2 positive pericytes were

Table 3.1 Terms and definitions of angiogenesis

Term	Definition
Vascular co-option	a non-angiogenic process whereby tumor cells hijack pre-existing blood-vessels[a]
Angiotropism/pericytic mimicry/ extravascular migratory metastasis	An embryologically derived mechanism of continuous extravascular tumor cell migration without entrance into vascular channels. (1) Tumor cells are recruited along the abluminal vascular surface instead of pericytes and replace pericytes; (2) Tumor cells migrate along the abluminal vascular surface[b]
Vascular (vasculogenic) mimicry	An angiogenic process whereby tumor cells organize themselves into vascular-like structures for the obtention of nutrients and oxygen independently of normal blood vessels or angiogenesis[c]

[a] Kuczynski EA, Vermeulen PB, Pezzella F, Kerbel RS, Reynolds AR. Vessel co-option in cancer. Nat Rev. Clin Oncol. 2019;16 (8):469–93

[b] Lugassy C, Kleinman HK, Vermeulen PB, Barnhill RL. Angiotropism, pericytic mimicry and extravascular migratory metastasis: an embryogenesis-derived program of tumor spread. Angiogenesis. 2020;23 (1):27–41

[c] Fernandez-Cortes M, Delgado-Bellido D, Oliver FJ. Vasculogenic Mimicry: Become an Endothelial Cell "But Not So Much". Front Oncol. 2019;9:803

reported to migrate into glioblastoma lesion and promote sprouting angiogenesis in mouse, but nestin negative pericytes did not implicated in the angiogenesis (Abbott et al. 2010). Therefore, pericyte heterogeneity might be considered in angiogenesis and pre-metastatic microvasular niche. In a mouse lung pericyte model, fibroblast growth factor 2 (FGF2) induced pericyte proliferation and migration through direct effects on FGF receptor 2 signaling and indirect effects of PDGFR-β signaling on pericytes (Hosaka et al. 2018). However, it is unknown whether these processes occur in brain metastatic lesions, especially *in vivo*.

Pericytes weakly but consistently express the epithelial-specific protein receptor tyrosine kinase inhibitor 2 (Tie2). Pericytes are transformed into a pro-migratory phenotype via Tie2 inhibition. Tie2 controls sprouting angiogenesis and vascular maturation. Tie2 downstream signaling in pericytes includes calpain, Akt, and fork-head box O3A signaling. Tie2 inhibition strongly promotes tumor growth (Teichert et al. 2017). Therefore, pericytes may either promote or suppress metastasis in terms of angiogenesis in the microenvironment.

3.4.4 Angiotropism/PM/EVMM and Pericytes

The angiotropism/PM/EVMM concept was first established in melanoma studies. PM was suggested to occur in eight patients with metastatic brain melanomas (Rodewald et al. 2019). Melanoma cells migrated along pre-existing vessels into the brain, exhibiting angiotropism/vascular cooption and PM (Lugassy et al. 2020). The pericyte-like spread of cancer cell lines (from lung cancer, breast cancer, renal cell carcinoma, and colon cancer) was reported to activate Yes-associated protein (YAP) and myocardin-related transcription factor via cell adhesion molecule L1 (L1CAM) mechanotransduction signaling for metastatic colonization. L1CAM activated YAP by engaging β1 integrin and integrin-linked kinase. Metastatic cell-initiated outgrowth was reported to occur via L1CAM-YAP signaling in the perivascular microenvironment (Er et al. 2018). Pericytes also used L1CAM for perivascular spreading. Thus, pericyte function during PM is considered to competitively inhibit metastasis (Lugassy et al. 2020).

Two comprehensive gene analyses were performed for melanoma PM. Microarray analyses of tissue from 66 patients with melanoma uncovered 128 differentially expressed genes (DEGs), and 15 genes were considered critical for angiotropism/PM (Lugassy et al. 2011). These 15 genes were classified according to their relationships with cell migration, cell motility, and neurotropism (Table 3.2). In 28 DEGs from the *in vitro* co-culture model of PM, 20 genes exhibited properties linked to (i) cell migration, (ii) cancer progression and metastasis, (iii) EMT, (iv) embryonic and/or cancer stem cell properties, (v) pericyte recruitment, and (vi) inflammation (Lugassy et al. 2013) (Table 3.2).

These cDNA data were not derived from pericytes, and thus, the direct implication of pericytes in angiotropism/PM/EVMM remains to be explored.

3.4.5 Immune Evasion in the Tumor Microenvironment and Pericytes

Immature pericytes in tumors can promote local immunosuppression in glioma and other malignant tumors (Gaceb and Paul 2018; Ochs et al. 2013; Paiva et al. 2018). Melanoma-derived vascular pericytes induced CD4-positive T cell dysfunction or anergy, thereby participating in the subversion of immunosurveillance associated with tumor progression compared with the findings in control kidney pericytes. The immunoregulatory function of pericyte appeared to be reinforced by IL-6 and required the intrinsic expression of RGS5 in the tumor microenvironment (Bose et al. 2013). The role of pericytes in the tumor microenvironment appears to involve the immunosuppression of T cell activity and promotion of tumor growth in glioma (Ochs et al. 2013). It remains unknown whether the same or similar process occurs in metastatic brain tumors.

Further studies are needed to apply immunotherapy for brain malignant tumors. In summary, pericytes may either promote or suppress the proliferation of metastatic tumor cells in context-dependent manners.

Table 3.2 The cDNA microarray differentially expressed gene analysis of Angiotropism/ Pericytic mimicry in melanoma studies

Author, Year	Evaluation	Control	Differentially expressed genes related to angiotropism/PM
Lugassy 2011[90]	Human angiotropic melanoma	Human non-angiotropic melanoma	**Neural crest migration**: <Ups>TCOF1, NEIL3, HMMR, <Downs>AHNAK, KCTD11, CEBPA, AQP3 **Cell migration of other malignant tumors with neural crest origin :**<Ups> ECT2, GLS, <Downs> AGAP2, **Cell motility and/or migration:** <Ups>DBF4, FNBP1L, <Downs> FGD3, F10 **Neurotropism:** <Ups>KIF14
Lugassy 2013[91]	Co-cultured endothelial and melanoma cells	Endothelial cell alone, and Melanoma cell alone	**Cell migration :** <Ups>CCL2, ICAM1, IL6, PDGFB. <Downs>RGNEF, RANBP9 **Cancer progression and metastasis:** <Ups>CCL2, ICAM1, SELE, TRAF1, IL6, SERPINB2, CXCL6, <Downs>BLID, MALT1, UPF1, PLAA, RGNEF, ZXDC **EMT :** <Ups>CCL2, IL6, ICAM1, PDGFB **Embryonic and/or cancer stem cell properties:** <Ups>CCL2, PDGFB, EVX1, <Downs>CFDP1, RANBP9 **Pericyte recruitment :** <Ups>PDGFB **Inflammation :** <Ups>CCL2, IL6, TRAF1, CXCL6, SELE, ICAM1, SERPINB2, SLC7A2, C2CD4B, PDGFB

In these studies, direct implication of pericytes still remains to be solved

Abbreviations: *AGAP2*, centaurin gamma 1. *AHNAK*, homo sapiens nucleoprotein AHNAK or desmoyokin. *AQP3*, aquaporin 3. *BLID*, BH3-like otif-containing cell death inducer. *C2CD4B*, C2 calcium dependent domain-containing protein 4B. *CCL2*, chemokine ligand 2. *CEBPA*, CCAAT/enhancer-binding protein alpha. *CFDP1*, craniofacial development protein 1. *CXCL6*, chemokine ligand 6. *DBF4*, activator of S-phase kinase, homo sapience DBP4 homologue (S.cerevisiae). Downs, down-regulated genes. *ECT2*, epithelial cell transforming sequence 2 ongogene. *EMT*, endothelial-mesenchymal transition. *EVX1*, even-skipped homeobox 1. F10, coagulation factor 10. *FGD3*, fyve, RhoGEF, and PH domain-containing protein 3. *FNBP1L*, formin-binding protein 1-like. *GLS*, glutaminase C. *HMMR*, hyaluronan-mediated motility receptor. *ICAM1*, intercellular adhesion molecule 1. *IL6*, interleukin 6. *INHBA*, inhibin beta A. *KCTD11*, potassium channel tetrameriation domain-containing protein 11. *KIF14*, kinecin family member 14. *MALT1*, mucosa-associated lymphoid tissue lymphoma translation gene 1. *NEIL3*, endonuclease VII-like 3. *PDGFB*, platelet derived growth factor beta. *PM*, pericytic mimicry. *PLAA*, phospholipase A2-activating protein. *RANBP9*, RAN binding protein 9. *RG9MTD6*, RNA (guanine-9-) methyltransferase. *RGNEF*, 190 k-Da guanine nucleotide exchange factor. *SELE*, selectin E. *SERPINB2*, serpin peptidase inhibitor, clade B, member2. *SLC7A2*, solute carrier family 7, member2. *TCOF1*, Treacher Collins syndrome 1. *UPF1*, UPF1 RNA helicase and ATPase. *Ups*, up-regulated genes. *TRAF1*, TNF receptor-associated factor 1. *ZNF254*, zinc finger protein 254. *ZXDC*, ZXD family zinc finger protein C

3.5 Pericyte and Regenerative Medicine

3.5.1 Pericytes Represent One Origin of Neural Stem/ Progenitor Cells After Ischemic Stroke

The origin of pericytes is debatable. Korn et al. reported that cranial neuroectodermal cells differentiated into vascular smooth muscle cells and pericytes (Korn et al. 2002). Yamanishi et al. also reported that neural crest-derived cells invaded the telencephalon from the mesenchyme together with endothelial cells and differentiated into pericytes in the brain vasculature (Yamanishi et al. 2012). In line with these findings, one of the origins of pericytes is the ectoderm. Another origin is the mesenchyme, and hemangioblasts were reported to differentiate into both pericytes and endothelial cells (Rolny et al. 2005; Rucker et al. 2000). Recently, Yamamoto et al. reported that mature macrophages recruited from yolk sac trans-differentiated into cerebrovascular pericytes in the early vascular development in the CNS (Yamamoto et al. 2017). Pericytes play an important role in cerebrovascular morphology. During development, a lack of pericytes is reported to lead to endothelial hyperplasia and the formation of microaneurysm (Hellstrom et al. 2001). Regarding function, pericytes represent one component of the BBB, and they play an important role in maintaining the permeability of the BBB (Armulik et al. 2010). After ischemic stroke, pericytes protect the BBB and decrease brain edema (Zhang et al. 2011). In addition, neovasculature covered with pericytes, termed mature vessels, exhibit decreased microglia/macrophage infiltration after ischemic stroke; therefore, these matured vessels affect anti-inflammation and functional recovery after stroke (Yamaguchi et al. 2018; Bell et al. 2010).

Recently, pericytes were reported to have other roles such as multipotent stem cells after stroke. During embryonic development, pericytes express nestin, a marker of stem cells. Nestin expression decreases gradually during embryonic development and finally disappears after birth (Nakano-Doi et al. 2016). Concerning neural stem cells, it is well known that the subventricular zone (SVZ) of the lateral ventricle and the subgranular zone of the dentate gyrus of the hippocampus are major locations at which neural stem cells exist in the adult rodent brain. During CNS development, neuroepithelial cells in the subependymal layer proliferate and differentiate into neuronal precursors, and then they migrate into the neocortex from the subependymal layer. This is consistent with the localization of neural stem cells. It was also reported that neural stem cells migrated from the subventricular zone to the peri-infarct area via the rostral migratory stream and reactive astrocytes guided neural stem cells into close proximity to the infarct core after ischemic stroke (Saha et al. 2013). However, all neural stem cells that play a role in brain repair after stroke do not arise from the SVZ. After ischemic stroke, pial cells generate nestin-positive cells in the pia mater at the site of stroke (Jung et al. 2011). Some of these nestin-positive cells express NG2 and PDGFR-β, which are markers of microvascular pericytes, and migrate into the infarct cortex. These cells, termed ischemic pericytes (iPCs), form a neurosphere-like cell cluster in response to basic fibroblast growth factor and epidermal growth factor *in vitro*, and they exhibit neural stem/progenitor cell activity (Nakagomi et al. 2011; Dore-Duffy et al. 2006). These cells differentiate into Tuj1-positive neurons in neural conditioned medium. Both rodent and human pericytes expressed neuronal lineage features. When normal human iPCs were cultured under oxygen/glucose deprivation, they also formed neurosphere-like cell clusters and expressed Sox-2, Tuji1, and MAP (Nakagomi et al. 2015). iPCs expressed reprogramming factors, including c-Myc, Klf4, and Sox-2, in polymerase chain reaction (PCR). These iPCs differentiate into both neural stem cells and mesenchymal angioblasts. When these iPCs were cultured in endothelial conditioned medium, they expressed Flk1 and CD31 via mesenchymal angioblasts, and they differentiated into mature endothelial cells. Furthermore, when iPCs were cultured in neurobasal medium, they expressed Iba-1. PDGFR-β and Iba-1 double-positive cells were observed in perivascular lesion in infarct areas after ischemic stroke. As these cells did not express CD68, a marker of perivascular macrophages, iPCs were also considered to differentiate into microglia with phagocytic functions (Kwon 1997; Sakuma et al. 2016). These facts indicated that pericytes acquired stemness through reprogramming and gained the capability to differentiate into various cells in response to ischemia, thereby playing an important role in endogenous brain repair following ischemic stroke.

3.5.2 *Pericytes in Brain Tumors*

In metastatic brain tumors, pericytes act both positively and negatively. When disseminated cancer cells (DCCs) formed a tumor mass in the brain, metastatic, they require perivascular localization, which provides oxygen, nutrition, and growth factors to promote their growth in the brain (Rafii et al. 2016; Morrison and Scadden 2014). As a first step of perivascular colonization by DCCs, the cells must compete with and dislodge pericytes. DCCs grow and spread through L1CAM, a cell adhesion molecule that is also used by pericytes for perivascular spread (Er et al. 2018). Pericytes were also reported to prevent brain metastasis by lung cancer by enhancing BBB and secreting mediators. Thus, pericytes play an important role in preventing cancer metastasis.

However, once cancer cells can colonize the brain, pericytes might promote tumor formation. Indeed, VEGF and PDGF-β produced by tumor cells increase the number of tumor vessels covered by pericytes (Wang et al. 2016; Buttigliero et al. 2016), and this angiogenesis facilitates tumor growth. Tumor growth is also promoted through vasculature co-option, indicating that tumors infiltrate normal tissue and adopt the pre-existing vasculature (Plate et al. 2012). This co-option is considered one of the causes of resistance to anti-angiogenesis treatment (Bridgeman et al. 2017). Because the vessels are pre-existing, anti-angiogenesis treatments such as bevacizumab are ineffective against these co-opted vessels. Pericytes promote tumor co-option via Cdc42 activation in glioblastoma multiforme (GBM). Contrarily, inhibition of Cdc42 activity decreases co-option and induces the transformation of pericytes into macrophage-like cell, which have anti-tumor effects (Caspani et al. 2014). This co-option was also observed in metastatic tumors, and thus, pericytes can be targeted to prevent blood vessel co-option.

In cancer recurrence, mesenchymal stem cells (MSCs) and pericytes were reported to migrate into tumors in response to inflammatory mediators produced by damaged tissue (Kidd et al. 2012). After radiotherapy in lung cancer, stromal cell derived factor-1α and PDGF-β levels were increased in tumors, and MSCs differentiated into pericytes (Wang et al. 2016). These pericytes contributed to vascular network formation, termed vasculogenesis, thereby leading to tumor growth (Folkman 2007). Contrarily, over-expression of PDGF-β in colorectal and pancreatic cancer cells inhibited tumor growth by increasing the coverage of endothelial cells by pericytes (McCarty et al. 2007). Thus, the role of pericytes might differ by tumor type.

3.5.3 Pericytes and Regenerative Medicine in Brain Tumor

As pericyte transplantation in amyloid model mice increased cerebral blood flow and decreased amyloid-beta levels in the treated hemisphere, cell therapy using pericytes has potential use in the treatment of Alzheimer's disease (Tachibana et al. 2018). Although one report examined the transplantation of pericytes in ischemic stroke (Youn et al. 2015), research only pericyte transplantation is limited. Many reports concerning stem cell transplantation revealed the enhancement of endogenous or transdifferentiated pericytes (Yamaguchi et al. 2018; Pisati et al. 2007; Garbuzova-Davis et al. 2017).

Stem cell transplantation therapy for brain tumor has two anti-tumor effects. The first anti-tumor effect is the inhibition of brain tumors by stem cells. A few reports described positive effects. In human-skin derived stem cell transplantation model for GBM, stem cells differentiated into pericytes. These cells upregulated TGF-β1 and downregulated VEGF, thereby inhibiting tumor invasion and angiogenesis and leading to decreased tumor growth (Pisati et al. 2007). This is consistent with a report that pericyte coverage inhibits the invasion of prostate cancer cells (Welen et al. 2009). In another report, the authors focused on the maintenance effect of MSCs on the BBB. They hypothesized that enhanced BBB integrity might decrease inflammation, inflammatory factor production, and tumor-associated macrophage numbers. These effects might decrease tumor growth by decreasing the number of blood vessels (Kaneko et al. 2015). The second anti-tumor effect involves the tumor-topic ability of stem cells. Ordinary, transplanted stem cells tend to localize at tumor sites. When stem cells with anti-tumor effects are generated through bioengineering techniques and transplanted in the brain, they localize at tumor site and exert anti-tumor effects (Bagci-Onder et al. 2015; Zhao et al. 2012; Mooney et al. 2018; Binello and Germano 2012). However, this strategy has some problems. Stem cell transplantation ordinarily enhances angiogenesis by increasing various trophic factor (Yamaguchi et al. 2018; Horie et al. 2011). In addition, stem cell carry a risk of tumorgenicity (Serakinci et al. 2004). It was also reported that pericytes can be converted from tumor-suppressing to tumor-promoting cells through contact-dependent interactions with GBM cells (Caspani et al. 2014). Transplanted stem cells have the potential to support tumor growth. Further investigation is needed.

3.6 Summary

Despite representing more than half of all brain tumors encountered clinically, brain metastases have been historically understudied. Meanwhile, the primary impediment to the effective treatment of brain metastases is the blood-brain barrier. Given that little is known regarding the role of pericytes in brain tumor formation, this review sought to clarify current knowledge on the biology of these cells. Our findings indicated that pericytes have a key role in the function of the blood-brain barrier, whereas tumor cells can induce pericytes to promote the development and progression of brain metastasis through pathological angiogenesis. Based on the findings of this review, pericytes could represent a key cellular target for the treatment and prevention of brain metastasis. These data provided new insights into the biology of pericytes and their potential roles in the pathogenesis of brain metastasis. These findings should spur additional studies to better clarify the mechanisms by which pericytes promote brain metastasis and stimulate the development of treatment strategies targeting these cells.

Acknowledgments We would like to express our gratitude to Masami Niwa (Emeritus Professor of Nagasaki University/PhamaCocell Company Ltd., Japan), Maria A. Deli (Biological Research Centre, Hungary), and William A. Banks (Veterans Affairs Puget Sound Health Care System and University of Washington, USA) for helpful discussions in preparing the manuscript. We wish to thank Kei Sato, Yuki Matsunaga, Yusuke Iki, Nobutaka Horie, and Tsuyoshi Izumo for providing insights and expertise that greatly assisted in the preparation of this manuscript, although they may not agree with all of the interpretations/conclusions of this paper.

Disclosure of Interests All authors declare they have no conflict of interest.

Ethical Approval This article does not contain any studies with human participants or animals performed by any of the authors.

Funding This study was partially funded by JSPS and HAS under the Japan-Hungary Research Cooperative Program (to Y.M. and S.N.), Grants-in-Aid for Scientific Research (Fostering Joint International Research) 15KK0349 (to Y.M.), Grants-in-Aid for Scientific Research (C) 17K10840 (to Y.M.), and (C) 17K10838 (to S.N.).

References

Abbott NJ, Ronnback L, Hansson E (2006) Astrocyte-endothelial interactions at the blood-brain barrier. Nat Rev Neurosci 7(1):41–53

Abbott NJ, Patabendige AA, Dolman DE, Yusof SR, Begley DJ (2010) Structure and function of the blood-brain barrier. Neurobiol Dis 37(1):13–25

Abramsson A, Lindblom P, Betsholtz C (2003) Endothelial and nonendothelial sources of PDGF-B regulate pericyte recruitment and influence vascular pattern formation in tumors. J Clin Invest 112(8):1142–1151

Armulik A, Genove G, Mae M, Nisancioglu MH, Wallgard E, Niaudet C et al (2010) Pericytes regulate the blood-brain barrier. Nature 468(7323):557–561

Bagci-Onder T, Du W, Figueiredo JL, Martinez-Quintanilla J, Shah K (2015) Targeting breast to brain metastatic tumours with death receptor ligand expressing therapeutic stem cells. Brain 138(Pt 6):1710–1721

Bagley RG, Weber W, Rouleau C, Teicher BA (2005) Pericytes and endothelial precursor cells: cellular interactions and contributions to malignancy. Cancer Res 65(21):9741–9750

Barlow KD, Sanders AM, Soker S, Ergun S, Metheny-Barlow LJ (2013) Pericytes on the tumor vasculature: jekyll or hyde? Cancer Microenviron 6(1):1–17

Bautch VL (2011) Stem cells and the vasculature. Nat Med 17(11):1437–1443

Bell RD, Winkler EA, Sagare AP, Singh I, LaRue B, Deane R et al (2010) Pericytes control key neurovascular functions and neuronal phenotype in the adult brain and during brain aging. Neuron 68(3):409–427

Berger M, Bergers G, Arnold B, Hammerling GJ, Ganss R (2005) Regulator of G-protein signaling-5 induction in pericytes coincides with active vessel remodeling during neovascularization. Blood 105(3):1094–1101

Bergers G, Song S (2005) The role of pericytes in blood-vessel formation and maintenance. Neuro Oncol 7(4):452–464

Betsholtz C (1995) Role of platelet-derived growth factors in mouse development. Int J Dev Biol 39(5):817–825

Binello E, Germano IM (2012) Stem cells as therapeutic vehicles for the treatment of high-grade gliomas. Neuro Oncol 14(3):256–265

Birbrair A, Zhang T, Wang ZM, Messi ML, Mintz A, Delbono O (2013) Type-1 pericytes participate in fibrous tissue deposition in aged skeletal muscle. Am J Physiol Cell Physiol 305(11):C1098–C1113

Birbrair A, Zhang T, Wang ZM, Messi ML, Olson JD, Mintz A et al (2014) Type-2 pericytes participate in normal and tumoral angiogenesis. Am J Physiol Cell Physiol 307(1):C25–C38

Boire A, Brastianos PK, Garzia L, Valiente M (2020) Brain metastasis. Nat Rev Cancer 20(1):4–11

Bose A, Barik S, Banerjee S, Ghosh T, Mallick A, Bhattacharyya Majumdar S et al (2013) Tumor-derived vascular pericytes anergize Th cells. J Immunol 191(2):971–981

Brandao M, Simon T, Critchley G, Giamas G (2019) Astrocytes, the rising stars of the glioblastoma microenvironment. Glia 67(5):779–790

Bridgeman VL, Vermeulen PB, Foo S, Bilecz A, Daley F, Kostaras E et al (2017) Vessel co-option is common in human lung metastases and mediates resistance to anti-angiogenic therapy in preclinical lung metastasis models. J Pathol 241(3):362–374

Brown LS, Foster CG, Courtney JM, King NE, Howells DW, Sutherland BA (2019) Pericytes and neurovascular function in the healthy and diseased brain. Front Cell Neurosci 13:282

Buttigliero C, Bertaglia V, Novello S (2016) Anti-angiogenetic therapies for central nervous system metastases from non-small cell lung cancer. Transl Lung Cancer Res 5(6):610–627

Carmeliet P, Jain RK (2011) Molecular mechanisms and clinical applications of angiogenesis. Nature 473(7347):298–307

Caspani EM, Crossley PH, Redondo-Garcia C, Martinez S (2014) Glioblastoma: a pathogenic crosstalk between tumor cells and pericytes. PLoS One 9(7):e101402

Chen Z, Xu XH, Hu J (2016) Role of pericytes in angiogenesis: focus on cancer angiogenesis and anti-angiogenic therapy. Neoplasma 63(2):173–182
Cody JJ, Scaturro P, Cantor AB, Yancey Gillespie G, Parker JN, Markert JM (2012) Preclinical evaluation of oncolytic deltagamma(1)34.5 herpes simplex virus expressing interleukin-12 for therapy of breast cancer brain metastases. Int J. Breast Cancer 2012:628697
Colombo MP, Trinchieri G (2002) Interleukin-12 in anti-tumor immunity and immunotherapy. Cytokine Growth Factor Rev 13(2):155–168
Dalkara T, Gursoy-Ozdemir Y, Yemisci M (2011) Brain microvascular pericytes in health and disease. Acta Neuropathol 122(1):1–9
De Palma M, Biziato D, Petrova TV (2017) Microenvironmental regulation of tumour angiogenesis. Nat Rev Cancer 17(8):457–474
Dias Moura Prazeres PH, Sena IFG, Borges IDT, de Azevedo PO, Andreotti JP, de Paiva AE et al (2017) Pericytes are heterogeneous in their origin within the same tissue. Dev Biol 427(1):6–11
Dore-Duffy P, Katychev A, Wang X, Van Buren E (2006) CNS microvascular pericytes exhibit multipotential stem cell activity. J Cerebral Blood Flow Metab: Off J Int Soc Cereb Blood Flow and Metab 26(5):613–624
Er EE, Valiente M, Ganesh K, Zou Y, Agrawal S, Hu J et al (2018) Pericyte-like spreading by disseminated cancer cells activates YAP and MRTF for metastatic colonization. Nat Cell Biol 20(8):966–978
Folkman J (2007) Angiogenesis: an organizing principle for drug discovery? Nat Rev Drug Discov 6(4):273–286
Fujimoto T, Nakagawa S, Morofuji Y, Watanabe D, Ujifuku K, Horie N et al (2020) Pericytes suppress brain metastasis from lung cancer in vitro. Cell Mol Neurobiol 40(1):113–121
Gaceb A, Paul G (2018) Pericyte Secretome. Adv Exp Med Biol 1109:139–163
Garbuzova-Davis S, Haller E, Lin R, Borlongan CV (2017) Intravenously transplanted human bone marrow endothelial progenitor cells engraft within brain capillaries, preserve mitochondrial morphology, and display pinocytotic activity toward blood-brain barrier repair in ischemic stroke rats. Stem Cells 35(5):1246–1258
Gautam J, Yao Y (2018) Roles of pericytes in stroke pathogenesis. Cell Transplant 27(12):1798–1808
Gril B, Evans L, Palmieri D, Steeg PS (2010) Translational research in brain metastasis is identifying molecular pathways that may lead to the development of new therapeutic strategies. Eur J Cancer 46(7):1204–1210
Hamano Y, Zeisberg M, Sugimoto H, Lively JC, Maeshima Y, Yang C et al (2003) Physiological levels of tumstatin, a fragment of collagen IV alpha3 chain, are generated by MMP-9 proteolysis and suppress angiogenesis via alphaV beta3 integrin. Cancer Cell 3(6):589–601
Hamzah J, Jugold M, Kiessling F, Rigby P, Manzur M, Marti HH et al (2008) Vascular normalization in Rgs5-deficient tumours promotes immune destruction. Nature 453(7193):410–414
Harrell CR, Simovic Markovic B, Fellabaum C, Arsenijevic A, Djonov V, Volarevic V (2018) Molecular mechanisms underlying therapeutic potential of pericytes. J Biomed Sci 25(1):21
Haseloff RF, Blasig IE, Bauer HC, Bauer H (2005) In search of the astrocytic factor(s) modulating blood-brain barrier functions in brain capillary endothelial cells in vitro. Cell Mol Neurobiol 25(1):25–39
Hellstrom M, Gerhardt H, Kalen M, Li X, Eriksson U, Wolburg H et al (2001) Lack of pericytes leads to endothelial hyperplasia and abnormal vascular morphogenesis. J Cell Biol 153(3):543–553
Henrik Heiland D, Ravi VM, Behringer SP, Frenking JH, Wurm J, Joseph K et al (2019) Tumor-associated reactive astrocytes aid the evolution of immunosuppressive environment in glioblastoma. Nat Commun 10(1):2541
Heyder C, Gloria-Maercker E, Entschladen F, Hatzmann W, Niggemann B, Zanker KS et al (2002) Realtime visualization of tumor cell/endothelial cell interactions during transmigration across the endothelial barrier. J Cancer Res Clin Oncol 128(10):533–538
Hollinger S, Hepler JR (2002) Cellular regulation of RGS proteins: modulators and integrators of G protein signaling. Pharmacol Rev 54(3):527–559

Holm A, Heumann T, Augustin HG (2018) Microvascular mural cell organotypic heterogeneity and functional plasticity. Trends Cell Biol 28(4):302–316

Horie N, Pereira MP, Niizuma K, Sun G, Keren-Gill H, Encarnacion A et al (2011) Transplanted stem cell-secreted vascular endothelial growth factor effects poststroke recovery, inflammation, and vascular repair. Stem Cells 29(2):274–285

Hosaka K, Yang Y, Seki T, Fischer C, Dubey O, Fredlund E et al (2016) Pericyte-fibroblast transition promotes tumor growth and metastasis. Proc Natl Acad Sci U S A 113(38):E5618–E5627

Hosaka K, Yang Y, Nakamura M, Andersson P, Yang X, Zhang Y et al (2018) Dual roles of endothelial FGF-2-FGFR1-PDGF-BB and perivascular FGF-2-FGFR2-PDGFRbeta signaling pathways in tumor vascular remodeling. Cell Discov 4:3

Husemann Y, Geigl JB, Schubert F, Musiani P, Meyer M, Burghart E et al (2008) Systemic spread is an early step in breast cancer. Cancer Cell 13(1):58–68

Jia W, Lu R, Martin TA, Jiang WG (2014) The role of claudin-5 in blood-brain barrier (BBB) and brain metastases (review). Mol Med Rep 9(3):779–785

Jung KH, Chu K, Lee ST, Bahn JJ, Jeon D, Kim JH et al (2011) Multipotent PDGFRbeta-expressing cells in the circulation of stroke patients. Neurobiol Dis 41(2):489–497

Kaneko Y, Tajiri N, Staples M, Reyes S, Lozano D, Sanberg PR et al (2015) Bone marrow-derived stem cell therapy for metastatic brain cancers. Cell Transplant 24(4):625–630

Kidd S, Spaeth E, Watson K, Burks J, Lu H, Klopp A et al (2012) Origins of the tumor microenvironment: quantitative assessment of adipose-derived and bone marrow-derived stroma. PLoS One 7(2):e30563

Kienast Y, von Baumgarten L, Fuhrmann M, Klinkert WE, Goldbrunner R, Herms J et al (2010) Real-time imaging reveals the single steps of brain metastasis formation. Nat Med 16(1):116–122

Korn J, Christ B, Kurz H (2002) Neuroectodermal origin of brain pericytes and vascular smooth muscle cells. J Comp Neurol 442(1):78–88

Kovac A, Erickson MA, Banks WA (2011) Brain microvascular pericytes are immunoactive in culture: cytokine, chemokine, nitric oxide, and LRP-1 expression in response to lipopolysaccharide. J Neuroinflammation 8:139

Kwon YK (1997) Expression of brain-derived neurotrophic factor mRNA stimulated by basic fibroblast growth factor and platelet-derived growth factor in rat hippocampal cell line. Mol Cells 7(3):320–325

Leontovich AA, Jalalirad M, Salisbury JL, Mills L, Haddox C, Schroeder M et al (2018) NOTCH3 expression is linked to breast cancer seeding and distant metastasis. Breast Cancer Res 20(1):105

Liebner S, Dijkhuizen RM, Reiss Y, Plate KH, Agalliu D, Constantin G (2018) Functional morphology of the blood-brain barrier in health and disease. Acta Neuropathol 135(3):311–336

Lindahl P, Johansson BR, Leveen P, Betsholtz C (1997) Pericyte loss and microaneurysm formation in PDGF-B-deficient mice. Science 277(5323):242–245

Lindblom P, Gerhardt H, Liebner S, Abramsson A, Enge M, Hellstrom M et al (2003) Endothelial PDGF-B retention is required for proper investment of pericytes in the microvessel wall. Genes Dev 17(15):1835–1840

Lockman PR, Mittapalli RK, Taskar KS, Rudraraju V, Gril B, Bohn KA et al (2010) Heterogeneous blood-tumor barrier permeability determines drug efficacy in experimental brain metastases of breast cancer. Clin Cancer Res 16(23):5664–5678

Lorger M, Felding-Habermann B (2010) Capturing changes in the brain microenvironment during initial steps of breast cancer brain metastasis. Am J Pathol 176(6):2958–2971

Lu-Emerson C, Duda DG, Emblem KE, Taylor JW, Gerstner ER, Loeffler JS et al (2015) Lessons from anti-vascular endothelial growth factor and anti-vascular endothelial growth factor receptor trials in patients with glioblastoma. J Clin Oncol 33(10):1197–1213

Lugassy C, Lazar V, Dessen P, van den Oord JJ, Winnepenninckx V, Spatz A et al (2011) Gene expression profiling of human angiotropic primary melanoma: selection of 15 differentially

expressed genes potentially involved in extravascular migratory metastasis. Eur J Cancer 47(8):1267–1275
Lugassy C, Wadehra M, Li X, Corselli M, Akhavan D, Binder SW et al (2013) Pilot study on "pericytic mimicry" and potential embryonic/stem cell properties of angiotropic melanoma cells interacting with the abluminal vascular surface. Cancer Microenviron 6(1):19–29
Lugassy C, Kleinman HK, Vermeulen PB, Barnhill RL (2020) Angiotropism, pericytic mimicry and extravascular migratory metastasis: an embryogenesis-derived program of tumor spread. Angiogenesis 23(1):27–41
Lyle LT, Lockman PR, Adkins CE, Mohammad AS, Sechrest E, Hua E et al (2016) Alterations in pericyte subpopulations are associated with elevated blood-tumor barrier permeability in experimental brain metastasis of breast cancer. Clin Cancer Res 22(21):5287–5299
Ma SC, Li Q, Peng JY, Zhouwen JL, Diao JF, Niu JX et al (2017) Claudin-5 regulates blood-brain barrier permeability by modifying brain microvascular endothelial cell proliferation, migration, and adhesion to prevent lung cancer metastasis. CNS Neurosci Ther 23(12):947–960
Machida T, Takata F, Matsumoto J, Takenoshita H, Kimura I, Yamauchi A et al (2015) Brain pericytes are the most thrombin-sensitive matrix metalloproteinase-9-releasing cell type constituting the blood-brain barrier in vitro. Neurosci Lett 599:109–114
Manzur M, Hamzah J, Ganss R (2009) Modulation of g protein signaling normalizes tumor vessels. Cancer Res 69(2):396–399
McCarty MF, Somcio RJ, Stoeltzing O, Wey J, Fan F, Liu W et al (2007) Overexpression of PDGF-BB decreases colorectal and pancreatic cancer growth by increasing tumor pericyte content. J Clin Invest 117(8):2114–2122
Mooney R, Hammad M, Batalla-Covello J, Abdul Majid A, Aboody KS (2018) Concise review: neural stem cell-mediated targeted cancer therapies. Stem Cells Transl Med 7(10):740–747
Morrison SJ, Scadden DT (2014) The bone marrow niche for haematopoietic stem cells. Nature 505(7483):327–334
Murgai M, Ju W, Eason M, Kline J, Beury DW, Kaczanowska S et al (2017) KLF4-dependent perivascular cell plasticity mediates pre-metastatic niche formation and metastasis. Nat Med 23(10):1176–1190
Nakagawa S, Aruga J (2019) Sphingosine 1-phosphate signaling is involved in impaired blood-brain barrier function in ischemia-reperfusion injury. Mol Neurobiol 57:1594–1606
Nakagawa S, Deli MA, Kawaguchi H, Shimizudani T, Shimono T, Kittel A et al (2009) A new blood-brain barrier model using primary rat brain endothelial cells, pericytes and astrocytes. Neurochem Int 54(3-4):253–263
Nakagomi T, Molnar Z, Nakano-Doi A, Taguchi A, Saino O, Kubo S et al (2011) Ischemia-induced neural stem/progenitor cells in the pia mater following cortical infarction. Stem Cells Dev 20(12):2037–2051
Nakagomi T, Kubo S, Nakano-Doi A, Sakuma R, Lu S, Narita A et al (2015) Brain vascular pericytes following ischemia have multipotential stem cell activity to differentiate into neural and vascular lineage cells. Stem cells (Dayton, Ohio) 33(6):1962–1974
Nakano-Doi A, Nakagomi T, Sakuma R, Takahashi A, Tanaka Y, Kawamura M et al (2016) Expression patterns and phenotypic changes regarding stemness in brainpericytes in health and disease. J Stem Cell Res & Therap 6(3):332
Nitta T, Hata M, Gotoh S, Seo Y, Sasaki H, Hashimoto N et al (2003) Size-selective loosening of the blood-brain barrier in claudin-5-deficient mice. J Cell Biol 161(3):653–660
Ochs K, Sahm F, Opitz CA, Lanz TV, Oezen I, Couraud PO et al (2013) Immature mesenchymal stem cell-like pericytes as mediators of immunosuppression in human malignant glioma. J Neuroimmunol 265(1-2):106–116
Ozen I, Roth M, Barbariga M, Gaceb A, Deierborg T, Genove G et al (2018) Loss of regulator of G-protein signaling 5 leads to neurovascular protection in stroke. Stroke 49(9):2182–2190
Paiva AE, Lousado L, Guerra DAP, Azevedo PO, Sena IFG, Andreotti JP et al (2018) Pericytes in the premetastatic niche. Cancer Res 78(11):2779–2786

Parkes I, Chintawar S, Cader MZ (2018) Neurovascular dysfunction in dementia – human cellular models and molecular mechanisms. Clin Sci (Lond) 132(3):399–418

Patel RR, Mehta MP (2007) Targeted therapy for brain metastases: improving the therapeutic ratio. Clin Cancer Res 13(6):1675–1683

Pecorino L (2016) Molecular biology of cancer : mechanisms, targets, and therapeutics, 4th edn. Oxford University Press, Oxford, p 375

Pisati F, Belicchi M, Acerbi F, Marchesi C, Giussani C, Gavina M et al (2007) Effect of human skin-derived stem cells on vessel architecture, tumor growth, and tumor invasion in brain tumor animal models. Cancer Res 67(7):3054–3063

Plate KH, Scholz A, Dumont DJ (2012) Tumor angiogenesis and anti-angiogenic therapy in malignant gliomas revisited. Acta Neuropathol 124(6):763–775

Quintero-Fabian S, Arreola R, Becerril-Villanueva E, Torres-Romero JC, Arana-Argaez V, Lara-Riegos J et al (2019) Role of matrix metalloproteinases in angiogenesis and cancer. Front Oncol 9:1370

Rafii S, Butler JM, Ding BS (2016) Angiocrine functions of organ-specific endothelial cells. Nature 529(7586):316–325

Ribatti D, Nico B, Crivellato E (2011) The role of pericytes in angiogenesis. Int J Dev Biol 55(3):261–268

Ribeiro AL, Okamoto OK (2015) Combined effects of pericytes in the tumor microenvironment. Stem Cells Int 2015:868475

Rodewald AK, Rushing EJ, Kirschenbaum D, Mangana J, Mittmann C, Moch H et al (2019) Eight autopsy cases of melanoma brain metastases showing angiotropism and pericytic mimicry. Implications for extravascular migratory metastasis. J Cutan Pathol 46(8):570–578

Rolny C, Lu L, Agren N, Nilsson I, Roe C, Webb GC et al (2005) Shb promotes blood vessel formation in embryoid bodies by augmenting vascular endothelial growth factor receptor-2 and platelet-derived growth factor receptor-beta signaling. Exp Cell Res 308(2):381–393

Roth M, Gaceb A, Enstrom A, Padel T, Genove G, Ozen I et al (2019) Regulator of G-protein signaling 5 regulates the shift from perivascular to parenchymal pericytes in the chronic phase after stroke. FASEB J 33(8):8990–8998

Rucker HK, Wynder HJ, Thomas WE (2000) Cellular mechanisms of CNS pericytes. Brain Res Bull 51(5):363–369

Saha B, Peron S, Murray K, Jaber M, Gaillard A (2013) Cortical lesion stimulates adult subventricular zone neural progenitor cell proliferation and migration to the site of injury. Stem Cell Res 11(3):965–977

Sakuma R, Kawahara M, Nakano-Doi A, Takahashi A, Tanaka Y, Narita A et al (2016) Brain pericytes serve as microglia-generating multipotent vascular stem cells following ischemic stroke. J Neuroinflammation 13(1):57

Sengillo JD, Winkler EA, Walker CT, Sullivan JS, Johnson M, Zlokovic BV (2013) Deficiency in mural vascular cells coincides with blood-brain barrier disruption in Alzheimer's disease. Brain Pathol 23(3):303–310

Serakinci N, Guldberg P, Burns JS, Abdallah B, Schrodder H, Jensen T et al (2004) Adult human mesenchymal stem cell as a target for neoplastic transformation. Oncogene 23(29):5095–5098

Shah AC, Price KH, Parker JN, Samuel SL, Meleth S, Cassady KA et al (2006) Serial passage through human glioma xenografts selects for a Deltagamma134.5 herpes simplex virus type 1 mutant that exhibits decreased neurotoxicity and prolongs survival of mice with experimental brain tumors. J Virol 80(15):7308–7315

Shen J, Ishii Y, Xu G, Dang TC, Hamashima T, Matsushima T et al (2012) PDGFR-beta as a positive regulator of tissue repair in a mouse model of focal cerebral ischemia. J Cereb Blood Flow Metab 32(2):353–367

Shumakovich MA, Mencio CP, Siglin JS (2017) Moriarty RA. Geller HM, Stroka KM. Astrocytes from the brain microenvironment alter migration and morphology of metastatic breast cancer cells. FASEB J

Sieczkiewicz GJ, Herman IM (2003) TGF-beta 1 signaling controls retinal pericyte contractile protein expression. Microvasc Res 66(3):190–196
Sims DE (2000) Diversity within pericytes. Clin Exp Pharmacol Physiol 27(10):842–846
Su X, Huang L, Qu Y, Xiao D, Mu D (2019) Pericytes in cerebrovascular diseases: an emerging therapeutic target. Front Cell Neurosci 13:519
Sun Y, Jin K, Xie L, Childs J, Mao XO, Logvinova A et al (2003) VEGF-induced neuroprotection, neurogenesis, and angiogenesis after focal cerebral ischemia. J Clin Invest 111(12):1843–1851
Sweeney MD, Ayyadurai S, Zlokovic BV (2016) Pericytes of the neurovascular unit: key functions and signaling pathways. Nat Neurosci 19(6):771–783
Tachibana M, Yamazaki Y, Liu CC, Bu G, Kanekiyo T (2018) Pericyte implantation in the brain enhances cerebral blood flow and reduces amyloid-beta pathology in amyloid model mice. Exp Neurol 300:13–21
Takata F, Sumi N, Nishioku T, Harada E, Wakigawa T, Shuto H et al (2008) Oncostatin M induces functional and structural impairment of blood-brain barriers comprised of rat brain capillary endothelial cells. Neurosci Lett 441(2):163–166
Teglasi V, Csury DT, Dezso K, Bugyik E, Szabo V, Szallasi Z et al (2019) Origin and distribution of connective tissue and pericytes impacting vascularization in brain metastases with different growth patterns. J Neuropathol Exp Neurol 78(4):326–339
Teichert M, Milde L, Holm A, Stanicek L, Gengenbacher N, Savant S et al (2017) Pericyte-expressed Tie2 controls angiogenesis and vessel maturation. Nat Commun 8:16106
Thijssen VL, Paulis YW, Nowak-Sliwinska P, Deumelandt KL, Hosaka K, Soetekouw PM et al (2018) Targeting PDGF-mediated recruitment of pericytes blocks vascular mimicry and tumor growth. J Pathol 246(4):447–458
Tietz S, Engelhardt B (2015) Brain barriers: crosstalk between complex tight junctions and adherens junctions. J Cell Biol 209(4):493–506
Uzunalli G, Dieterly AM, Kemet CM, Weng HY, Soepriatna AH, Goergen CJ et al (2019) Dynamic transition of the blood-brain barrier in the development of non-small cell lung cancer brain metastases. Oncotarget 10(59):6334–6348
Valiente M, Obenauf AC, Jin X, Chen Q, Zhang XH, Lee DJ et al (2014) Serpins promote cancer cell survival and vascular co-option in brain metastasis. Cell 156(5):1002–1016
van Hinsbergh VW, Koolwijk P (2008) Endothelial sprouting and angiogenesis: matrix metalloproteinases in the lead. Cardiovasc Res 78(2):203–212
Vanlandewijck M, He L, Mae MA, Andrae J, Ando K, Del Gaudio F et al (2018) A molecular atlas of cell types and zonation in the brain vasculature. Nature 554(7693):475–480
Wang HH, Cui YL, Zaorsky NG, Lan J, Deng L, Zeng XL et al (2016) Mesenchymal stem cells generate pericytes to promote tumor recurrence via vasculogenesis after stereotactic body radiation therapy. Cancer Lett 375(2):349–359
Wasilewski D, Priego N, Fustero-Torre C, Valiente M (2017) Reactive astrocytes in brain metastasis. Front Oncol 7:298
Weidle UH, Birzele F, Kollmorgen G, Ruger R (2017) The multiple roles of exosomes in metastasis. Cancer Genomics Proteomics 14(1):1–15
Welen K, Jennbacken K, Tesan T, Damber JE (2009) Pericyte coverage decreases invasion of tumour cells into blood vessels in prostate cancer xenografts. Prostate Cancer Prostatic Dis 12(1):41–46
Wilhelm I, Fazakas C, Molnar K, Vegh AG, Hasko J, Krizbai IA (2018) Foe or friend? Janus-faces of the neurovascular unit in the formation of brain metastases. J Cereb Blood Flow Metab 38(4):563–587
Yamaguchi S, Horie N, Satoh K, Ishikawa T, Mori T, Maeda H et al (2018) Age of donor of human mesenchymal stem cells affects structural and functional recovery after cell therapy following ischaemic stroke. J Cereb Blood Flow Metab 38(7):1199–1212
Yamamoto S, Muramatsu M, Azuma E, Ikutani M, Nagai Y, Sagara H et al (2017) A subset of cerebrovascular pericytes originates from mature macrophages in the very early phase of vascular development in CNS. Sci Rep 7(1):3855

Yamanishi E, Takahashi M, Saga Y, Osumi N (2012) Penetration and differentiation of cephalic neural crest-derived cells in the developing mouse telencephalon. Dev Growth Differ 54(9):785–800
Yamazaki T, Mukouyama YS (2018) Tissue specific origin, development, and pathological perspectives of pericytes. Front Cardiovasc Med 5:78
Yong VW (2005) Metalloproteinases: mediators of pathology and regeneration in the CNS. Nat Rev Neurosci 6(12):931–944
Yoshimasu T, Sakurai T, Oura S, Hirai I, Tanino H, Kokawa Y et al (2004) Increased expression of integrin alpha3beta1 in highly brain metastatic subclone of a human non-small cell lung cancer cell line. Cancer Sci 95(2):142–148
Youn SW, Jung KH, Chu K, Lee JY, Lee ST, Bahn JJ et al (2015) Feasibility and safety of intra-arterial pericyte progenitor cell delivery following mannitol-induced transient blood-brain barrier opening in a canine model. Cell Transplant 24(8):1469–1479
Yuan X, Wu Q, Wang P, Jing Y, Yao H, Tang Y et al (2019) Exosomes derived from pericytes improve microcirculation and protect blood-spinal cord barrier after spinal cord injury in mice. Front Neurosci 13:319
Zhang J, Takahashi HK, Liu K, Wake H, Liu R, Maruo T et al (2011) Anti-high mobility group box-1 monoclonal antibody protects the blood-brain barrier from ischemia-induced disruption in rats. Stroke 42(5):1420–1428
Zhao D, Najbauer J, Annala AJ, Garcia E, Metz MZ, Gutova M et al (2012) Human neural stem cell tropism to metastatic breast cancer. Stem Cells 30(2):314–325
Zhao Z, Nelson AR, Betsholtz C, Zlokovic BV (2015) Establishment and dysfunction of the blood-brain barrier. Cell 163(5):1064–1078
Zonneville J, Safina A, Truskinovsky AM, Arteaga CL, Bakin AV (2018) TGF-beta signaling promotes tumor vasculature by enhancing the pericyte-endothelium association. BMC Cancer 18(1):670

Chapter 4
Effects of Cytomegalovirus on Pericytes

Donald J. Alcendor

Abstract

Introduction

Human cytomegalovirus (HCMV) is a member of Betaherpesvirus family and is the leading infectious cause of neurosensory hearing loss, vision loss and neurocognitive disability among congenitally infected children. HCMV is a ubiquitous central nervous system (CNS) pathogen that causes significant morbidity and mortality in individuals that are immune compromised. Even more, HCMV is an important opportunistic pathogen that can cause life threatening disease in HIV patients and transplant recipients. HCMV has broad tropism for multiple cell types including cells of the vascular barrier systems. My aim is to explore HCMV infectivity in human vascular pericyte populations with implications for its role in human vascular diseases.

Methods

Human primary low passaged pericyte populations were examined for HCMV infectivity along with controls by means of phase microscopy, immunofluorescence, immunohistochemistry, electron microscopy, real-time reverse-transcription polymerase chain reaction (RT-PCR), and quantitative RT-PCR (qRT-PCR), the proinflammatory cytokines assay, and HCMV-GFP recombinant virus.

Results

I have discovered that human vascular pericytes from multiple barrier systems that includes the blood brain barrier (BBB), inner blood retinal barrier (IRRB), glomerular barrier, and pericytes of vascular systems of the placenta, and adipose tissue are fully permissive for HCMV lytic replication. I also observed consistent induction of proinflammatory and angiogenic cytokines after HCMV infection in all of these pericyte populations. Even more, I have examined murine retinal pericyte and found that they are also permissive for mouse cytomegalovirus (MCMV) lytic replication.

D. J. Alcendor (✉)
Center for AIDS Health Disparities Research, Meharry Medical College, Nashville, TN, USA
e-mail: dalcendor@mmc.edu

A. Birbrair (ed.), *Biology of Pericytes – Recent Advances*, Stem Cell Biology and Regenerative Medicine 68, https://doi.org/10.1007/978-3-030-62129-2_4

Conclusions
The mechanisms associated with HCMV infection and the increased risks for the development of vascular diseases requires further investigation. Studies have shown that people exposed to HCMV infection had higher risk for vascular disease. There are antivirals that can effectively inhibit HCMV replication thus therapeutic modalities designed to specifically protect pericytes populations from HCMV infection could impact vascular disease outcomes especially among the elderly and immune compromise patients.

Keywords Pericytes · HCMV · Cytomegalovirus · Infection · Inflammation · Brain · Placenta · Kidney · Ocular disease · Adipose · Liver · Stellate cells · Neurovascular unit · Latency · Neurodegenerative disease · Virus replication

4.1 Introduction

Cytomegalovirus is the most common congenital virus infection in the world. HCMV congenital infections in newborn infants ranges from 0.5% to 0.7% in developed countries and from 1 to 2% in developing countries (Kenneson and Cannon 2007). Approximately 13% of newborns with congenital HCMV are symptomatic at birth and have clinical presentations that includes microcephaly, intrauterine growth restriction, seizures, brain calcifications, hepatosplenomegaly, thrombocytopenia and chorioretinitis (Dollard et al. 2007). Congenital human cytomegalovirus (HCMV) infections can result in abnormalities including intrauterine growth restriction (IUGR), vision loss, neurocognitive disability, motor deficits, seizures, and hearing loss (Nigro and Adler 2011; Stagno et al. 1986). Forty percent of mothers with primary HCMV infection during gestation transmit the infection to their babies (Boppana et al. 1992, 2005). Even more, 58% of transplacental transmission of HCMV occur in women who are seropositive with non-primary maternal infections (Kylat et al. 2006; Boppana et al. 2005). Annually, about 1 out of 200 babies in the US are born with congenital HCMV (Pass and Anderson 2014; Pass et al. 2006; Enders et al. 2011; Wang et al. 2011). Preventing congenital HCMV is a public health priority (Rawlinson et al. 2017). HCMV not only causes life-threatening disease in immunocompromised individuals (Lichtner et al. 2015; Beam et al. 2014), but also HCMV-associated pathologies that lead to long-term health risks represents an important health disparity in underserved communities. Higher infection rates are observed among non-Hispanic Blacks and Mexican Americans than among non-Hispanic Whites (Colugnati et al. 2007). Higher frequencies of new HCMV infections were found among non-Hispanic Blacks and Mexican Americans compared to Whites, and primary infections among adolescent girls 12–19 years of age were 50× more likely in seronegative non-Hispanic Blacks and 15× more likely in seronegative Mexican Americans than in non-Hispanic Whites (Colugnati et al. 2007). Higher rates of primary maternal infections among ethnic minorities resulted in increased rates of congenital disease, perinatal morbidity and higher health care expense.

The human blood-brain barrier (BBB) consists of brain microvascular endothelial cells, brain vascular pericytes, and astrocytes; together they are known as the neurovascular unit (NVU). Pericytes of the BBB play an essential role in a range of microvascular functions, including angiogenesis, vascular remodeling, regression, and stabilization, as well as generation and maintenance of the BBB (Bergers and Song 2005; Armulik et al. 2010, 2011; Hamilton et al. 2010). My studies have revealed that primary human pericytes from vascular compartments of the brain, retina, glomerulus, placenta, and adipose tissue are all permissive for HCMV lytic replication. Even more, I have observed that the pericyte component in vascular compartments of the brain, retina, glomerulus (mesangial cells/pericyte of the glomerulus) and placental vascular pericytes are the most permissive cell type for HCMV infection when compared to other related cells types (Alcendor et al. 2012; Wilkerson et al. 2015; Popik et al. 2019; Aronoff et al. 2017). I also find consistent upregulation of proinflammatory cytokines in these pericyte populations post HCMV infection when compared control cells (Alcendor et al. 2012; Wilkerson et al. 2015; Popik et al. 2019; Aronoff et al. 2017). This permissiveness for HCMV infection occurs across species as demonstrated when I used murine retinal pericytes and mouse cytomegalovirus (MCMV). The connection between HCMV and vascular disease has been well documented. The molecular mechanisms that governs HCMV universal tropism for human pericytes populations had not been appreciated. Therefore the impact of HCMV infection on pericyte populations and its role in the development and progression of vascular disease requires further investigation. Insights obtained from these studies could provide new information for developing novel therapeutic approaches for stroke and other disorders with a neurovascular vascular component, including Alzheimer's disease and Parkinson Disease.

4.2 Cytomegalovirus Biology

It was Hugo Ribbert who in 1881 discovered the first evidence of "cytomegalia" as inclusions in human kidney and parathyroid gland cells (Ribbert 1904). Many years later it was collaborations between Smith, Rowe, and Weller in 1956 and 1957 that the virus was first isolated and referred as cytomegalovirus (Smith 1956; Rowe et al. 1956; Weller et al. 1957). Human cytomegalovirus or human herpesvirus type-5 (HHV-5) is a member of the human Herpesvirus family in the order Herpesvirales, in the family Herpesviridae, and in the subfamily Betaherpesvirinae (Brito and Pinney 2020). HCMV, has a lipid bilayer enveloped and a protective viral tegument protein layer with an icosahedral capsid (Fig. 4.1a) and an electron dense viral core (Fig. 4.1b) (Yu et al. 2017). HCMV has a double-linear stranded DNA genome of about 235 kb (Martí-Carreras and Maes 2019).

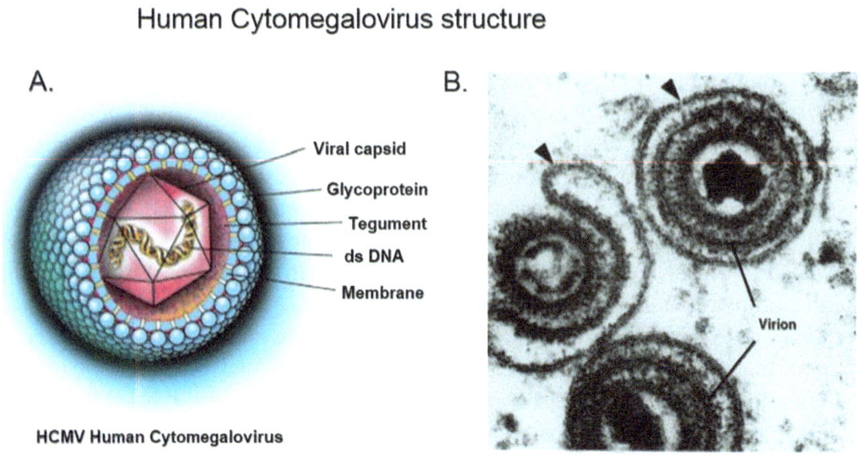

Fig. 4.1 HCMV virion structure. (**a**) Features of the HCMV virion structure showing viral capsid, glycoprotein spikes, viral tegument protein, DNA genome, and viral membrane (**b**) Scanning electron micrograph showing the HCMV virion structure with black arrows pointing to the lipid bilayer membrane

The seroprevalence of HCMV world-wide is between 45% and 100% in adult populations (Cannon et al. 2010). HCMV can cause primary infections in immunocompetent individuals in the form of a mononucleosis-like syndrome (Alberola et al. 2000). Most often these infections are asymptomatic but the virus can establish life-long latency and can reactivate in immunosuppressed individuals (Heald-Sargent et al. 2020). Even more immunocompetent individuals will often shed virus intermittently throughout the life time without clinical disease (Vancíková and Dvorák 2001). The primary target cells of HCMV are monocytes, lymphocytes, and epithelial cells, and the virus establishes latency as circularized episomes inside the nuclei of bone marrow progenitor cells that are CD33+ and CD34+ as well as peripheral blood mononuclear cells (Schottstedt et al. 2010; Collins-McMillen et al. 2018; Sinclair and Sissons 2006). However, HCMV has been shown to infect fibroblasts; epithelial, endothelial, stromal cells, and smooth muscle cells (Haspot et al. 2012); and adipocytes (Bouwman et al. 2008). Our recent studies suggest that vascular pericytes are primary target cells for HCMV and likely serve as amplification reservoirs for HCMV replication in multiple vascular bed. HCMV is a ubiquitous opportunistic pathogen and most notably HCMV can cause severe life-threatening disease in the HIV patient that goes undiagnosed or who is untreated for their HIV disease, the transplant patient under iatrogenic immunuosuppression, and a child of an infected mother that acquires a primary infection early in pregnancy (Boeckh and Geballe 2011). Viral latency and reactivation from latency along genetic strain variation that occurs among clinical strains of HCMV represents a major obstacle for the development of an effective anti-HCMV vaccine (Chen et al. 2019).

HCMV has been implicated in the development of atherosclerosis leading to coronary heart disease, malignant gliomas, systemic autoimmune diseases, type 1 diabetes, type 2 diabetes, Alzheimer's disease, rheumatoid arthritis, systemic lupus erythematosus, and contributing to frailty via chronic inflammation (Blankenberg et al. 2001; Farias et al. 2019; Pawlik et al. 2003; Pak et al. 1988; Chen et al. 2012; Nell et al. 2013; Rothe et al. 2016; Berman and Belmont 2017; Wang et al. 2010). The role of HCMV in these chronic diseases have not been fully determined and need further study (Janahi et al. 2018).

4.3 Pericytes

Pericytes, were first describes by Eberth and Rouget and were also known as Roget cells or mural cells in the 1870s. However the name "pericyte" was introduced by Zimmermann in 1923 (Eberth 1871; Rouget 1873; Zimmermann 1923; Attwell et al. 2016). Pericytes were described as cells that were abluminal to endothelial cells that were later shown to have contractile properties due to their cytoplasmic extensions that wrap around the endothelial cells lining the capillaries (Fig. 4.2a-1, a-2) and venules throughout the body (Sims 1986; Bergers and Song 2005; Armulik et al. 2010, 2011; Hamilton et al. 2010; Rustenhoven et al. 2017). Primary pericytes can be cultivated to confluency and stained to confirm biomarker characterization (Fig. 4.2b-1, b-2, b-3). There are no unique antigenic biomarker for pericytes however they can be identified by their expression of the platelet-derived growth factor receptor-beta (PDGFRb) and of the neural/glial antigen 2 (NG2 proteoglycan) a co-receptor for PDGF (Fig. 4.2b-4) (Smyth et al. 2018; Stallcup 2018). The highest density of vascular pericytes in humans have been observed in the brain and retina (Sims 1986). There are functional differences among pericytes depending on their origin, morphology and their organ-derived vascular bed. Pericytes and microvascular endothelial cells share a common membrane (Birbrair 2018; Abbott et al. 2010) (Fig. 4.3a). The human blood-brain barrier (BBB) consists of brain microvascular endothelial cells, brain vascular pericytes, and astrocytes; together they are known as the neurovascular unit (NVU) (Brown et al. 2019) (Fig. 4.3b). Perciytes are pluripotent cells that are an essential component of the blood-brain-barrier and play major role in the BBB development, stability, angiogenesis, immunoregulation, function, and maintenance of the BBB (Fig. 4.3c). The physical contact and paracrine signaling between pericytes and endothelial cells are essential for new blood vessel formation, maturation, and maintenance. During cellular invasion after traumatic brain injury pericytes are among the first cell to invade the primary lesion in models of brain and spinal injury (Göritz et al. 2011; Birbrair et al. 2018; Dias et al. 2018; Hesp et al. 2018) in support of vascular remodeling.

Recent studies have revealed that pericyte deficiency in the CNS can cause a breakdown of the BBB and lead to other degenerative changes in the brain. Most

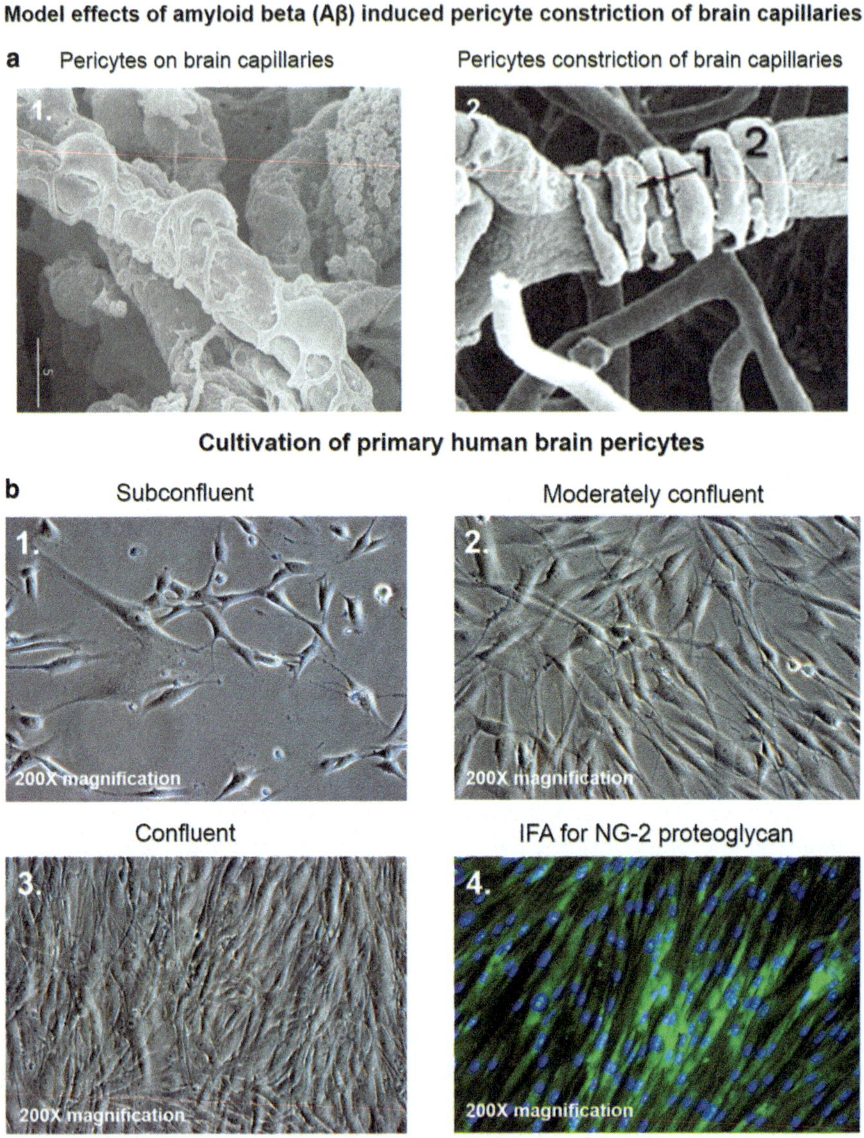

Fig. 4.2 Brain vascular pericytes. (**a**) Pericytes as they appear on rat brain capillaries (2a-1) Rat brain capillaries undergoing constriction by vascular pericytes (2a-2). (**b**) Primary human brain pericytes subconfluent (2b-1); Primary human pericytes moderately confluent (2b-2); Confluent primary human brain pericytes (2b-3); Primary human brain pericytes stained with a monoclonal antibody for NG-2 proteoglycan. Total magnification is 200×

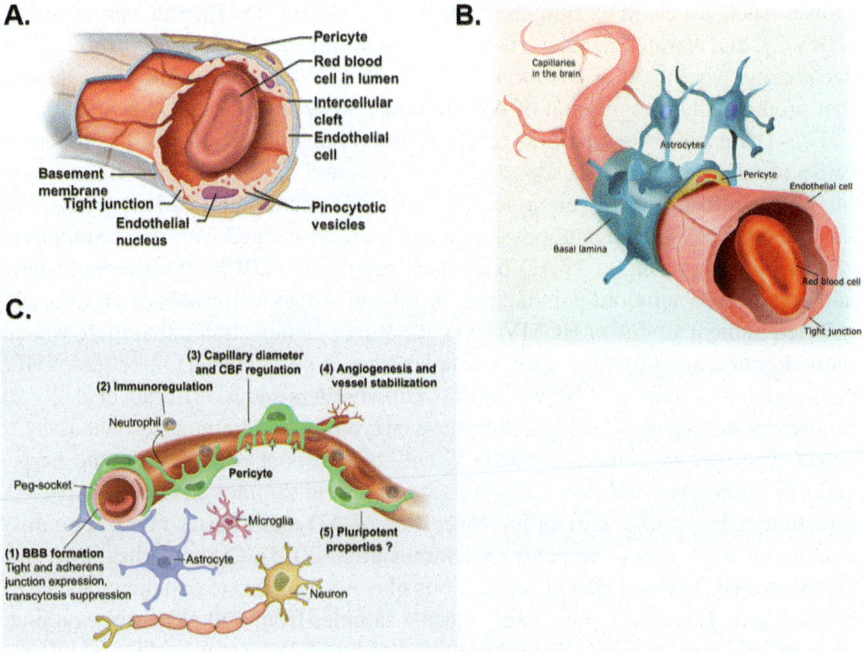

Fig. 4.3 Pericytes and the neurovascular unit (NVU). Brain capillaries featuring pericytes. (**a**) Cross-section of a brain capillary illustrating the capillary basement membrane, as well as the endothelial cells and brain pericytes abluminal to endothelial cells sharing a contiguous basement membrane. (**b**) Neurovascular unit (NVU) of the blood-brain barrier (BBB) featuring capillary endothelial cells with tight junction, pericytes and astrocytes. (**c**) The NVU and other brain parenchymal cells, including neurons and microglia. Important functions of pericytes are also shown listed 1–5

recently, pericytes dysfunction has been observed in Alzheimer's disease (AD) pathology (Salmina et al. 2019; Yamazaki and Kanekiyo 2017; Liesz 2019; Cai et al. 2018). Studies reveal that Amyloid-β (Aβ) induces neurovascular dysfunction that leads to functional changes in the brain microvascular resulting in the impairment of capillaries to respond to neuronal activity (Nortley et al. 2019). Studies by Nortley et al., suggest that brain vascular pericytes link Aβ to vascular dysfunction in AD (Nortley et al. 2019). Using human brain slices and in a mouse model they reported that oxidative stress caused by toxicity of Aβ causes constriction of capillaries via the generation of reactive species (ROS) by the induction of NOX4 (reduced nicotinamide adenine dinucleotide phosphate oxidase 4). ROS then triggers the release of endothelin-1, which interacts with ETA receptors on pericytes to cause pericyte contraction and subsequent capillary constriction (Nortley et al. 2019). It is clear that AD is a multifactorial disease and the implication HCMV in AD pathogenesis is controversial. Infection due to reactivation of latent neurotropic viruses such as HCMV, Human herpesvirus 1 (HHV-1), Human herpesvirus 2 (HHV-2), and Varicella zoster virus, have been linked to AD neuropathology for decades but whether viral infections could be considered causative in AD or a factor that promotes the progression of AD continues to be investigated (Sochocka et al. 2017). There have been several recent reports that link HCMV as an infectious cause of AD. Laurain et al., reported virological and immunological characteristic of HCMV infection that were associated with AD (Lurain et al. 2013). They demonstrated HCMV serum antibody levels that were associated with the development of neurofibrillary tangles in AD patients (Lurain et al. 2013). In this same study, authors showed amyloid-β induction in human foreskin fibroblasts (HFF cells) infected clinical strains of HCMV. Even more they demonstrated that there was no induction of amyloid-β after herpes simplex virus type 1 (HSV-1) infection of HFF cells and no association of HSV-1 serum antibody titers and AD (Lurain et al. 2013). Studies by Barnes et al., using solid-phase enzyme-linked immunosorbent assay to detect type-specific immunoglobulin G antibody responses to HCMV and herpes simplex virus type 1 (HSV-1) in 849 archived serum samples showed that HCMV infection is associated with an increased risk of AD and a faster rate of cognitive decline in older diverse populations (Barnes et al. 2015). Other studies have suggested that HCMV and HSV-1 interactions play a role in AD (Lövheim et al. 2018). In a study by Loveheim et al., using plasma samples from 360 AD cases examined both HCMV and HSV IgG and IgM antibodies by ELISA assay and found interaction between HCMV and HSV1 to be associated significantly with AD development (Lövheim et al. 2018). However, the role of HCMV, HSV-1/2, and other neurotropic viruses in AD pathobiology is controversial and the mechanisms to support these associations as causative or supportive of progressive disease will require further investigations. The permissiveness of brain pericytes for HCMV infection and the role pericyte loss and dysfunction in AD associated vascular pathology could be linked and deserves further study. As pericyte death reliably indicates very mild cognitive impairments (Nation et al. 2019), it is likely that pericyte death is not only a symptom of AD but may also be a potential cause of AD. Hypoxia influences amyloid precursor protein (APP) mechanisms that lead to an increased production of b- and g-secretase and therefore increased Aβ production (Zlokovic 2011). This suggests that pericyte loss may be an early event in the progression of AD, as pericyte loss reduces cerebral blood flow, thus creating potential hypoxic areas throughout the brain.

4.4 Cytomegalovirus and Vascular Disease

Several serological and molecular-biological studies have shown an association between HCMV infection of endothelial cells and in the development of atherosclerosis (AS) (Gkrania-Klotsas et al. 2012; Muhlestein et al. 2000; Simanek et al. 2011; Spyridopoulos et al. 2016). A meta-analysis of data retrieved from electronic databases that involved 30 studies, 3328 cases and 2090 controls showed that HCMV infection is significantly associated with an increased risk for atherosclerosis. A meta-analysis conducted by Lv et al., involving 68 studies, from 24 countries (12,027 cases and 15,386 controls) suggests that HCMV infection is associated with an increased risk for vascular diseases (Lv et al. 2017). The study found that people exposed to HCMV infection had higher risk for vascular diseases (OR 1.70 [95% CI 1.43–2.03]) (Lv et al. 2017). HCMV infection has been shown to be a risk factor for the development of renal vascular disease. Most recently, Lee et al., have shown the high burden of HCMV results in poor vascular health in renal transplant patients compared to healthy controls (Lee et al. 2019). After Examining carotid intima-media thickness, FMD, eGFR and plasma levels of HCMV antibodies and the expression of ICAM-1, VCAM-1, P-selectin, sIFNαR2, sTNFR1, sCD14 and CRP they predicted that high viral burden was associated with poor endothelial health and vascular damage in renal transplant patients when compared to controls (Lee et al. 2019). In addition, a study by Huang et al., involving 200 patients diagnosed with stroke and 200 controls found a correlation between HCMV and stroke (Huang et al. 2012). HCMV seropositivity was higher in the stroke patients than in controls (55.0% vs. 23.5%; $P < 0.0001$). Huang et al., also observed that the presence of HCMV DNA increased the risk of stroke (Huang et al. 2012). The mechanisms associated with HCMV infection and the increased risks for the development of vascular diseases requires further investigation. Even more HCMV infection upregulates the endothelin receptor type B protein (ETBR) expression and mRNA expression in both endothelial cells and smooth muscle cells (Yaiw et al. 2015). ETBR, a G protein-coupled receptor that mediates the vascular effects of ET-1 a potent vasoconstrictor (Yaiw et al. 2015). HCMV may have a role in cardiovascular disease via upregulation of ETBR. ET-1 is also known to cause capillary constriction via ETA receptor interactions on pericytes.

4.4.1 Cytomegalovirus Infection of Pericytes in the Neurovascular Unit

The blood-brain barrier (BBB) interfaces the peripheral circulation and the central nervous system (CNS) allowing nutrients into the CNS and preventing blood-borne pathogens from harming the brain. This barrier is an elaborate network of tight junctions (TJ) between capillary endothelial cells that lack fenestrae and have a reduced capacity for pinocytosis (Ballabh et al. 2004; Engelhardt 2003). The TJ of the capillary endothelium is supported by astrocytic endfeet and pericytes. Cerebral vascular pericytes (CNS pericytes) have been shown to enhance TJ barrier function, stimulate expression of TJ proteins and reduce the paracellular permeability of the capillary endothelium (Garberg et al. 2005; Ramsauer et al. 2002; Shepro and Morel 1993; Tsukita et al. 2001). Pericytes are adult multipotent, contractile and migratory stem cells (Balabanov and Dore-Duffy 1998; Dore-Duffy 2008) that surround capillaries and actively communicate with other cells of the neurovasculature, including endothelial cells, astrocytes and neurons. Completely surrounded by a basal lamina, they also contribute to the deposition of the basal lamina during vascular development and angiogenesis (Dore-Duffy and Cleary 2011). Although a critical cellular component for the development and function of the BBB, the role of pericytes in HCMV infection and dissemination has largely been ignored (Lai and Kuo 2005). Rather, to date, astrocytes and brain microvascular endothelial cells (BMVEC) cells have been implicated as cell types that support HCMV dissemination at the blood-brain barrier level (Cheeran et al. 2009). Recent studies have revealed that pericyte deficiency in the CNS can cause a breakdown of the BBB and lead to other degenerative changes in the brain (Cai et al. 2017). Ribbert (1904) demonstrated murine CMV (MCMV) infection of pericytes in brown and white adipose tissue of young adult infected mice. These mice later developed viral-induced inflammatory lesions in peripancreatic and salivary gland adipose tissues. However it was Alcendor et al., that was the first demonstrate HCMV infection of primary human brain pericytes (Alcendor et al. 2012). Using a clinical isolate of HCMV (SBCMV), microscopy of infected pericytes showed virion production and typical cytomegalic cytopathology (Fig. 4.4a1, a-2, a-3). This finding was confirmed by the expression of major immediate early and late virion proteins and by the presence of HCMV mRNA (Fig. 4.4a4, a5) and finally by electron microscopy (Fig. 4.4a-6). Brain pericytes were fully permissive for CMV lytic replication after 96 h in culture compared to human astrocytes or human brain microvascular endothelial cells (BMVEC) (Fig. 4.4b-1, b-2, b-3, b-4, b-5, b-6). However, temporal transcriptional expression of pp65 virion protein after SBCMV infection was lower than that seen with the HCMV Towne laboratory strain (Alcendor et al. 2012). Using RT-PCR and dual-labeled immunofluorescence, proinflammatory cytokines CXCL8/IL-8, CXCL11/ITAC, and CCL5/Rantes were upregulated in SBCMV-infected cells, as were tumor necrosis factor-alpha (TNF-alpha), interleukin-1 beta (IL-1beta), and interleukin-6 (IL-6) (Alcendor et al. 2012). Pericytes exposed to SBCMV elicited higher levels of IL-6 compared to both mock-infected as well as heat-killed virus controls (Alcendor et al. 2012). A 6.6-fold induction of IL-6 and no induction TNF-alpha was observed in SBCMV-infected cell supernatants at 24 h

Fig. 4.4 (continued) pericytes, and (3) pericytes 72 h after infection with SBCMV. Immunofluorescence staining of SBCMV-infected pericytes for (4) HCMV MIE protein and (5) pp65 late protein. (6) TEM of SBCMV-infected pericytes showing HCMV virions in the cytoplasm (see arrow). With the exception of the TEM, images were taken on a Nikon TE2000S microscope (200× magnification). *HCMV* human cytomegalovirus, *MIE* major immediate early protein, *SBCMV* primary HCMV isolate from a patient, *TEM* transmission electron microscopy. (**b**) Time course analysis of human cytomegalovius-GFP (HCMV-GFP) infection of BBB (blood-brain barrier) cells. For comparison purposes, (a) the top panel includes phase contrast images of human mock and infected brain microvascular endothelial cells, brain vascular pericytes and astrocytes. The bottom panel shows phase contrast images of infected brain microvascular endothelial cells, brain pericytes and astrocytes with a fluorescent overlay showing HCMV-GFP-positive cells. Total magnification is 200×

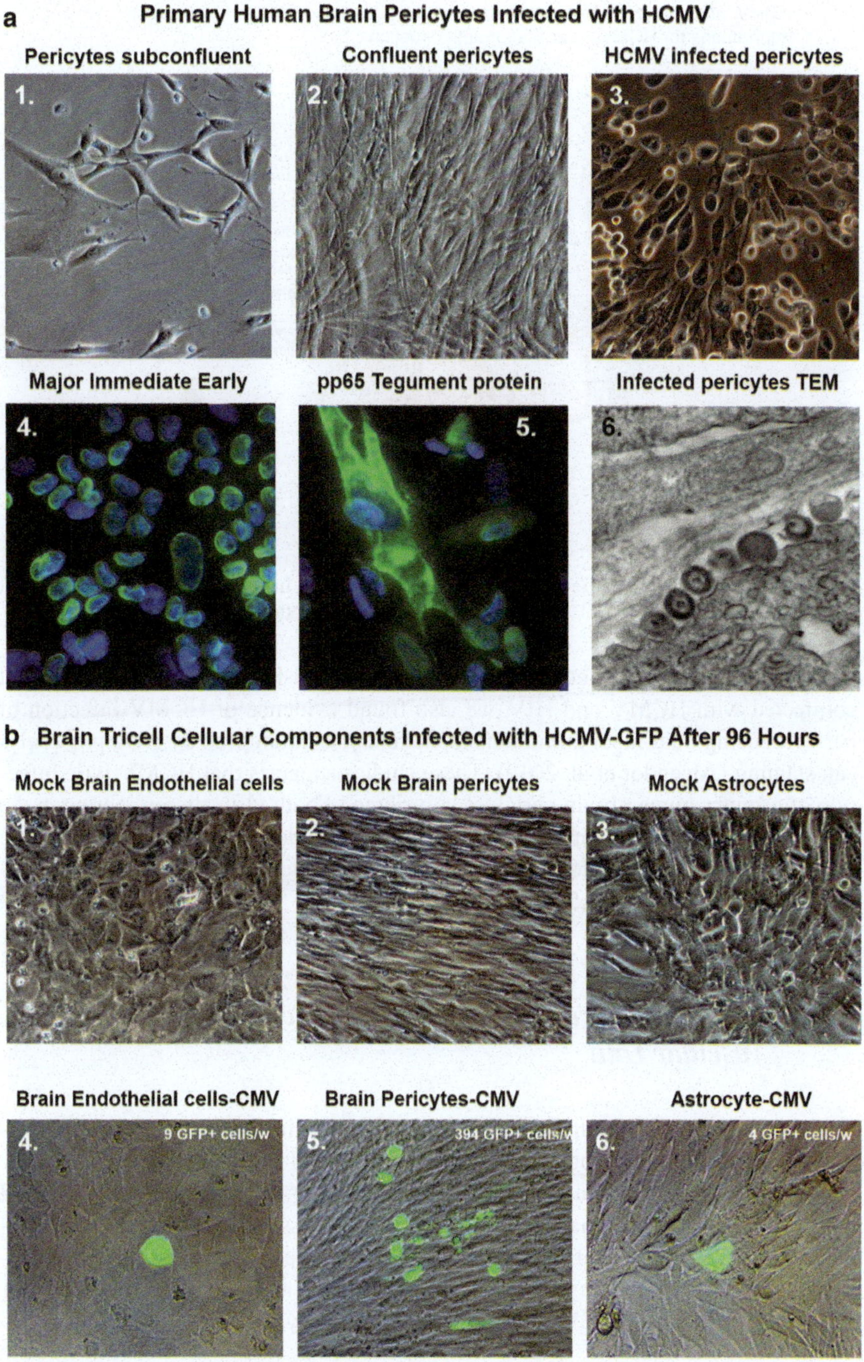

Fig. 4.4 (**a**) Primary brain vascular pericytes. Phase contrast images of: (1) an uninfected subconfluent monolayer of primary brain vascular pericytes, (2) a confluent monolayer of brain vascular

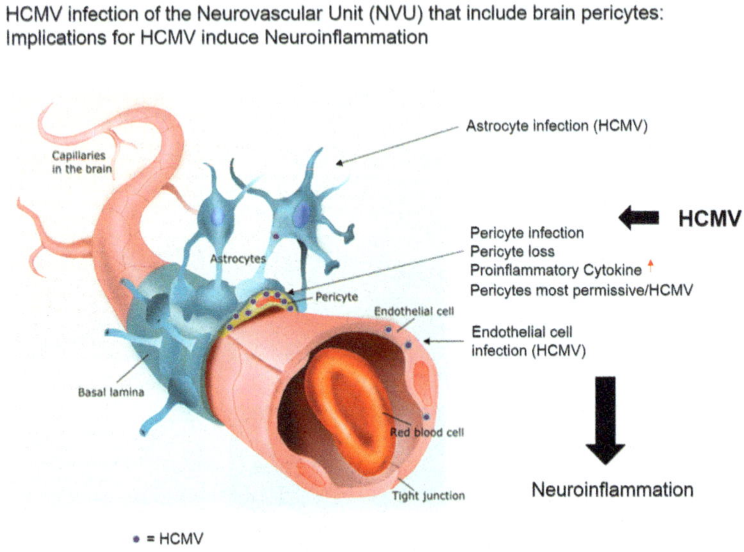

Fig. 4.5 Model of HCMV infection and viral dissemination in the NVU. HCMV infection of cellular components of the NVU and the effects of HCMV infection in pericytes

post infection (Alcendor et al. 2012). Using archival brain tissue from a patient coinfected with HCMV and HIV, we also found evidence of HCMV infection of pericytes using dual-label immunohistochemistry, as monitored by NG2 proteoglycan staining (Alcendor et al. 2012). These studies suggests that HCMV lytic infection of primary human brain pericytes contribute to both virus dissemination in the CNS as well as neuroinflammation. A hypothetical model of HCMV infection of the NVU shows viral replication preference for brain pericytes when compared astrocytes and BMVEC (Fig. 4.5).

4.4.2 Cytomegalovirus Infection of Pericytes in Retinal Vascular Unit

The inner blood-retinal barrier (IBRB) consists of retinal microvascular endothelial cells covered with tightly associated pericytes and Müller cells; together, they make up the retinal vascular unit (RVU) (Wilkerson et al. 2015). My study showed that normal primary human retinal pericytes expressed the biomarker neuron-glial antigen 2 and that retinal pericytes are highly permissive for HCMV infection as demonstrated by cytomegalic cytopathology and expressed of the major immediate protein (MIE) and the late phosphorylated envelop protein 65

(Fig. 4.6a-1, a-2, a-3, a-4, a-5) and shown to support lytic replication of a recombinant GFP virus (Fig. 4.6a-6) (Wilkerson et al. 2015). qRT-PCR analysis showed full lytic replication of HCMV in retinal pericytes. Pericytes were the most permissive cell type for HCMV replication when compared to retinal endothelial cells and Möller cells (Fig. 4.6b1, b-2, b-3, b-4, b-5, b-6) Pericytes exposed to SBCMV expressed higher levels of vascular endothelial cell growth factor mRNA compared to controls (Wilkerson et al. 2015). Luminex analysis of supernatants from SBCMV-infected retinal pericytes had increased levels of macrophage inflammatory protein-1α, beta-2 microglobulin (B2-m), matrix metalloproteinase-3 and -9 (MMP3/9), and lower levels of IL-6 and IL-8 compared to controls. At 24 h post infection, pericytes expressed higher levels of IL-8, TIMP-1 (tissue inhibitor of metalloproteinase-1), and RANTES (regulated upon activation normal T cell-expressed and presumably secreted) but lower levels of MMP9 (Wilkerson et al. 2015). Time course analysis showed that both brain and retinal pericytes were more permissive for HCMV infection than other cellular components of the BBB (blood-brain barrier) and IBRB. Using a Tricell culture model of the IBRB (retinal endothelial, pericytes, Müller cells), retinal pericytes were most permissive for SBCMV infection (Wilkerson et al. 2015). SBCMV infection of this IBRB Tricell mixture for 96 h that resulted in increased levels of IL-6, MMP9, and stem cell factor with a concomitant decrease in granulocyte-macrophage colony-stimulating factor and TNF-alpha (Wilkerson et al. 2015). In retinal pericytes, HCMV induces proinflammatory and angiogenic cytokines. In the IBRB, pericytes likely serve as an amplification reservoir which contributes to retinal inflammation and angiogenesis. Even more, I examined the permissiveness of normal murine retinal pericytes expressing alpha actin smooth muscle (Fig. 4.7a–c) for mouse cytomegalovirus (MCMV) infection. At 72 h post infection I observed pericytes being fully permissive for MCMV lytic replication as demonstrated by expression of the late MCMV M55 late protein (Fig. 4.7d) and lytic replication of a recombinant MCMV expressing green fluorescent protein (GFP) (Fig. 4.7e).

4.4.3 Cytomegalovirus Infection of Mesangial Cells: Pericytes of the Glomerular Vascular Unit

Mesangial cells are considered specialized pericytes of the renal glomerulus (Yamanaka 1988; Diaz-Flores et al. 2009; Lindahl et al. 1998; Smith et al. 2012). The glomerular vascular unit (GVU) consists of glomerular endothelial cells, podocytes, and mesangial cells (Popik et al. 2019). The molecular crosstalk between mesangial cells, podocytes, and glomerular endothelial cells is essential for glomerular filtration. Mesangial cells form a supporting framework that maintains the structural integrity of the glomerular tuft that includes the glomerular capillaries

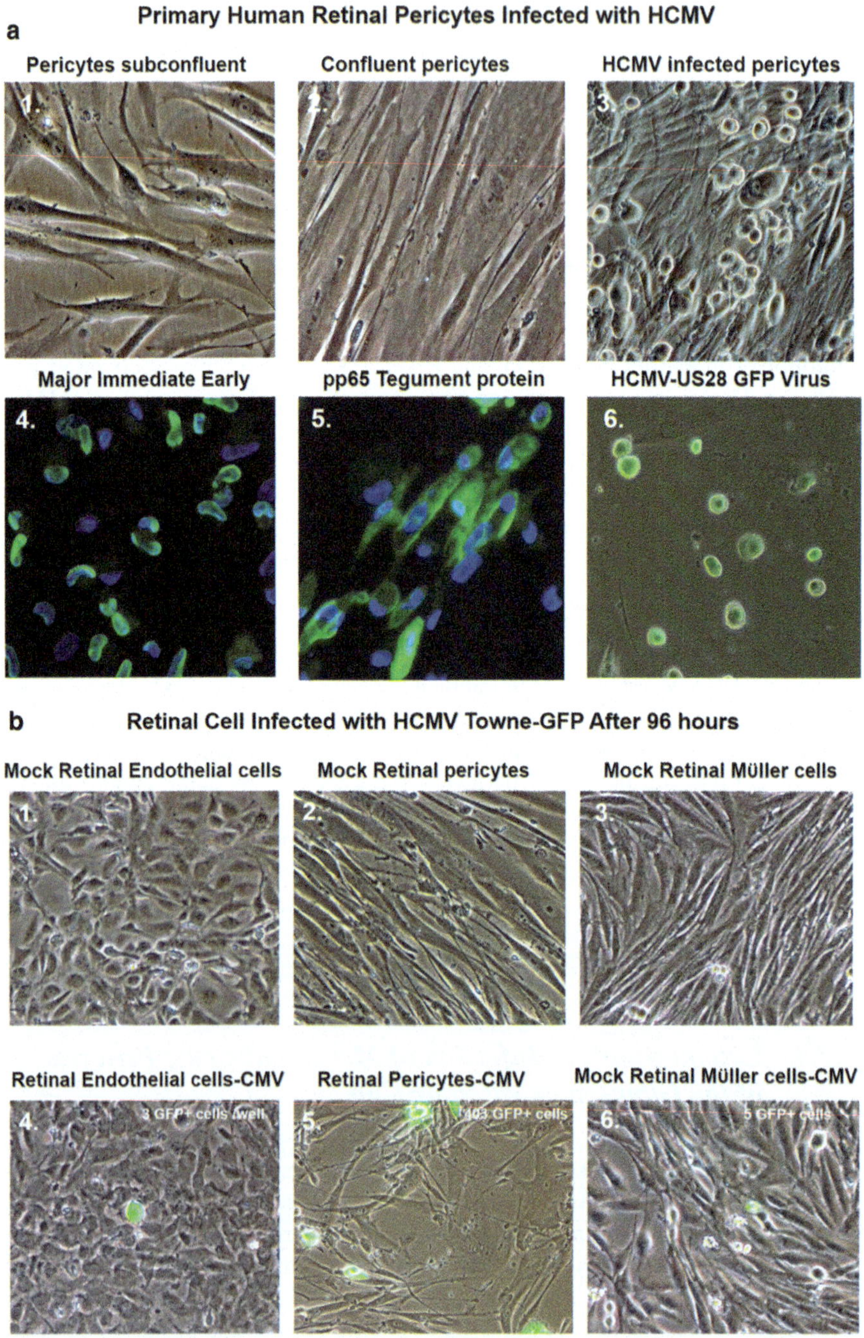

Fig. 4.6 Human cytomegalovirus (HCMV) infectivity of primary human retinal pericytes. (**a**) Phase contrast and fluorescent images of retinal pericytes (1) uninfected subconfluent

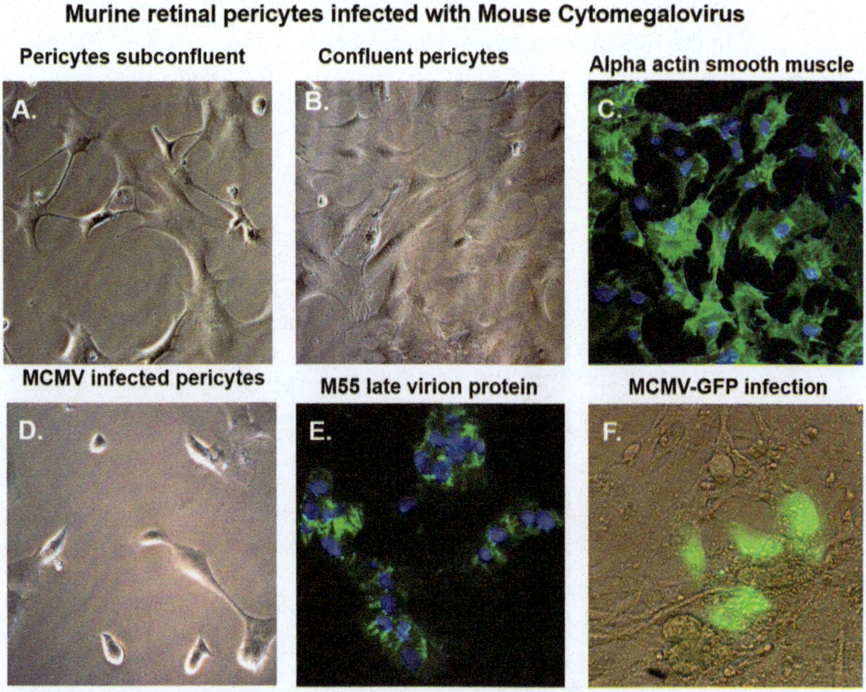

Fig. 4.7 MCMV infection of mouse retinal pericytes. Cultivated mouse pericytes infected with mouse cytomegalovirus. (**a**) Pericytes subconfluent; (**b**) Pericytes confluent; (**c**) Pericytes staining positive for alpha actin smooth muscle (green FITC/blue DAPI); (**d**) Phase image of pericytes infected with MCMV; (**e**) Pericytes staining positive for the MCMV M55 late protein; (**f**) Pericytes infected with a GFP-positive MCMV (virus infected appear green)

(Wilkerson et al. 2015). Mesangial cells affect glomerular hemodynamics by altering glomerular vascular resistance (Wilkerson et al. 2015). To date, mesangial cells, and their contribution to HCMV infection in the glomerulus is poorly understood. I showed that mesangial cells and glomerular endothelial cells but not podocytes were permissive for both lab adapted and clinical strains of HCMV (Fig. 4.8a) (Popik et al. 2019). A hypothetical model is shown to support this finding (Fig. 4.8b).

Fig. 4.6 (continued) monolayer of retinal pericytes, (2) confluent monolayer of retinal pericytes and (3) pericytes 96 h after infection with the clinical strain SBCMV. Immunofluorescence staining of SBCMV-infected retinal pericytes for (4) major immediate protein (MIE) protein and (5) pp65 late stage protein. (6) Phase fluorescent overlay image of human retinal pericytes infected with a recombinant HCMV virus expressing GFP. All images were taken on a Nikon TE2000S microscope (200× magnification); (**b**) Time course analysis of human cytomegalovirus-GFP (HCMV-GFP) infection of IBRB (inner blood-retinal barrier) cells. (a) Top panel: phase contrast images of human mock and infected retinal microvascular endothelial cells, retinal pericytes and Müller cells. Bottom panel: phase contrast images of infected retinal microvascular endothelial cells, retinal pericytes and Müller cells with a fluorescent overlay showing HCMV-GFP-positive cells. Magnification 200×

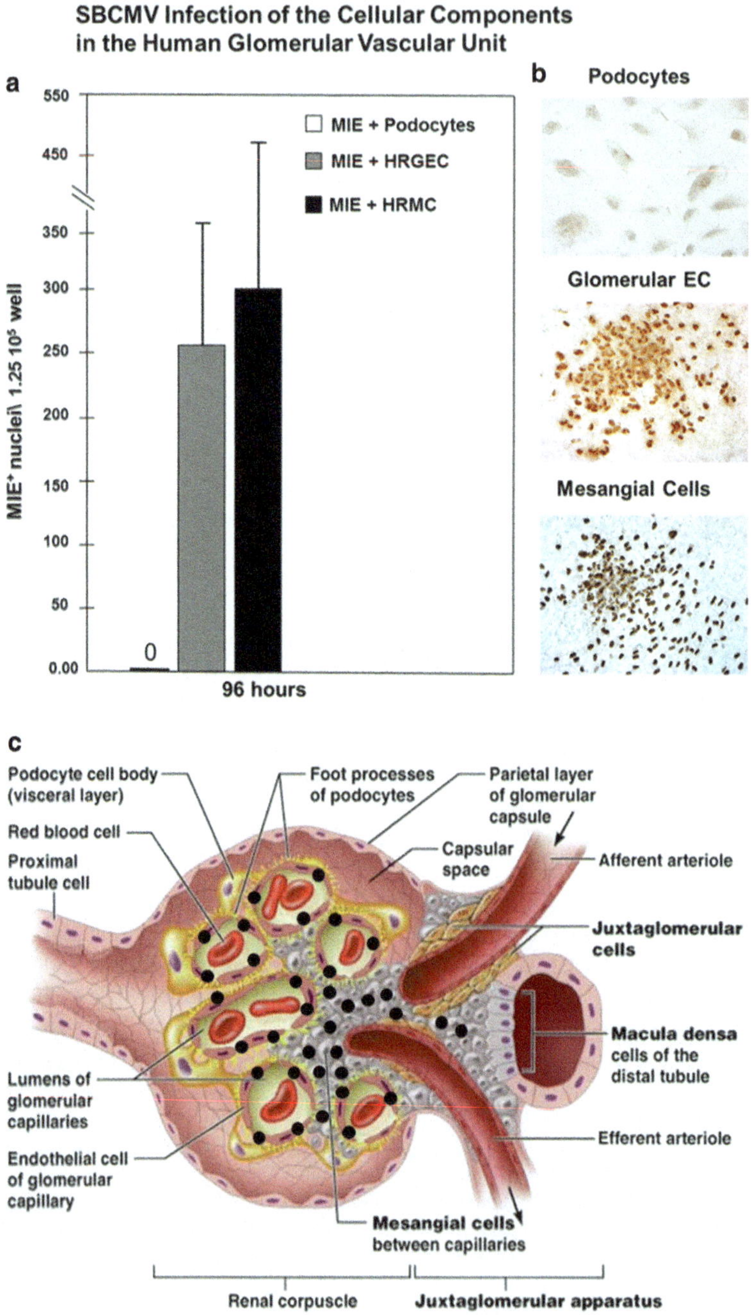

Fig. 4.8 Infection with the clinical strain (SBCMV) of GVU (Glomerular Vascular Unit) cells 96 h after infection. (**a**) A graph showing the number of infected SBCMV positive podocytes (open bars), glomerular endothelial cells (gray bars), and mesangial (black bars) per 4 × 106 total cells

Luminex analysis revealed dysregulation proinflammatory cytokines expressed by mesangial cells exposed to the SBCMV clinical strain (Popik et al. 2019). HCMV upregulation of angiogenic and proinflammatory cytokines could contribute to glomerular inflammation (Popik et al. 2019). Briefly, mesangial cells exposed to SBCMV for 96 h revealed an increase in the expression of B2-m (beta-2 microglobulin) ferritin, complement C3, IL-6, IL-7, IL-8, RANTES, VEGF, and MMP-3 compared to mock infected controls (Popik et al. 2019). Taken together upregulation of these factor are known to greatly impact the development of kidney disease. However, because these data represent in vitro studies in primary cells infected at low multiplicities of infection this will require further investigation. Although I was able to demonstrate HCMV infection in pericytes in renal tissue from a transplant patient supporting in vivo for HCMV infection of pericytes in renal tissue (Popik et al. 2019).

4.4.4 Cytomegalovirus Infection of Pericytes in at the Placental Vascular Barrier

Placental pericytes are essential for endothelial cell proliferation as well as for placental microvasculature stability and integrity but have largely been ignored in placenta biology (Price et al. 1990). Pericytes are also critical for placental vascular development and angiogenesis. A blood-placental barrier model consisting of trophoblasts and placental pericytes has been established. HCMV placental pathogenesis models that include placental pericytes have not been reported. The signaling mechanisms between placental pericytes, cytotrophoblast, and villous fibroblasts are largely unknown (Alcendor et al. 2012).

I was among the first (to my knowledge) to report the infectivity of human placental pericytes for HCMV, their potential role in viral dissemination in placental tissue, and the implications for HCMV-associated congenital disease (Aronoff et al. 2017). I first tracked HCMV in placental tissue obtained from a child with HCMV inclusion disease by dual labeled immunohistochemistry and showed overtime that the virus infects cytotrophoblasts and spreads into villous during dissemination (Fig. 4.9a). My findings show that placental pericytes strongly support HCMV replication, inducing proinflammatory and angiogenic cytokines that could contribute to viral dissemination, placenta inflammation, and dysregulation of placental angiogenesis. I showed that primary human placenta pericytes were more permissive for HCMV infectivity than either primary human cytotrophoblasts or villous fibroblasts (Fig. 4.9b, c). I also

Fig. 4.8 (continued) 96 h after infection. (**b**) Right panel: Immunohistochemical stained images of SBCMV infected podocytes, glomerular endothelial cells, and mesangial cells showing SBCMV positive cells (stained brown). All images were taken on a Nikon TE2000S microscope mounted with a charge-coupled device (CCD) camera at 200× magnification. (**c**) Hypothetical model of HCMV dissemination in the glomerulus of the human kidney showing HCMV infection of mesangial cells and glomerular endothelial cells but not podocytes

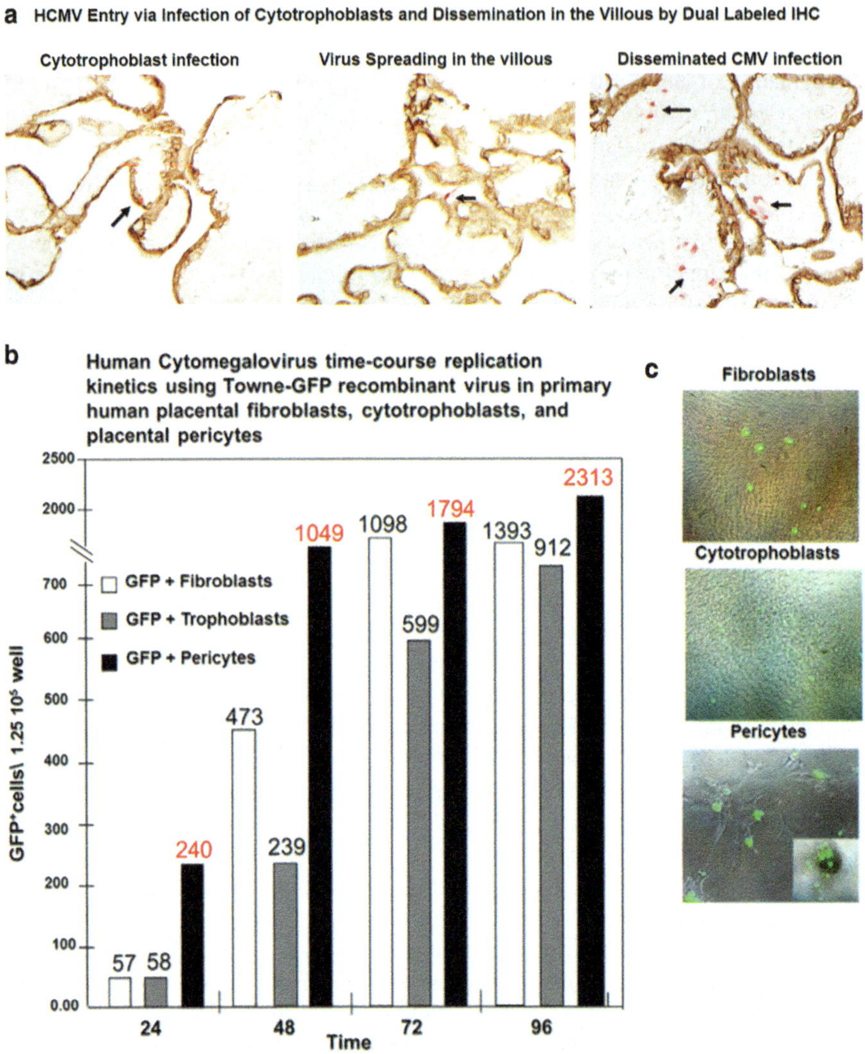

Fig. 4.9 HCMV dissemination in the placenta that includes placental pericytes. (**a**) HCMV entry into placenta tissue demonstrated by dual labeled IHC (trophoblasts, brown color, HCMV red color); (**b**) HCMV replication kinetics examined using a GFP recombinant virus in human placental fibroblasts, cytotrophoblasts and placental pericytes. Infections were performed in chamber slides in triplicate and the average number GFP positive cells are plotted in the graph; (**c**) HCMV-GFP positives cells after infection (insert for pericytes shows evidence of giant cell formation)

demonstrate the HCMV induction of giant cell formation only occurs infected pericytes (Fig. 4.9c see insert). Using a mixture of placental pericytes, cytotrophoblasts and villous fibroblasts, I find dysregulation of cytokines with a unique profile when compared with pericytes alone at early and late times post infection (Aronoff et al. 2017). I hypothesize that human placental pericytes are the most permissive cell type in the placenta for HCMV infection and serve as amplification reservoirs for HCMV dissemination in placental tissue that is governed by virus host-cell receptor entry.

4.4.5 Cytomegalovirus Infection of Pericytes in Adipose Vasculature

The relevance of HCMV infection of pericytes in adipose vasculature is not apparent however I observed both full lytic replication in human adipose pericytes (obtained from Paula Dore Duffy from Wayne State University) data unpublished. To my knowledge there is only one report of mouse cytomegalovirus (MCMV) infection of mouse pericytes in salivary gland adipose vasculature tissue (Price et al. 1990). After MCMV infection of young adult mice they observed MCMV replication in adipocytes, fibroblasts, endothelial cells and pericytes in both brown and white adipose tissues from salivary glands as demonstrated by immunoperoxidase staining and electron microscopy (Price et al. 1990). However I shown human adipose tissue pericytes are fully permissive for HCMV lytic replication as demonstrated by infecting adipose pericytes with Toledo HCMV (lab strain) when compared to the mock infected control (Fig. 4.10). This is the first unpublished report of HCMV infection of human adipose vascular pericytes. This finding supports the notion that all human vascular pericytes are permissive for HCMV. Even

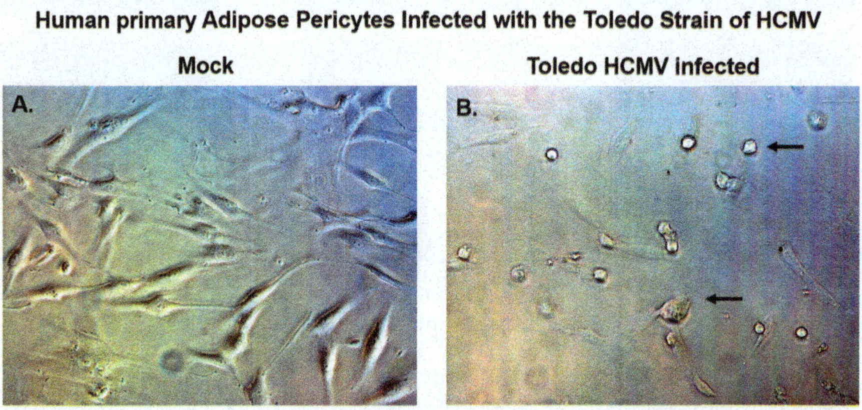

Fig. 4.10 HCMV infection of human adipose pericytes. (**a**) Mock infected primary human adipose pericytes cultivated in vitro; (**b**) Human adipose pericytes cultivated in vitro and infected with the Toledo strain of HCMV. Cells showing cytopathology are identified with black arrows

more, it suggest that adipose tissue could serve as a reservoir for HCMV replication and dissemination which is a novel concept that will require further investigation.

4.5 Discussion and Conclusions

These studies described published evidence from my laboratory of HCMV lytic replication in human pericyte populations of the brain, retinal, placental, glomerular (mesangial cells), as well as unpublished evidence of HCMV infection and lytic replication in pericytes from human adipose tissue (Fig. 4.11). These studies relied

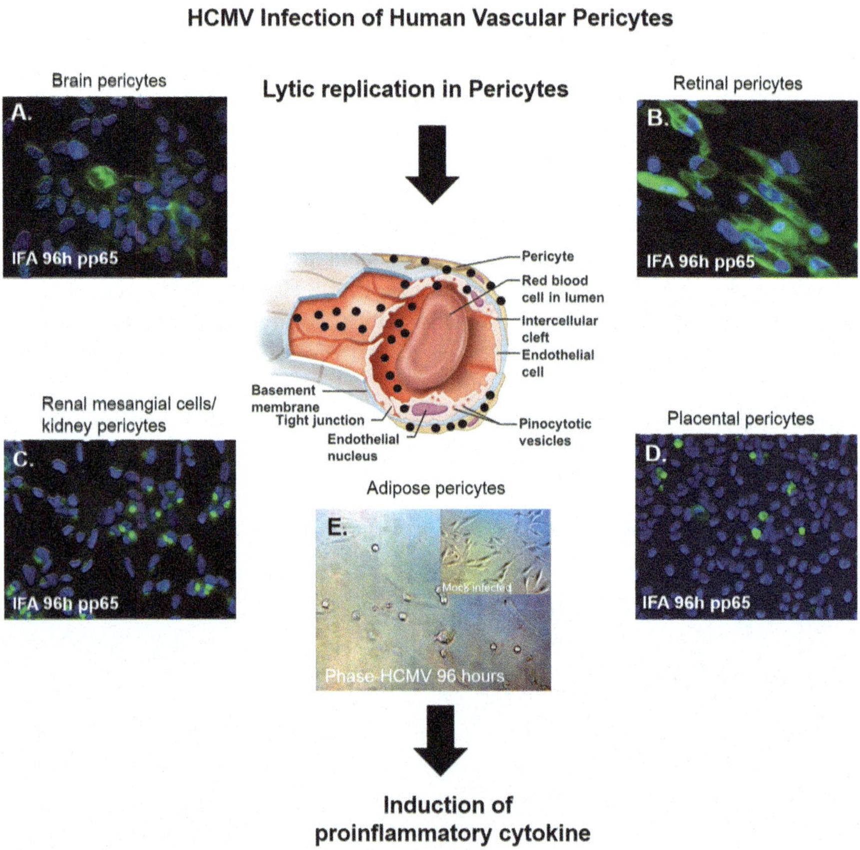

Fig. 4.11 Human vascular pericytes currently known to be permissive for HCMV lytic replication. Human vascular pericytes shown to be permissive for HCMV infection. (**a**) Brain pericyte; (**b**) Retinal pericytes; (**c**) Renal mesangial cells (kidney pericytes); (**d**) Placental pericytes; (**e**) Adipose pericytes (phase image). All cell types with the exception of adipose pericytes were stained with an antibody to the HCMV pp65 protein late protein. All images were taken at 200× magnification

mainly on in vitro data obtained from experiments using primary cells at low passage and infection at low multiplicity. Pericyte population were largely found to be the most permissive cell type for HCMV infection within neurovascular unit, the retinal vascular unit, placenta vascular barrier, and the glomerular vascular unit. In these vascular barrier systems, HCMV infection resulted in the induction of proinflammatory and angiogenic cytokines that likely contributes to vascular inflammation. Hence, vascular pericytes may represent a global reservoir for HCMV that can mobilized during immunosuppression. In addition, results from these studies support the notion that all pericytes populations, regardless of tissue origin, are permissive for HCMV infection, and that loss of HCMV immune surveillance or intermittent viral shedding over time could contribute to pericyte loss or dysfunction, microvascular instability, and subclinical progressive vascular disease. Pericytes have also been implication cerebral vascular diseases like AD However the role of HCMV, HSV-1/2, in AD remains highly controversial and requires further study.

Acknowledgements The author would like to thank Waldemar Popik and Atanu Khatua for past collaborations.

Dedication Special dedication to Dr. James E.K. Hildreth MD, PhD.

Compliance with Ethical Standards

Funding: D.J.A. was supported by the Meharry Zika Startup Grant and a Research Centers in Minority Institutions (RCMI) program grant (U54MD007586-01).

Disclosure of Interests: All authors declare they have no conflict of interest.

Ethical approval: *Studies involving humans.*
All procedures performed in studies involving human participants were in accordance with the ethical standards of the institutional and/or national research committee and with the 1964 Helsinki declaration and its later amendments or comparable ethical standards. Approval was granted by the Institutional Review Boards (IRB) of Meharry Medical College, Johns Hopkins University Medical Center, and Vanderbilt University Medical Center. This includes all de-identified specimens from patients on slides.

Ethical approval: *Studies involving animals.*
This article does not contain any studies with animals performed by any of the authors.

Informed Consent: Informed consent for participation and publication was obtained from all individual participants included in the study.

References

Abbott NJ, Patabendige AA, Dolman DE, Yusof SR, Begley DJ (2010) Structure and function of the blood-brain barrier. Neurobiol Dis 1:13–25

Alberola J, Tamarit A, Igual R, Navarro D (2000) Early neutralizing and glycoprotein B (gB)-specific antibody responses to human cytomegalovirus (HCMV) in immunocompetent individuals with distinct clinical presentations of primary HCMV infection. J Clin Virol 2:113–122

Alcendor DJ, Charest AM, Zhu WQ, Vigil HE, Knobel SM (2012) Infection and upregulation of proinflammatory cytokines in human brain vascular pericytes by human cytomegalovirus. J Neuroinflamm 9:95
Armulik A, Genové G, Mäe M, Nisancioglu MH, Wallgard E, Niaudet C, He L, Norlin J, Lindblom P, Strittmatter K et al (2010) Pericytes regulate the blood-brain barrier. Nature 468:557–561
Armulik A, Genove G, Betsholtz C (2011) Pericytes: developmental, physiological, and pathological perspectives, problems, and promises. Dev Cell 21:193–215
Aronoff DM, Correa H, Rogers LM, Arav-Boger R, Alcendor DJ (2017) A Placental pericytes and cytomegalovirus infectivity: implications for HCMV placental pathology and congenital disease. Am J Reprod Immunol 2017:78
Attwell D, Mishra A, Hall CN, O'Farrell FM, Dalkara T (2016) What is a pericyte? J Cereb Blood Flow Metab 2:451–455
Balabanov R, Dore-Duffy P (1998) Role of the CNS microvascular pericyte in the blood-brain barrier. J Neurosci Res 6:637–644
Ballabh P, Braun A, Nedergaard M (2004) The blood-brain barrier: an overview: structure, regulation, and clinical implications. Neurobiol Dis 1:1–13
Barnes LL, Capuano AW, Aiello AE, Turner AD, Yolken RH, Torrey EF, Bennett DA (2015) Cytomegalovirus infection and risk of Alzheimer disease in older black and white individuals. J Infect Dis 2:230–237
Beam E, Dioverti V, Razonable RR (2014) Emerging cytomegalovirus management strategies after solid organ transplantation: challenges and opportunities. Curr Infect Dis Rep 9:419
Bergers G, Song S (2005) The role of pericytes in blood-vessel formation and maintenance. Neuro-Oncology 7:452–464
Berman N, Belmont HM (2017) Disseminated cytomegalovirus infection complicating active treatment of systemic lupus erythematosus: an emerging problem. Lupus 4:431–434
Birbrair A (2018) Pericyte biology: development, homeostasis, and disease. Adv Exp Med Biol 1109:1–3
Birbrair A, Zhang T, Files DC, Mannava S, Smith T, Wang ZM et al (2018) Type-1 pericytes accumulate after tissue injury and produce collagen in an organ-dependent manner. Stem Cell Res Ther 5:122
Blankenberg S, Rupprecht HJ, Bickel C, Espinola-Klein C, Rippin G, Hafner G, Ossendorf M, Steinhagen K, Meyer J (2001) Cytomegalovirus infection with interleukin-6 response predicts cardiac mortality in patients with coronary artery disease. Circulation 24:2915–2921
Boeckh M, Geballe AP (2011) Cytomegalovirus: pathogen, paradigm, and puzzle. J Clin Investig 121:1673–1680
Boppana SB, Pass RF, Britt WJ, Stagno S, Alford CA (1992) Symptomatic congenital cytomegalovirus infection: neonatal morbidity and mortality. Pediatr Infect Dis J 11:93–99
Boppana SB, Fowler KB, Pass RF, Rivera LB, Bradford RD, Lakeman FD, Britt WJ (2005) Congenital cytomegalovirus infection: association between virus burden in infancy and hearing loss. J Pediatr 146:817–823
Bouwman JJ, Visseren FL, Bouter KP, Diepersloot RJ (2008) Infection-induced inflammatory response of adipocytes in vitro. Int J Obes 32:892–901
Brito AF, Pinney JW (2020) The evolution of protein domain repertoires: shedding light on the origins of the Herpesviridae family. Virus Evol 6(1):veaa001. https://doi.org/10.1093/ve/veaa001
Brown LS, Foster CG, Courtney JM, King NE, Howells DW, Sutherland BA (2019) Pericytes and neurovascular function in the healthy and diseased brain. Front Cell Neurosci 13:282
Cai W, Liu H, Zhao J, Chen LY, Chen J, Lu Z, Hu X (2017) Pericytes in brain injury and repair after ischemic stroke. Transl Stroke Res 2:107–121
Cai Z, Qiao PF, Wan CQ, Cai M, Zhou NK, Li Q (2018) Role of blood-brain barrier in Alzheimer's disease. J Alzheimers Dis 4:1223–1234
Cannon MJ, Schmid DS, Hyde TB (2010) Review of cytomegalovirus seroprevalence and demographic characteristics associated with infection. Rev Med Virol 20:202–213

Cheeran MC, Lokensgard JR, Schleiss MR (2009) Neuropathogenesis of congenital cytomegalovirus infection: disease mechanisms and prospects for intervention. Clin Microbiol Rev 1:99–126
Chen S, de Craen AJ, Raz Y, Derhovanessian E, Vossen AC, Westendorp RG, Pawelec G, Maier AB (2012) Cytomegalovirus seropositivity is associated with glucose regulation in the oldest old. Results from the Leiden 85-plus Study. Immun Ageing 1:18
Chen SJ, Wang SC, Chen YC (2019) Antiviral agents as therapeutic strategies against cytomegalovirus infections. Viruses 12(1):23
Collins-McMillen D, Buehler J, Peppenelli M, Goodrum F (2018) Molecular determinants and the regulation of human cytomegalovirus latency and reactivation. Viruses 10(8):20. pii: E444
Colugnati FA, Staras SA, Dollard SC, Cannon MJ (2007) Incidence of cytomegalovirus infection among the general population and pregnant women in the United States. BMC Infect Dis 7:71
Dias DO, Kim H, Holl D, Werne Solnestam B, Lundeberg J, Carlén M et al (2018) Reducingpericytederivedscarringpromotesrecoveryafterspinal cordinjury. Cell 173:153.e22–165.e22
Diaz-Flores L, Gutierrez R, Madrid JF, Varela H, Valladares F, Acosta E, Martin-Vasallo P, Díaz-Flores JL (2009) Pericytes. Morphofunction, interactions and pathology in a quiescent and activated mesenchymal cell niche. Histol Histopathol 24:909–969
Dollard SC, Grosse SD, Ross DS (2007) New estimates of the prevalence of neurological and sensory sequelae and mortality associated with congenital cytomegalovirus infection. Rev Med Virol 5:355–363
Dore-Duffy P (2008) Pericytes: pluripotent cells of the blood brain barrier. Curr Pharm Des 16:1581–1593
Dore-Duffy P, Cleary K (2011) Morphology and properties of pericytes. Methods Mol Biol 686:49–68
Eberth CJ (1871) Handbuch der Lehre von der Gewegen des Menschen und der Tiere, vol 1. Engelmann, Leipzig
Enders G, Daiminger A, Bäder U, Exler S, Enders M (2011) Intrauterine transmission and clinical outcome of 248 pregnancies with primary cytomegalovirus infection in relation to gestational age. J Clin Virol 52:244–256
Engelhardt B (2003) Development of the blood-brain barrier. Cell Tissue Res 1:119–129
Farias KPRA, Moreli ML, Floriano VG, da Costa VG (2019) Evidence based on a meta-analysis of human cytomegalovirus infection in glioma. Arch Virol 164(5):1249–1257
Garberg P, Ball M, Borg N, Cecchelli R, Fenart L, Hurst RD, Lindmark T, Mabondzo A, Nilsson JE, Raub TJ, Stanimirovic D, Terasaki T, Oberg JO (2005) In vitro models for the blood–brain barrier. Toxicol In Vitro 19:299–334
Gkrania-Klotsas E, Langenberg C, Sharp SJ, Luben R, Khaw KT, Wareham NJ (2012) Higher immunoglobulin G antibody levels against cytomegalovirus are associated with incident ischemic heart disease in the population-based EPIC-Norfolk cohort. J Infect Dis 2012(206):1897–1903
Göritz C, Dias DO, Tomilin N, Barbacid M, Shupliakov O, Frisén J (2011) A pericyte origin of spinal cord scar tissue. Science 333:238–242
Hamilton NB, Attwell D, Hall CN (2010) Pericyte-mediated regulation of capillary diameter: a component of neurovascular coupling in health and disease. Front Neuroenerg 2:1–14
Haspot F et al (2012) Human cytomegalovirus entry into dendritic cells occurs via a macropinocytosis-like pathway in a pH independent and cholesterol-dependent manner. PLoS One 7:e34795
Heald-Sargent TA, Forte E, Liu X, Thorp EB, Abecassis MM, Zhang ZJ, Hummel MA (2020) New insights into the molecular mechanisms and immune control of cytomegalovirus reactivation. Transplantation 104:e118–e124. https://doi.org/10.1097/TP.0000000000003138
Hesp ZC, Yoseph RY, Suzuki R, Jukkola P, Wilson C, Nishiyama A et al (2018) Proliferating NG2-cell-dependent angiogenesis and scar formation alter axon growth and functional recovery after spinal cord injury in mice. J Neurosci 38:1366–1382
Huang ZR, Yu LP, Yang XC, Zhang F, Chen YR, Feng F, Qian XS, Cai J (2012) Human cytomegalovirus linked to stroke in a Chinese population. CNS Neurosci Ther 6:457–460

Janahi EMA, Das S, Bhattacharya SN, Haque S, Akhter N, Jawed A, Wahid M, Mandal RK, Lohani M, Areeshi MY, Ramachandran VG, Almalki S, Dar SA (2018) Cytomegalovirus aggravates the autoimmune phenomenon in systemic autoimmune diseases. Microb Pathog 120:132–139
Kenneson A, Cannon MJ (2007) Review and meta-analysis of the epidemiology of congenital cytomegalovirus (CMV) infection. Rev Med Virol 4:253–276
Kylat RI, Kelly EN, Ford-Jones EL (2006) Clinical findings and adverse outcomes in neonates with symptomatic congenital cytomegalovirus (SCCMV) infection. Eur J Pediatrics 165:773–778
Lai CH, Kuo KH (2005) The critical component to establish in vitro BBB model: Pericyte. Brain Res Brain Res Rev 2:258–265
Lee S, Brook E, Affandi J, Howson P, Tanudjaja SA, Dhaliwal S, Irish A, Price P (2019) A high burden of cytomegalovirus marks poor vascular health in transplant recipients more clearly than in the general population. Clin Transl Immunol 2:e1043
Lichtner M, Cicconi P, Vita S, Cozzi-Lepri A, Galli M, Lo Caputo S, Saracino A, De Luca A, Moioli M, Maggiolo F, Marchetti G, Vullo V, d'Arminio Monforte A, ICONA Foundation Study (2015) Cytomegalovirus coinfection is associated with an increased risk of severe non-AIDS-defining events in a large cohort of HIV-infected patients. J Infect Dis 2:178–186
Liesz A (2019) The vascular side of Alzheimer's disease. Science 6450:223–224
Lindahl P, Hellstrom M, Kalen M, Karlsson L, Pekny M, Pekna M, Soriano P, Betsholtz C (1998) Paracrine PDGF-B/PDGF-Rbeta signaling controls mesangial cell development in kidney glomeruli. Development 125:3313–3322
Lövheim H, Olsson J, Weidung B, Johansson A, Eriksson S, Hallmans G, Elgh F (2018) Interaction between cytomegalovirus and herpes simplex virus type 1 associated with the risk of Alzheimer's disease development. J Alzheimers Dis 3:939–945
Lurain NS, Hanson BA, Martinson J, Leurgans SE, Landay AL, Bennett DA, Schneider JA (2013) Virological and immunological characteristics of human cytomegalovirus infection associated with Alzheimer disease. J Infect Dis 4:564–572
Lv YL, Han FF, Gong LL, Liu H, Ma J, Yu WY, Wan ZR, Jia YJ, Zhang W, Shi M, Liu LH (2017) Human cytomegalovirus infection and vascular disease risk: a meta-analysis. Virus Res 227:124–134
Martí-Carreras J, Maes P (2019) Human cytomegalovirus genomics and transcriptomics through the lens of next-generation sequencing: revision and future challenges. Virus Genes 2:138–164
Muhlestein JB, Horne BD, Carlquist JF, Madsen TE, Bair TL, Pearson RR, Anderson JL (2000) Cytomegalovirus seropositivity and C-reactive protein have independent and combined predictive value for mortality in patients with angiographically demonstrated coronary artery disease. Circulation 102:1917–1923
Nation DA, Ho JK, Dutt S, Han SD, MHC L (2019) Alzheimer's disease neuroimaging initiative neuropsychological decline improves prediction of dementia beyond Alzheime's disease biomarker and mild cognitive impairment diagnoses. J Alzheimers Dis 4:1171–1182
Nell S, Lurain BA, Hanson JM, Leurgans SE, Landay AL, Bennett DA, Schneider JA (2013) Virological and immunological characteristics of human cytomegalovirus infection associated with Alzheimer disease. J Infect Dis 4:564–572
Nigro G, Adler SP (2011) Cytomegalovirus infections during pregnancy. Curr Opin Obstet Gynecol 23:123–128
Nortley R, Korte N, Izquierdo P, Hirunpattarasilp C, Mishra A, Jaunmuktane Z, Kyrargyri V, Pfeiffer T, Khennouf L, Madry C, Gong H, Richard-Loendt A, Huang W, Saito T, Saido TC, Brandner S, Sethi H, Attwell D (2019) Amyloid β oligomers constrict human capillaries in Alzheimer's disease via signaling to pericytes. Science 19(6450):365
Pak CY, Eun HM, McArthur RG, Yoon JW (1988) Association of cytomegalovirus infection with autoimmune type 1 diabetes. Lancet 8601:1–4
Pass RF, Anderson B (2014) Mother-to-child transmission of cytomegalovirus and prevention of congenital infection. J Pediatr Infect Dis Soc Suppl 1:S2–S6

Pass RF, Fowler KB, Boppana SB, Britt WJ, Stagno S (2006) Congenital cytomegalovirus infection following first trimester maternal infection: symptoms at birth and outcome. J Clin Virol 35:216–220

Pawlik A, Ostanek L, Brzosko I, Brzosko M, Masiuk M, Machalinski B, Gawronska-Szklarz B (2003) The expansion of CD4+CD28- T cells in patients with rheumatoid arthritis. Arthritis Res Ther 4:R210–R213

Popik W, Correa H, Khatua A, Aronoff DM, Alcendor DJ (2019) Mesangial cells, specialized renal pericytes and cytomegalovirus infectivity: implications for HCMV pathology in the glomerular vascular unit and post-transplant renal disease. J Transl Sci 2019:5

Price P, Eddy KS, Papadimitriou JM, Robertson TA, Shellam GR (1990) Cytomegalovirus infection of adipose tissues induces steatitis in adult mice. Int J Exp Pathol 4:557–571

Ramsauer M, Krause D, Dermietzel R (2002) Angiogenesis of the blood–brain barrier in vitro and the function of cerebral pericytes. FASEB J 16:1274–1276

Rawlinson WD, Boppana SB, Fowler KB, Kimberlin DW, Lazzarotto T, Alain S, Daly K, Doutré S, Gibson L, Giles ML, Greenlee J, Hamilton ST, Harrison GJ, Hui L, Jones CA, Palasanthiran P, Schleiss MR, Shand AW, van Zuylen WJ (2017) Congenital cytomegalovirus infection in pregnancy and the neonate: consensus recommendations for prevention, diagnosis, and therapy. Lancet Infect Dis 6:e177–e188

Ribbert H (1904) Ueber protozoenartige Zellen in der Niere eines syphilitischen Neugeborenen und in der Parotis von Kindern. Zbl All Pathol 15:945–948

Rothe K, Quandt D, Schubert K, Rossol M, Klingner M, Jasinski-Bergner S, Scholz R, Seliger B, Pierer M, Baerwald C, Wagner U (2016) Latent cytomegalovirus infection in rheumatoid arthritis and increased frequencies of cytolytic LIR-1+CD8+ T cells. Arthritis Rheumatol 2:337–346

Rouget C (1873) Me'moire sur le de'veloppement, la structure et les propriete's physiologiques des capillaires sanguins et lymphatiques. Arch Physiol Norm Path 5:603–663

Rowe WP, Hartley JW, Waterman S, Turner HC, Huebner RJ (1956) Cytopathogenic agents resembling human salivary gland virus recovered from tissue cultures of human adenoids. Proc Soc Exp Biol 92:418–424

Rustenhoven J, Jansson D, Smyth LC, Dragunow M (2017) Brain pericytes as mediators of neuroinflammation. Trends Pharmacol Sci 3:291–304

Salmina AB, Komleva YK, Lopatina OL, Birbrair A (2019) Pericytes in Alzheimer's disease: novel clues to cerebral amyloid angiopathy pathogenesis. Adv Exp Med Biol 1147:147–166

Schottstedt V, Blümel J, Burger R et al (2010) Human cytomegalovirus (HCMV)-revised. Transfus Med Hemother 2010(37):365–375

Shepro D, Morel NM (1993) Pericyte physiology. FASEB J 7:1031–1038

Simanek AM, Dowd JB, Pawelec G, Melzer D, Dutta A, Aiello AE (2011) Seropositivity to cytomegalovirus, inflammation, all-cause and cardiovascular disease-related mortality in the United States. PLoS One 6:e16103

Sims DE (1986) The pericyte—a review. Tissue Cell 18:153–174

Sinclair J, Sissons P (2006) Latency and reactivation of human cytomegalovirus. J Gen Virol 87(Pt 7):1763–1779

Smith MG (1956) Propagation in tissue cultures of a cytopathogenic virus from human salivary gland virus (SGV) disease. Proc Soc Exp Biol Med 92:424–430

Smith SW, Chand S, Savage CO (2012) Biology of the renal pericyte. Nephrol Dial Transplant 27:2149–2155

Smyth LCD, Rustenhoven J, Scotter EL, Schweder P, Faull RLM, Park TIH, Dragunow M (2018) Markers for human brain pericytes and smooth muscle cells. J Chem Neuroanat 92:48–60

Sochocka M, Zwolińska K, Leszek J (2017) The infectious etiology of Alzheimer's disease. Curr Neuropharmacol 7:996–1009

Spyridopoulos I et al (2016) CMV seropositivity and T-cell senescence predict increased cardiovascular mortality in octogenarians: results fromthe Newcastle 85+ study. Aging Cell 15:389–392

Stagno S, Pass RF, Cloud G, Britt WJ, Henderson RE, Walton PD, Veren DA, Page F, Alford C (1986) Primary cytomegalovirus infection in pregnancy. Incidence, transmission to fetus, and clinical outcome. JAMA 14:1904–1808

Stallcup WB (2018) The NG2 proteoglycan in pericyte biology. Adv Exp Med Biol 1109:5–19

Tsukita S, Furuse M, Itoh M (2001) Multifunctional strands in tight junctions. Nat Rev Mol Cell Biol 2:285–293

Vancíková Z, Dvorák P (2001) Cytomegalovirus infection in immunocompetent and immunocompromised individuals--a review. Curr Drug Targets Immune Endocr Metabol Disord 2:179–187

Wang GC, Kao WH, Murakami P, Xue QL, Chiou RB, Detrick B, McDyer JF, Semba RD, Casolaro V, Walston JD, Fried LP (2010) Cytomegalovirus infection and the risk of mortality and frailty in older women: a prospective observational cohort study. Am J Epidemiol 10:1144–1152

Wang C, Zhang X, Bialek S, Cannon MJ (2011) Attribution of congenital cytomegalovirus infection to primary versus non-primary maternal infection. Clin Infect Dis 2:e11–e13

Weller TH, MacAuley JC, Craig JM, Wirth P (1957) Isolation of intranuclear inclusion producing agents from infants with illnesses resembling cytomegalic inclusion disease. Exp Biol Med 94:4–12

Wilkerson I, Laban J, Mitchell J, Sheibani N, Alcendor DJ (2015) Retinal pericytes and cytomegalovirus infectivity: Implications for both CMV induced retinopathy and congenital ocular disease. J Neuroinflamm 12:2

Yaiw KC, Mohammad AA, Costa H, Taher C, Badrnya S, Assinger A, Wilhelmi V, Ananthaseshan S, Estekizadeh A, Davoudi B, Ovchinnikova O, Shlyakhto E, Rafnsson A, Khan Z, Butler L, Rahbar A, Pernow J, Söderberg-Nauclér C (2015) Human cytomegalovirus up-regulates endothelin receptor type B: implication for vasculopathies? Open Forum Infect Dis 4:ofv155

Yamanaka N (1988) Development of the glomerular mesangium. Pediatr Nephrol 2:85–91

Yamazaki Y, Kanekiyo T (2017) Blood-brain barrier dysfunction and the pathogenesis of Alzheimer's disease. Int J Mol Sci 18(9):E1965. pii: E1965

Yu X, Jih J, Jiang J, Zhou ZH (2017) Atomic structure of the human cytomegalovirus capsid with its securing tegument layer of pp150. Science 356:6345. pii: eaam6892

Zimmermann KW (1923) Der feinere Bau der Blutkapillaren. Z Anat Entwicklungsgesch 68:29–109

Zlokovic BV (2011) Neurovascular pathways to neurodegeneration in Alzheimer's disease and other disorders. Nat Rev Neurosci 12:723–738

Chapter 5
Pericytes in Retinal Ischemia

Luis Alarcon-Martinez, Gulce Kureli, and Turgay Dalkara

Abstract

Introduction

Recent research has disclosed involvement of pericytes in neurovascular coupling of retinal capillaries independent of the arteriole-mediated functional hyperemia. This view is supported by studies showing expression of the contractile protein α-smooth muscle actin in capillary pericytes in low but sufficient quantities to control capillary diameter. During retinal ischemia, injured pericytes persistently contract and disrupt microcirculation as well as impairing blood-retinal barrier function. Studies revealing the link between retinal ischemia and pericyte dysfunction are reviewed in this chapter.

Methods

Journals and databases (Pubmed and Google Scholar) were surveyed with the keywords "Retina", "Pericyte" "Microcirculation" "Capillary" and "Ischemia". Related works were reviewed.

Results

During retinal ischemia, focal microvascular constrictions emerge at pericyte locations, suggesting that contracted pericytes constrict the capillaries and disrupt microcirculation as previously reported in cerebral and cardiac ischemia. In addition to ex vivo studies, the colocalization of capillary constrictions with pericytes was also observed in vivo in retinas of NG2-DsRed transgenic mice. Pericyte contractions persisted after recanalisation, causing incomplete reperfusion of the retina. Studies with calcium-sensitive fluoroprobe Fluo-4 and genetically-encoded calcium indicator, GCaMP6, showed that a sustained intrapericytic calcium elevation mediated persistent constrictions. Calcium increase and constrictions were coincident

L. Alarcon-Martinez
Department of Neuroscience and Centre de Recherche du Centre Hospitalier de l'Université de Montréal, Université de Montréal, Montréal, QC, Canada

G. Kureli · T. Dalkara (✉)
Institute of Neurological Sciences and Psychiatry, Hacettepe University, Ankara, Turkey
e-mail: tdalkara@hacettepe.edu.tr

A. Birbrair (ed.), *Biology of Pericytes – Recent Advances*, Stem Cell Biology and Regenerative Medicine 68, https://doi.org/10.1007/978-3-030-62129-2_5

with loss of peri-microvascular glycogen, suggesting that glycogen can support microvascular calcium homeostasis within the first hour of ischemia by providing glucose. Indeed, inhibition of glycogen utilization accelerated the emergence of ischemic microvascular constrictions. Intravitreal injection of adenosine reversed capillary constrictions possibly by relaxing pericytes, encouraging that incomplete reperfusion can be reverted. Ischemic pericyte injury also contributes to increased blood-retina barrier permeability and causes delayed pericyte loss and angiogenesis, which have been extensively characterized in diabetic retinopathy.

Conclusions
Recent research has documented the crucial role of pericytes in retinal ischemia. Studies showing reversal of these pathological events create exciting opportunities for treatment of retinal diseases of ischemic origin.

5.1 Pericytes Regulate Microcirculatory Blood Flow in the Retina

Retina is one of the tissues with high metabolic demand (Yu et al. 2019a). It consists of several layers rich in either cell soma (i.e. nuclear) or synapses (i.e. plexiform) and is supported by two vascular systems: the choroid – which mainly feeds the photoreceptors in the outer retina, and the retinal vasculature – which supports the inner retina, encompassing horizontal, bipolar, amacrine, and retinal ganglion cells. Unlike choroid, retinal vasculature autoregulates and responds to flickering light stimuli (Bill 1975; Grunwald et al. 1984; Riva et al. 1997; Srienc et al. 2010), suggesting a capacity to regulate blood flow in response to changing metabolic demand as documented for the brain vasculature (Hall et al. 2014; Rungta et al. 2018). Neurons consume significant amount of energy supplied mainly by ATP to reestablish the neuronal and astrocytic membrane ion gradients after neuronal firing (Attwell and Laughlin 2001). Restoration of energy stores requires an increased blood flow to match the enhanced oxygen and glucose demand (Attwell et al. 2010; Newman 2013; Yu et al. 2019a). This is achieved by using a complex communication signaling between neurons, astrocytes and vessels, which is named neurovascular coupling (Roy and Sherrington 1890). The physiological consequence of this coupling is termed functional hyperemia (Attwell et al. 2010; Newman 2013).

The structure supporting functional hyperemia is the neurovascular unit, which is formed by neurons, glial cells, endothelia, smooth muscle cells and pericytes (Attwell et al. 2010; Newman 2013; Dalkara and Alarcon-Martinez 2015). Although it is known that each component of the neurovascular unit contributes to neurovascular coupling to varying degrees along the vascular tree, the underlying mechanisms are not completely understood. Active neurons/interneurons may produce vasoactive mediators that can directly or indirectly modify the intracellular calcium of smooth muscle cells and pericytes, varying the tension of contractile proteins, and, consequently, the diameter of the vessels that they surround (Alarcon-Martinez et al. 2019; Puro 2007, 2012; Rungta et al. 2018). Such a mechanism involving neuron-glia-vessel communication in the retina has recently been disclosed (Biesecker et al. 2016). Briefly, active neurons release ATP, which binds to the purinergic receptors on Müller cells – the most abundant glial cell in the retina that extends fine processes around microvessels and into the plexiform layers. Activation of purinergic receptors leads to calcium release in Müller cells from internal stores, which induces synthesis of vasoactive arachidonic acid metabolites such as prostaglandin E2 and epoxyeicosatrienoic acids, or 20-hydroxyeicosatetraenoic acid (Attwell et al. 2010; Biesecker et al. 2016; Metea and Newman 2006; Mishra et al. 2016). Due to the close proximity of the Müller cell endfeet to the vessels, the metabolites discharged over smooth muscle cells and pericytes can modify their intracellular calcium and, hence, the vascular diameter (Alarcon-Martinez et al. 2019; Puro 2007, 2012). The activation of Müller cells, however, exclusively leads to dilation of the capillaries in the intermediate plexus but not of the arterioles (Biesecker et al. 2016; Kornfield and Newman 2014), suggesting presence of a pericyte-mediated neurovascular coupling independent of the arteriole-mediated functional hyperemia. This is in contrast to previous studies (e.g. Hill et al. 2015), which proposed that capillary pericytes were devoid of α-smooth muscle actin (α-SMA) – a protein responsible for the vascular constriction, hence, the arterioles ensheathed with smooth muscle cells and pre-capillary arterioles wrapped by pericytes, both expressing large amounts of α-SMA, were the sole regulators of the blood flow. However, recently, it has been shown that pericytes located on retinal capillaries of the three plexuses do contain α-SMA in low but sufficient quantities to allow them to change their shape and, consequently, control capillary diameter (Alarcon-Martinez et al. 2018). Of note, in histological sections, the detection of small amounts of α-SMA in capillary pericytes is challenging because of the rapid depolymerization of filamentous α-SMA during slow tissue fixation with commonly used fixatives such as aldehydes. In contrast, rapid fixation methods consistently reveal the low amounts of α-SMA even in pericytes on high order distal capillaries (Alarcon-Martinez et al. 2018) as originally reported by Herman and D'Amore (1985). Moreover, Alarcon-Martinez et al. (2018) were able to suppress α-SMA expression in downstream capillary pericytes but not in upstream pericytes with short interfering RNA. This is in accord with the histological findings suggesting presence of a smaller amount of α-SMA in downstream mid-capillary pericytes compared to the α-SMA-rich wrapping-type upstream pericytes as well as the vascular smooth muscle cells. These findings were confirmed by single-cell RNA sequence of the cells of brain vasculature in a study that provides an extensive online database for RNA expression in pericytes, smooth muscle cells, endothelial cells, glial cells and other cell types in the brain vasculature (Vanlandewijck et al. 2018). As expected, a great amount of α-SMA mRNA was located in smooth muscle cells and a low quantity in pericytes (Zeisel et al. 2018; http://mousebrain.org/genesearch.html; please also see the on-line section of Vanlandewijck et al. 2018; http://betsholtzlab.org/VascularSingle Cells/database.html; Acta2 expression in pericytes = 14.73 average counts). Recently, Yu et al. confirmed the expression of α-SMA in retinal capillaries of human macula region (Yu et al. 2019b).

Evidence accumulates for the retinal pericytes having the basic machinery to constrict and dilate capillaries and contribute to the regulation of functional hyperemia as previously reported for the brain (Hall et al. 2014; Hamilton et al. 2010; Rungta et al. 2018). In addition to expressing receptors for vasoactive mediators (Hamilton et al. 2010; Peppiatt et al. 2006; Puro 2007), their capability to regulate blood flow is also suggested by their morphology characterized by processes enwrapping microvessels as well as by immunohistochemical and ultrastructural detection of contractile proteins (Shepro and Morel 1993; Nehls and Drenckhahn 1991; Bandopadhyay et al. 2001; Allt and Lawrenson 2001; Das et al. 1988; Fujimoto and Singer 1987; Herman and D'Amore 1985; Joyce et al. 1985a, 1985b; Kelley et al. 1987; Wallow and Burnside 1980). In vitro studies on retina or on isolated microvessels or cultured pericytes have consistently showed that retinal pericytes are capable of contracting or relaxing when stimulated with vasoactive mediators or physiological means (Kelley et al. 1987; Peppiatt et al. 2006; Puro 2007, 2012). The above findings strongly suggest that retinal pericytes are capable of regulating the microcirculatory blood flow. However, improvement in the imaging techniques, allowing visualizing single elements of the neurovascular unit in the retina of live animals are still needed to gain further insight. Multi-photon laser scanning microscopy used with genetically encoded fluorescent tagged animals is an ideal technique as previously used for the in vivo recording of mitochondria in retinal ganglion cells (Takihara et al. 2015).

5.2 Retinal Pericytes Are Vulnerable to Ischemia

Pericytes are vulnerable cells involved in pathophysiology of several neurological diseases such as ischemia, epilepsy, Alzheimer's disease, diabetic retinopathy and spreading depolarization (Cogan et al. 1961; Hall et al. 2014; Khennouf et al. 2018; Kisler et al. 2017; Leal-Campanario et al. 2017; Puro 2012; Sweeney et al. 2016, 2018; Winkler et al. 2014; Yemisci et al. 2009). During retinal ischemia, focal microvascular constrictions emerge at pericyte locations (Alarcon-Martinez et al. 2019), suggesting that contracted pericytes may constrict the capillaries and disrupt microcirculation as previously reported in cerebral or cardiac ischemia (Dalkara et al. 2016; Hall et al. 2014; O'Farrell et al. 2017; Yemisci et al. 2009). In addition to ex vivo whole-mount retinas immunolabeled with pericyte markers, the colocalization of capillary constrictions with pericytes was also observed in vivo in retinas of NG2-DsRed transgenic mice expressing a red fluorescent protein in pericytes by adaptive optics scanning light ophthalmoscopy (AOSLO)(Alarcon-Martinez et al. 2019) (Fig. 5.1).

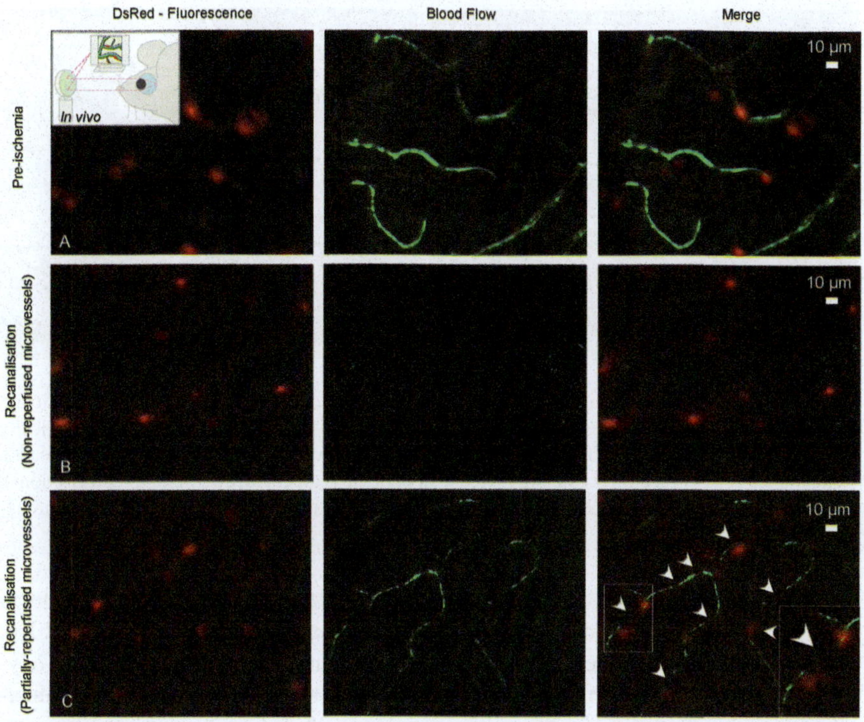

Fig. 5.1 In vivo demonstration of reduced capillary blood flow at pericyte locations after ischemia. (A-C) Retinal pericytes and capillary blood flow were identified in live animals with AOSLO by using transgenic mice expressing the red fluorescent protein, DsRed specifically in pericytes (NG2-DsRed). Retinal blood flow was recorded by monitoring the erythrocyte motion (pseudo-colored in green). (A) Before ischemia, microvessels exhibiting an uninterrupted blood flow. (B) After recanalisation, movement of erythrocytes was extremely reduced in some microvessels (no-reflow). (C) Some microvessels showed an incomplete reperfusion (black segments, arrowheads), due to capillary constrictions by pericytes. Scale bars = 10 μm. With permission of BioMed central (BMC) as publisher (Alarcon-Martinez et al. 2019)

5.3 Mechanisms of Ischemia-Induced Pericyte Contractions

The cellular mechanisms underlying ischemia-induced pericyte contraction in the retina are being disclosed. Pericyte contraction is physiologically regulated by intracellular Ca^{2+} concentrations (Hamilton et al. 2010; Kamouchi et al. 2004; Wu et al. 2003). Accordingly, a sustained intracellular calcium increase in pericytes during ischemia can prompt persistent vascular constriction as illustrated with intravitreally injected calcium-sensitive fluoroprobe Fluo-4 as well as in the mouse retina expressing the genetically encoded calcium indicator, GCaMP6 (Alarcon-Martinez et al. 2019) (Fig. 5.2). An uncontrolled rise in intracellular calcium may originate from multiple sources, including voltage-gated calcium channels, non-specific cation channels, purinergic receptors, Na^+/Ca^{2+} exchangers and calcium-induced

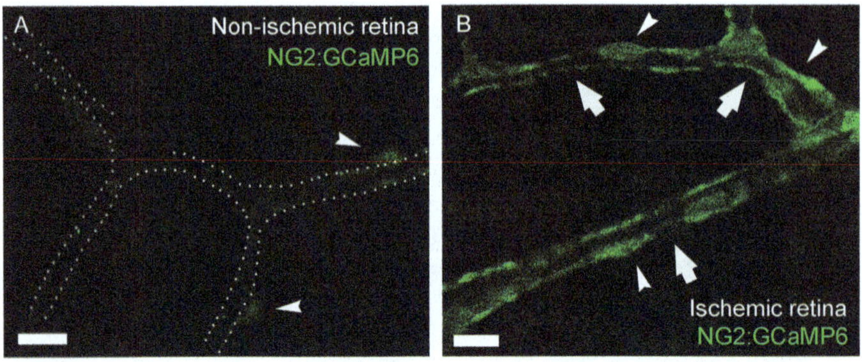

Fig. 5.2 Intracellular calcium elevation promotes pericyte contraction and capillary constriction during ischemia. (A-B) Representative examples of the mouse retinas expressing the genetically encoded calcium indicator GCaMP6 specifically in pericytes (NG2:GCaMP6) (arrowheads). Ischemia leads to intracellular calcium increase in pericytes (green, arrows point to constrictions). Scale bars in A-B = 10 μm. With permission of BioMed central (BMC) as publisher (Alarcon-Martinez et al. 2019)

calcium release from the endoplasmic reticulum (Hamilton et al. 2010; Kamouchi et al. 2007; Nakaizumi and Puro 2011; Nakamura et al. 2009; Yamanishi et al. 2006; Yemisci et al. 2009). Moreover, calcium overload can be facilitated by reactive oxygen species (ROS) (Kamouchi et al. 2007; Nakamura et al. 2009), which are generated in high quantities on the microvascular wall during ischemia-reperfusion (Gursoy-Ozdemir et al. 2004, 2012). Confirming that ischemia-induced retinal pericyte contractions are mediated by α-SMA activation with intracellular calcium rise, microvascular constrictions were prevented with calcium antagonist amlodipine or by silencing α-SMA protein expression with small RNA interference (Alarcon-Martinez et al. 2019). Interestingly, the intracellular calcium rise in pericytes as well as capillary constrictions appear 1 h after ischemia induced by thrombotic occlusion of the central retina artery, suggesting that an energy source, perhaps glycogen located in Müller cells, can sustain pericyte metabolism during the first hour of ischemia when glucose transport from blood is severely limited (Alarcon-Martinez et al. 2019). Glucose derived from the Müller glia reservoir of glycogen can rapidly be mobilized to the pericyte underneath possibly by way of transporters and gap junctions (Suzuki et al. 2009) – the latter of which are rapidly induced during ischemia (Danesh-Meyer et al. 2012, 2016; Kerr et al. 2012). Although evidence is not complete for the presence of functional gap junctions between glial endfeet and pericytes, gap junctions are well known to connect glial perivascular endfeet, the endothelial cells with each other and the endothelia with pericytes. This communication pathways may allow propagation of calcium signaling through gap junctions during ischemia and directly contribute to pericyte calcium rise or indirectly via release of vasoconstrictive mediators from glial endfeet onto pericytes because carbenoxolone, a gap junction blocker or a selective peptide inhibitor of Connexin-43 (Cx43) gap junctions lowered the ischemia-induced calcium increase in retinal pericytes (Alarcon-Martinez et al. 2019). Retinal ischemia led to a rapid increase in Cx43 hemichannels on Müller cell endfeet covering pericytes (Alarcon-Martinez et al. 2019). Carbenoxolone also reduced the ischemia-induced capillary constrictions. Interestingly, pericyte calcium increase and capillary constrictions were coincident with loss of peri-microvascular glycogen staining by periodic acid-Schiff (Alarcon-Martinez et al. 2019), which suggests that glycogen supported microvascular calcium homeostasis within the first hour of ischemic insult by providing glucose (Kuwabara and Cogan 1961; Wasilewa et al. 1976). Further supporting this hypothesis, inhibition of glycogen degradation with 1,4-dideoxy-1,4-imino-d-arabinitol (DAB) accelerated the emergence of microvascular constrictions during ischemia (Brown 2004; Gurer et al. 2009; Kasischke et al. 2004; Rossi et al. 2007). In line with these findings, Müller glia have been shown to be resistant to glucose deprivation or hypoxia owing to their glycogen stores (Toft-Kehler et al. 2017).

In addition to the above reviewed acute effects, the delayed adverse effects of ischemia on retinal microvasculature and pericytes have been shown in several models. For instance, elevation of the intraocular pressure as a retinal ischemia/reperfusion injury model leads to retinal capillary degeneration in mice and rats (Zheng et al. 2007). Pericyte loss was evident 2 weeks after ischemia-reperfusion injury (Nakahara et al. 2015), suggesting that some of the permanently contracted pericytes died in rigor (Hall et al. 2014). Pericytes are detached from microvessels in the peri-infarct tissue and promote angiogenesis and neurogenesis, both of which are reportedly improve the stroke outcome in rodent models (Liu et al. 2012), however, this issue is not well characterized after retinal ischemia as we will discuss below.

5.4 Incomplete Microcirculatory Reflow After Central Retinal Artery Recanalisation

The no-reflow phenomenon, which refers to incomplete tissue reperfusion after re-opening of an occluded artery or reinstitution of blood flow following cardiac arrest was first reported by Ames et al. (1968) after global and by Crowell and Olsson (1972) after focal cerebral ischemia. Recently, studies with imaging techniques resolving single capillary blood flow in the intact mouse brain confirmed the presence of this phenomenon in vivo (Dziennis et al. 2015; Hill et al. 2015; Lee et al. 2016; Yemisci et al. 2009). Clots formed by leukocytes, platelets, fibrin and, more recently, ischemia-induced persistent contraction of microvascular pericytes were suggested as the source of the no-reflow phenomenon (Hall et al. 2014; Hallenbeck et al. 1986; Yemisci et al. 2009; Zhang et al. 1999; del Zoppo et al. 1991, 2011).

Similar to the brain, the occlusion of the central retinal artery in the mouse induces nodal microvascular constrictions coincident with pericyte locations, which generally do not recover after recanalisation (Alarcon-Martinez et al. 2019). The persistent constrictions lead to an impaired reflow, which intensifies as a function of

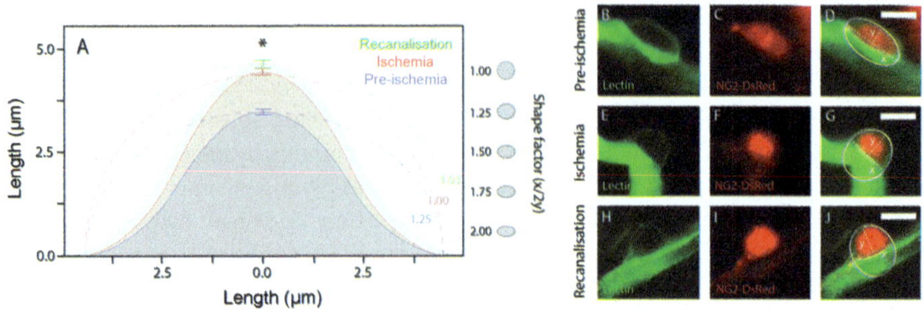

Fig. 5.3 Pericyte morphology changes to a contractile phenotype during ischemia and recanalisation. (A) Ex vivo analysis of DsRed+ pericytes after labeling pericyte basement membrane with lectin (green) confirms that their soma protruded away during ischemia and recanalisation as suggested by their shape factor. (B-J) Images show pericyte shape before ischemia (B-D), during ischemia (E-G) and recanalization (H-J). Scale bars = 5 µm. With permission of BioMed central (BMC) as publisher (Alarcon-Martinez et al. 2019)

the duration and severity of ischemia (Alarcon-Martinez et al. 2019; Hill et al. 2015; Yemisci et al. 2009; Lee et al. 2016). In a recent in vivo study, the authors used AOSLO to image the retinal blood flow and genetically encoded fluorescent-tagged pericytes in NG2-DsRed transgenic mice (Alarcon-Martinez et al. 2019). During ischemia, pericyte somata on microvessels rounded-up, a morphological hallmark of pericyte contraction (Murphy and Wagner 1994; Schallek et al. 2013; Takahashi et al. 1997), which persisted after recanalization (Fig. 5.3). These pericyte morphological changes corresponded to decreases in capillary diameter, causing erythrocyte entrapment. Remarkably, intravitreal injection of the potent pericyte relaxant adenosine prevented (pre-treatment) or reversed the pericyte contractions (when administered one hour after ischemia), encouraging that impaired reperfusion can be reverted before pericytes die in rigor hours later (Hall et al. 2014; Li and Puro 2001). In line with these observations, systemic administration of adenosine-squalene nanoparticles improved cerebral microcirculation after recanalisation in mouse stroke models possibly by reversing pericyte contraction by slowly releasing adenosine in endothelial cells (Gaudin et al. 2014).

ROS formation on vessel wall may also be involved in the ischemia-induced pericyte contraction and impede reperfusion after ischemia. Indeed, S-PBN (2-sulfophenyl-N-tert-butylnitrone), a sulfonyl derivative of the ROS scavenger PBN, although blood-brain barrier (BBB)-impermeable, provided neuroprotection by improving microcirculatory reperfusion and, hence, secondarily reducing parenchymal ROS generation in a mouse stroke model (Taskiran-Sag et al. 2018). Consequently, reducing microcirculatory no-reflow appears to be an important target to improve the outcome of recanalization in the brain and retina (Dalkara and Arsava 2012; Dalkara 2019; Alarcon-Martinez et al. 2018, 2019) and, to support as well as sustain neuroprotection (Gursoy-Ozdemir et al. 2012; Erdener and Dalkara 2019).

5.5 Pericytes and Blood-Retina Barrier Leakiness in Retinal Ischemia

The existence of a blood-retina barrier (BRB) isolating retinal microenvironment from circulation is essential for the retina to properly function (Kaur et al. 2008; Cunha-Vaz 2010). BRB is composed of an outer barrier – located at the retinal pigment epithelium; and an inner barrier – located at the inner retinal vasculature. Inner BRB, which is more selective than the outer BRB, is mainly formed by the tight junctions between endothelial cells, a continuous (non-fenestrated) basal lamina, pericytes/smooth muscle cells, and neighboring astrocytes and Müller cells. The inner BRB constitutes a barrier between retinal tissue and the lumen of the retinal vessels.

The importance of pericytes for retinal vasculature can be inferred from the fact that pericyte:endothelial cell ratio reaches the highest number in the body by 1:1 in the retina (Frank et al. 1987; Shepro and Morel 1993). Pericyte coverage is fundamental for the formation and maintenance of inter-endothelial junctions as tight and adherens junctions that are essential for an intact barrier function. During development of the retinal vasculature, the interaction between endothelial cells, pericytes and astrocytes results in increased expression of the tight junction protein ZO-1 (Kim et al. 2009). Several studies also point to the PDGFB/PDGFR-β pathway as an important axis for the formation and maintenance of a functional BBB/BRB (Armulik et al. 2010; Daneman et al. 2010; Trost et al. 2016). Blockage of pericyte PDGFR-β signaling by monoclonal antibodies inhibited recruitment of retinal pericytes to the vasculature and resulted in abnormal vascular development, leading to plasma leakage, edema and hemorrhages (Uemura et al. 2002). In transgenic mice lacking the retention motif of PDGFB, newborns developed abnormal vascular sprouting and decreased pericyte coverage. This was followed by abnormal vascular plexus formation, detachment of pericytes, anomalous development of nuclear and photoreceptor layers and fibrosis (Lindblom et al. 2003). In adult animals, however, the impairment of PDGFB/PDGFR-β signaling pathway does not directly destabilize BRB but BRB becomes vulnerable to any additional noxious stimulus (Park et al. 2017). However, these studies are limited by pericyte heterogeneity because not all pericytes are positive for PDGFR-β (Santos et al. 2018). Due to the heterogeneity in protein expression of pericytes (Armulik et al. 2011; Santos et al. 2018), there is not a pan-pericyte marker to unambiguously identify all pericytes. Pericytes also express neural-glial antigen 2/chondroitin sulphate proteoglycan 4 (NG2/CSPG4), aminopeptidase N (CD13) and α-SMA to varying degrees (for review, see Armulik et al. 2011). This heterogeneity suggests the presence of subpopulations of pericytes with different functions (Armulik et al. 2011; Santos et al. 2018). Pericytes are highly plastic cells such that even specific pericyte subsets may assume distinct roles between organs (Birbrair et al. 2013, 2014a, 2014b). In addition to the heterogeneity of pericytes for PDGFR-β expression, there are other retinal cells than pericytes signaling through PDGFB/PDGFR-β pathway, further complicating the interpretation of the results obtained by modulating this pathway (Santos et al. 2018).

In acute brain ischemia, pericyte injury and, in the long term, pericyte death in addition to factors such as matrix metalloproteinase activation leads to BBB destabilization (Underly et al. 2016). Similarly to cerebral ischemia, hypoxia/ischemia induction in animal models leads to an increased permeability of BRB in the retina as well (Wilson et al. 1995; Kaur et al. 2007). Several studies pointed to common pathophysiological mechanisms in BBB and BRB permeability increase induced by hypoxia/ischemia such as increased vascular endothelial growth factor (VEGF) and nitric oxide production (Mayhan 1999; Dobrogowska et al. 1998; Kaur et al. 2007) and decreased expression of tight junction proteins (Fischer et al. 2002; Wang et al. 2007). Although a contribution from dysfunctional pericytes is highly likely, further studies are required to firmly establish a link between pericyte dysfunction and junctional protein expression in retinal ischemia. The relevance of pericyte loss/dysfunction to BRB leakiness in retinal ischemia has been most clearly documented in the pathophysiology of diabetic retinopathy (Beltramo and Porta 2013), as we will discuss below.

5.6 Pericytes and Angio- and Neuro-genesis Following Retinal Ischemia

Pericyte investment is a significant factor in vessel formation, sprouting, maturation and stabilization. Neovascularization and angiogenesis rely highly on the interaction between endothelial cells and pericytes during development (Gerhardt and Betsholtz 2003). This interaction is mediated by several signaling pathways as well as direct cell-to-cell contact (Armulik et al. 2011) (Fig. 5.4). PDGFB/PDGFR-β signaling pathway is a significant player in pericyte recruitment, proliferation and survival (Betsholtz 2004). Defective signaling of this pathway leads to pericyte loss, abnormal vascular formation and microaneurysms (Lindahl et al. 1997). TGF-β/TGFβR2 signaling, which is also crucial for angiogenesis, mediates inhibition of endothelial cell proliferation by pericytes (Caporarello et al. 2019). Ang1 released by pericytes acts on its receptor Tie2 on endothelial cells to promote vessel remodeling, maturation and stabilization. Dysfunction in Ang1/Tie2 signaling pathway between pericytes and endothelial cells makes the vasculature vulnerable to vascular stress, leading to abnormal and excessive angiogenesis (Jeansson et al. 2011). In a retinal vascular mural cell depletion model, providing external recombinant Ang1 rescued retinal edema and hemorrhages (Uemura et al. 2002), suggesting a potential treatment approach for diabetic retinopathy. It was shown that Tie2 expression is not specific to endothelial cells and deletion of Tie2 in pericytes delays postnatal retinal angiogenesis (Teichert et al. 2017). This signaling pathway is also important for the maintenance of the inner BRB, where Ang1 inhibits and Ang2 promotes the degradation of endothelial junctional proteins (Díaz-Coránguez et al. 2017). An increased ratio of Ang2 over Ang1 is seen in diabetic retinopathy, which may allow VEGF-induced retinal vascular permeability increase (Patel et al. 2005). VEGF released by

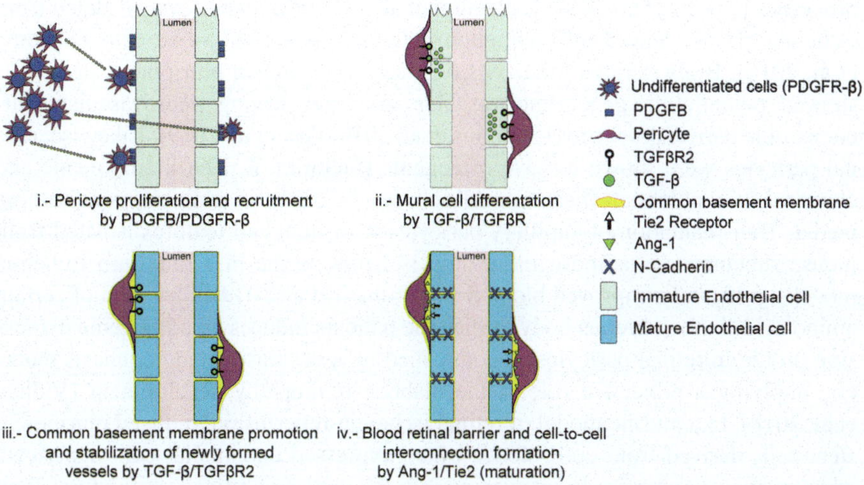

Fig. 5.4 The importance of pericytes in angiogenesis. i) PDGFB/PDGFR-β axis controls the recruitment of undifferentiated mesenchymal cells/pericytes to newly formed vessels. ii) TGF-β/TGFβR2 axis leads to differentiation of mural cells. iii) The TGF-β/TGFβR2 axis regulates formation of the endothelial cell/pericyte common basement membrane and stabilizes new vessels. iv) Ang-1/Tie2 axis promotes blood-retina barrier formation

perivascular cells acting on VEGFR-2 receptors on endothelial cells, promotes endothelial cell proliferation, survival and stabilization. Expression of VEGFR-1 by pericytes sequester excessive VEGF, preventing endothelial proliferation in already stabilized, mature vessels (Eilken et al. 2017).

The above-summarized signaling pathways mediate the ischemia-induced angiogenesis. In middle cerebral artery occlusion models and pericyte cultures, ischemia-induced upregulation of PDGFB/PDGFR-β signaling was associated with increased pericyte migratory capacity toward newly formed vessels (Nakamura et al. 2016) along with an increased expression of the neurotrophic factors (Arimura et al. 2012). Pericytes reportedly migrate from microvessels in peri-infarct areas in the brain, contributing to angiogenesis and repair, but also posing a possible risk of a leaky BBB (Kamouchi et al. 2012; Liu et al. 2012). Similarly, exposure of bovine retinal pericytes to in vitro hypoxia promoted their migration and induced VEGF production and Ang1/Tie2 mRNA upregulation in these cells (Yamagishi et al. 1999; Park et al. 2003). Retinal-pericyte-conditioned medium prevented excessive endothelial cell proliferation in an in vitro model of ischemia (Kondo et al. 2005). NG2 proteoglycan expressed by retinal pericytes may also be involved in ischemia-induced neovascularization, as its knockout resulted in decreased pericyte and endothelial cell proliferation, and increased ectopic vascularization into vitreous (Ozerdem and Stallcup 2004). These findings await further support from in vivo retinal ischemia studies.

Besides their role in angiogenesis, central nervous system pericytes exhibit vasculogenic and neurogenic potential after ischemia owing to their stem cell-like properties (Karow et al. 2014; Nakagomi et al. 2015). Brain pericytes isolated from ischemic brains showed differentiation potential to microglia-like cells (Sakuma et al. 2016). Brain pericyte-like cells obtained from human pluripotent stem cells showed potential in generating vascular structures having neurovascular unit-compatible barrier properties (Stebbins et al. 2019; Faal et al. 2019). Likewise, retinal pericytes were shown to have osteogenic (Schor et al. 1990), adipogenic and chondrogenic potential (Farrington-Rock et al. 2004) in suitable conditioning media. Transplantation of capillary pericytes with stem cell features isolated from mouse subcutaneous adipose tissue into ischemic tissue in a hindlimb ischemia model significantly improved blood flow (Yoshida et al. 2019). Elevation of certain amino acids in the vitreous body of diabetic patients induces an adipogenic expression profile in retinal pericytes, accompanied by a decrease in angiogenesis markers, implying a protective role in the diabetic retinopathy development (Vidhya et al. 2018). In a murine model of retinal ischemia-reperfusion, induced pluripotent stem cells derived from cord blood, which expressed endothelial cell or pericyte markers showed promise in integrating into the damaged retinal vasculature (Park et al. 2014a). Given their remarkable plasticity, whether retinal pericytes can locally contribute to neurogenesis after ischemia warrants further investigation.

5.7 Diabetic Retinopathy

Diabetic retinopathy is a serious microvascular complication causing ischemic as well as hemorrhagic retinal injury, which makes it the leading cause of blindness in middle-aged adults (Wong et al. 2018). The main risk factor of diabetic retinopathy is hyperglycemia. It is classified as proliferative or non-proliferative depending on the presence of neovascularization. Proliferative diabetic retinopathy presents dysfunctional vascularization with microaneurysms, vascular leakage, and non-perfused (ischemic) areas of the retina. The main treatment is the prevention of neovascularization by intraocular injections of anti-VEGF and laser photocoagulation therapy (Wong et al. 2018). Conversely, non-proliferative diabetic retinopathy does not present with neovascularization. However, vasculature is also dysfunctional, causing microaneurysms, hemorrhages, non-perfused (ischemic) areas, and microvascular abnormalities (Wong et al. 2018). The alterations in the retinal vasculature during diabetic retinopathy may progress in the following order: formation of microaneurysms and venular dilation, increased vascular permeability and microhemorrhages, occlusion of retinal vessels and ischemia resulting in small infarctions in the nerve fiber layer, and finally neoangiogenesis in proliferative diabetic retinopathy (Motiejūnaitė and Kazlauskas 2008).

As previously described, pericytes are essential for the BRB integrity. The initial pathophysiological event in the development of diabetic retinopathy is pericyte dropout (Hammes et al. 2002). Although the exact mechanism underlying pericyte loss is not clear, hyperglycemia-dependent alterations in intracellular signaling and metabolic pathways may be responsible for the pericyte dropout. Pericyte-mediated control of capillary dilation and constriction and spread of this vasomotor response along the microvasculature were reportedly impaired in retinas obtained from streptozotocin-treated hyperglycemic mice (Ivanova et al. 2017). Cui et al. (2006) also found that hyperglycemia promoted ROS production and apoptosis in cultured bovine retinal capillary pericytes. Similarly, hyperglycemia was shown to activate NF-κB in retinal pericytes and trigger expression of pro-apoptotic molecules both in donated diabetic eyes and in vitro experiments (Romeo et al. 2002). Hyperglycemia is also reported to prompt Ang2-dependent apoptosis in pericytes (Park et al. 2014b) and induce a decrease in pro-survival signals for pericytes by phosphorylating PDGFR-β (Geraldes et al. 2009). Likewise, anti-PDGFR-β antibodies induced pericyte loss and replicated the hallmarks of diabetic retinopathy in the mouse, including vascular hyperpermeability, edema and neoangiogenesis (Ogura et al. 2017). In addition, the loss of pericytes rendered endothelial cells more susceptible to VEGF-A released by infiltrating inflammatory cells (Ogura et al. 2017). Thus, it appears such that the pericyte dysfunction or dropout induced by diabetes is followed by endothelial cell loss and an overall vascular regression, consequently leading to hypoxia and ischemia, which contribute to initiation of neovascularization, a hallmark feature of diabetic retinopathy (Hammes et al. 2010). Overall, the damage to pericytes inflicted by diabetes mellitus sets a clinical example showing the importance of pericytes for a healthy BRB functioning and retinal vascularization.

5.8 Future Directions

The important role of pericytes in retinal microvascular physiology, especially in regulation of the BRB and microcirculation has been well documented. Pericytes' role in several pathological conditions such as retinal ischemia and subsequent no-reflow as well as diabetic retinopathy has also been disclosed. Now, it is an exciting time for translating these discoveries to pharmacological interventions that can restore disrupted microcirculation in retinal diseases. Also, clarification of reservations regarding whether or not retinal microvascular pericytes are contractile may prompt new lines of research that can further our understanding of retinal microcirculation and neurovascular coupling in health and disease.

Acknowledgements Dr. Luis Alarcon-Martinez prepared Fig. 5.4. We thank our colleagues who contributed to the work from which figures are reused.

Compliance with Ethical Standards

Funding: Previously published research papers from our laboratory (Alarcon-Martinez et al. 2018 and 2019), which were cited here and their figures were partly reused, were supported in part by a grant to Drs. Luis Alarcon-Martinez and Turgay Dalkara from the seventh Framework Programme EU Marie Curie Actions – Co-funded Brain Circulation Scheme, and The Scientific and Technological Research Council of Turkey (TÜBİTAK; project number: 112C013). Further funding is given in references, Alarcon-Martinez et al. 2018 and 2019. Dr. Turgay Dalkara's research is supported by The Turkish Academy of Sciences.

Disclosure of Interests: None.

Ethical Approval: For previously published research from our laboratory, all applicable international, national, and/or institutional guidelines for the care and use of animals were followed. Details of ethics committee and permit numbers are given in references, Alarcon-Martinez et al. 2018 and 2019.

References

Alarcon-Martinez L, Yilmaz-Ozcan S, Yemisci M, Schallek J, Kılıç K, Can A, Di Polo A, Dalkara T (2018) Capillary pericytes express α-smooth muscle actin, which requires prevention of filamentous-actin depolymerization for detection. eLife 7

Alarcon-Martinez L, Yilmaz-Ozcan S, Yemisci M, Schallek J, Kılıç K, Villafranca-Baughman D, Can A, Di Polo A, Dalkara T (2019) Retinal ischemia induces α-SMA-mediated capillary pericyte contraction coincident with perivascular glycogen depletion. Acta Neuropathol Commun 7

Allt G, Lawrenson JG (2001) Pericytes: cell biology and pathology. Cells Tissues Organs 169:1–11

Ames A, Wright RL, Kowada M, Thurston JM, Majno G (1968) Cerebral ischemia. II. The no-reflow phenomenon. Am J Pathol 52:437–453

Arimura K, Ago T, Kamouchi M, Nakamura K, Ishitsuka K, Kuroda J, Sugimori H, Ooboshi H, Sasaki T, Kitazono T (2012) PDGF receptor β signaling in pericytes following ischemic brain injury. Curr Neurovasc Res 9:1–9

Armulik A, Genové G, Mäe M, Nisancioglu MH, Wallgard E, Niaudet C, He L, Norlin J, Lindblom P, Strittmatter K, Johansson BR, Betsholtz C (2010) Pericytes regulate the blood–brain barrier. Nature 468(7323):557–561

Armulik A, Genové G, Betsholtz C (2011) Pericytes: developmental, physiological, and pathological perspectives, problems, and promises. Dev Cell 21(2):193–215

Attwell D, Laughlin SB (2001) An energy budget for signaling in the grey matter of the brain. J Cereb Blood Flow Metab 21:1133–1145

Attwell D, Buchan AM, Charpak S, Lauritzen M, Macvicar BA, Newman EA (2010) Glial and neuronal control of brain blood flow. Nature 468:232–243

Bandopadhyay R, Orte C, Lawrenson JG, Reid AR, De Silva S, Allt G (2001) Contractile proteins in pericytes at the blood-brain and blood-retinal barriers. J Neurocytol 30:35–44

Beltramo E, Porta M (2013) Pericyte loss in diabetic retinopathy: mechanisms and consequences. Curr Med Chem 20(26):3218–3225

Betsholtz C (2004) Insight into the physiological functions of PDGF through genetic studies in mice. Cytokine Growth Factor Rev 15(4):215–228

Biesecker KR, Srienc AI, Shimoda AM, Agarwal A, Bergles DE, Kofuji P, Newman EA (2016) Glial cell calcium signaling mediates capillary regulation of blood flow in the retina. J Neurosci 36:9435–9445

Bill A (1975) Blood circulation and fluid dynamics in the eye. Physiol Rev 55:383–417

Birbrair A, Zhang T, Wang Z-M, Messi ML, Enikolopov GN, Mintz A, Delbono O (2013) Role of pericytes in skeletal muscle regeneration and fat accumulation. Stem Cells Dev 22:2298–2314

Birbrair A, Zhang T, Wang Z-M, Messi ML, Mintz A, Delbono O (2014a) Pericytes: multitasking cells in the regeneration of injured, diseased, and aged skeletal muscle. Front Aging Neurosci 6:245

Birbrair A, Zhang T, Wang Z-M, Messi ML, Olson JD, Mintz A, Delbono O (2014b) Type-2 pericytes participate in normal and tumoral angiogenesis. Am J Physiol Cell Physiol 307:C25–C38

Brown AM (2004) Brain glycogen re-awakened. J Neurochem 89:537–552

Caporarello N, D'Angeli F, Cambria MT, Candido S, Giallongo C, Salmeri M, Lombardo C, Longo A, Giurdanella G, Anfuso CD, Lupo G (2019) Pericytes in microvessels: from "mural" function to brain and retina regeneration. Int J Mol Sci 20(24):6351

Cogan DG, Toussaint D, Kuwabara T (1961) Retinal vascular patterns. IV. Diabetic retinopathy. Arch Ophthalmol 66:366–378
Crowell RM, Olsson Y (1972) Impaired microvascular filling after focal cerebral ischemia in monkeys. J Neurosurg 36:303–309
Cui Y, Xu X, Bi H, Zhu Q, Wu J, Xia X, Ren Q, Ho PC (2006) Expression modification of uncoupling proteins and MnSOD in retinal endothelial cells and pericytes induced by high glucose: the role of reactive oxygen species in diabetic retinopathy. Exp Eye Res 83:807–816
Cunha-Vaz J (2010) Blood–retinal barrier. Encyclopedia of the eye, pp 209–215
Dalkara T (2019) Pericytes. Stroke 50:2985–2991
Dalkara T, Alarcon-Martinez L (2015) Cerebral microvascular pericytes and neurogliovascular signaling in health and disease. Brain Res 1623:3–17
Dalkara T, Arsava EM (2012) Can restoring incomplete microcirculatory reperfusion improve stroke outcome after thrombolysis? J Cereb Blood Flow Metab 32:2091–2099
Dalkara T, Alarcon-Martinez L, Yemisci M (2016) Role of pericytes in neurovascular unit and stroke. In: Chen J, Zhang JH, Hu X (eds) Non-neuronal mechanisms of brain damage and repair after stroke. Springer International Publishing, Cham, pp 25–43
Daneman R, Zhou L, Kebede AA, Barres BA (2010) Pericytes are required for blood–brain barrier integrity during embryogenesis. Nature 468(7323):562–566
Danesh-Meyer HV, Kerr NM, Zhang J, Eady EK, O'Carroll SJ, Nicholson LFB, Johnson CS, Green CR (2012) Connexin43 mimetic peptide reduces vascular leak and retinal ganglion cell death following retinal ischaemia. Brain J Neurol 135:506–520
Danesh-Meyer HV, Zhang J, Acosta ML, Rupenthal ID, Green CR (2016) Connexin43 in retinal injury and disease. Prog Retin Eye Res 51:41–68
Das A, Frank RN, Weber ML, Kennedy A, Reidy CA, Mancini MA (1988) ATP causes retinal pericytes to contract in vitro. Exp Eye Res 46:349–362
del Zoppo GJ, Schmid-Schönbein GW, Mori E, Copeland BR, Chang CM (1991) Polymorphonuclear leukocytes occlude capillaries following middle cerebral artery occlusion and reperfusion in baboons. Stroke 22:1276–1283
del Zoppo GJ, Sharp FR, Heiss W-D, Albers GW (2011) Heterogeneity in the penumbra. J Cereb Blood Flow Metab 31:1836–1851
Díaz-Coránguez M, Ramos C, Antonetti DA (2017) The inner blood-retinal barrier: cellular basis and development. Vis Res 139:123–137
Dobrogowska D, Lossinsky A, Tarnawski M, Vorbrodt A (1998) J Neurocytol 27:163–173
Dziennis S, Qin J, Shi L, Wang RK (2015) Macro-to-micro cortical vascular imaging underlies regional differences in ischemic brain.. Scientific Reports
Eilken HM, Diéguez-Hurtado R, Schmidt I, Nakayama M, Jeong H-W, Arf H, Adams S, Ferrara N, Adams RH (2017) Pericytes regulate VEGF-induced endothelial sprouting through VEGFR1. Nat Commun 8(1)
Erdener SE, Dalkara T (2019) Small vessels are a big problem in neurodegeneration and neuroprotection. Front Neurol 16(10):889
Faal T, Phan DT, Davtyan H, Scarfone VM, Varady E, Blurton-Jones M, Hughes CC, Inlay MA (2019) Induction of mesoderm and neural crest-derived pericytes from human pluripotent stem cells to study blood-brain barrier interactions. Stem Cell Rep 12:451–460
Farrington-Rock C, Crofts N, Doherty M, Ashton B, Griffin-Jones C, Canfield A (2004) Chondrogenic and adipogenic potential of microvascular pericytes. Circulation 110(15):2226–2232
Fischer S, Wobben M, Marti H, Renz D, Schaper W (2002) Hypoxia-induced Hyperpermeability in brain microvessel endothelial cells involves VEGF-mediated changes in the expression of zonula occludens-1. Microvasc Res 63:70–80
Frank RN, Dutta S, Mancini MA (1987) Pericyte coverage is greater in the retinal than in the cerebral capillaries of the rat. Invest Ophthalmol Vis Sci 28(7):1086–1091
Fujimoto T, Singer SJ (1987) Immunocytochemical studies of desmin and vimentin in pericapillary cells of chicken. J Histochem Cytochem 35:1105–1115

Gaudin A, Yemisci M, Eroglu H, Lepetre-Mouelhi S, Turkoglu OF, Dönmez-Demir B, Caban S, Sargon MF, Garcia-Argote S, Pieters G, Loreau O, Rousseau B, Tagit O, Hildebrandt N, Le Dantec Y, Mougin J, Valetti S, Chacun H, Nicolas V, Desmaële D, Andrieux K, Capan Y, Dalkara T, Couvreur P (2014) Squalenoyl adenosine nanoparticles provide neuroprotection after stroke and spinal cord injury. Nat Nanotechnol 9:1054–1062

Geraldes P, Hiraoka-Yamamoto J, Matsumoto M, Clermont A, Leitges M, Marette A, Aiello LP, Kern TS, King GL (2009) Activation of PKC-delta and SHP-1 by hyperglycemia causes vascular cell apoptosis and diabetic retinopathy. Nat Med 15:1298–1306

Gerhardt H, Betsholtz C (2003) Endothelial-pericyte interactions in angiogenesis. Cell Tissue Res 314(1):15–23

Grunwald JE, Riva CE, Stone RA, Keates EU, Petrig BL (1984) Retinal auto-regulation in open-angle glaucoma. Ophthalmology 91:1690–1694

Gurer G, Gursoy-Ozdemir Y, Erdemli E, Can A, Dalkara T (2009) Astrocytes are more resistant to focal cerebral ischemia than neurons and die by a delayed necrosis. Brain Pathol Zurich Switz 19:630–641

Gursoy-Ozdemir Y, Can A, Dalkara T (2004) Reperfusion-induced oxidative/nitrative injury to neurovascular unit after focal cerebral ischemia. Stroke 35:1449–1453

Gursoy-Ozdemir Y, Yemisci M, Dalkara T (2012) Microvascular protection is essential for successful neuroprotection in stroke. J Neurochem 123(Suppl 2):2–11

Hall CN, Reynell C, Gesslein B, Hamilton NB, Mishra A, Sutherland BA, O'Farrell FM, Buchan AM, Lauritzen M, Attwell D (2014) Capillary pericytes regulate cerebral blood flow in health and disease. Nature 508:55–60

Hallenbeck JM, Dutka AJ, Tanishima T, Kochanek PM, Kumaroo KK, Thompson CB, Obrenovitch TP, Contreras TJ (1986) Polymorphonuclear leukocyte accumulation in brain regions with low blood flow during the early postischemic period. Stroke 17:246–253

Hamilton NB, Attwell D, Hall CN (2010) Pericyte-mediated regulation of capillary diameter: a component of neurovascular coupling in health and disease. Front Neuroenerg 2

Hammes H-P, Lin J, Renner O, Shani M, Lundqvist A, Betsholtz C, Brownlee M, Deutsch U (2002) Pericytes and the pathogenesis of diabetic retinopathy. Diabetes 51(10):3107–3112

Hammes H-P, Feng Y, Pfister F, Brownlee M (2010) Diabetic retinopathy: targeting Vasoregression. Diabetes 60(1):9–16

Herman IM, D'Amore PA (1985) Microvascular pericytes contain muscle and non-muscle actins. J Cell Biol 101:43–52

Hill RA, Tong L, Yuan P, Murikinati S, Gupta S, Grutzendler J (2015) Regional blood flow in the Normal and ischemic brain is controlled by arteriolar smooth muscle cell contractility and not by capillary pericytes. Neuron 87:95–110

Ivanova E, Kovacs-Oller T, Sagdullaev BT (2017) Vascular Pericyte Impairment and Connexin43 Gap Junction Deficit Contribute to Vasomotor Decline in Diabetic Retinopathy. J Neurosci 37(32):7580–7894

Jeansson M, Gawlik A, Anderson G, Li C, Kerjaschki D, Henkelman M, Quaggin SE (2011) Angiopoietin-1 is essential in mouse vasculature during development and in response to injury. J Clin Investig 121(6):2278–2289

Joyce NC, Haire MF, Palade GE (1985a) Contractile proteins in pericytes. I. Immunoperoxidase localization of tropomyosin. J Cell Biol 100:1379–1386

Joyce NC, Haire MF, Palade GE (1985b) Contractile proteins in pericytes. II. Immunocytochemical evidence for the presence of two isomyosins in graded concentrations. J Cell Biol 100:1387–1395

Kamouchi M, Kitazono T, Ago T, Wakisaka M, Ooboshi H, Ibayashi S, Iida M (2004) Calcium influx pathways in rat CNS pericytes. Brain Res Mol Brain Res 126:114–120

Kamouchi M, Kitazono T, Ago T, Wakisaka M, Kuroda J, Nakamura K, Hagiwara N, Ooboshi H, Ibayashi S, Iida M (2007) Hydrogen peroxide-induced Ca2+ responses in CNS pericytes. Neurosci Lett 416:12–16

Kamouchi M, Ago T, Kitazono T (2012) Brain pericytes: emerging concepts and functional roles in brain homeostasis. Cell Mol Neurobiol 31:175–193

Karow M, Schichor C, Beckervordersandforth R, Berninger B (2014) Lineage-reprogramming of pericyte-derived cells of the adult human brain into induced neurons. J Vis Exp (87)
Kasischke KA, Vishwasrao HD, Fisher PJ, Zipfel WR, Webb WW (2004) Neural activity triggers neuronal oxidative metabolism followed by astrocytic glycolysis. Science 305:99–103
Kaur C, Sivakumar V, Yong Z, Lu J, Foulds W, Ling E (2007) Blood–retinal barrier disruption and ultrastructural changes in the hypoxic retina in adult rats: the beneficial effect of melatonin administration. J Pathol 212:429–439
Kaur C, Foulds W, Ling E (2008) Blood–retinal barrier in hypoxic ischaemic conditions: basic concepts, clinical features and management. Prog Retin Eye Res 27(6):622–647
Kelley C, D'Amore P, Hechtman HB, Shepro D (1987) Microvascular pericyte contractility in vitro: comparison with other cells of the vascular wall. J Cell Biol 104:483–490
Kerr NM, Johnson CS, Zhang J, Eady EK, Green CR, Danesh-Meyer HV (2012) High pressure-induced retinal ischaemia reperfusion causes upregulation of gap junction protein connexin43 prior to retinal ganglion cell loss. Exp Neurol 234:144–152
Khennouf L, Gesslein B, Brazhe A, Octeau JC, Kutuzov N, Khakh BS, Lauritzen M (2018) Active role of capillary pericytes during stimulation-induced activity and spreading depolarization. Brain J Neurol 141:2032–2046
Kim JH, Kim JH, Yu YS, Kim DH, Kim K-W (2009) Recruitment of pericytes and astrocytes is closely related to the formation of tight junction in developing retinal vessels. J Neurosci Res 87(3):653–659
Kisler K, Nelson AR, Montagne A, Zlokovic BV (2017) Cerebral blood flow regulation and neurovascular dysfunction in Alzheimer disease. Nat Rev Neurosci 18:419–434
Kondo T, Hosoya K-I, Hori S, Tomi M, Ohtsuki S, Terasaki T (2005) PKC/MAPK signaling suppression by retinal pericyte conditioned medium prevents retinal endothelial cell proliferation. J Cell Physiol 203:378–386
Kornfield TE, Newman EA (2014) Regulation of blood flow in the retinal trilaminar vascular network. J Neurosci 34:11504–11513
Kuwabara T, Cogan D (1961) Retinal glycogen. Trans Am Ophthalmol Soc 59:106–110
Leal-Campanario R, Alarcon-Martinez L, Rieiro H, Martinez-Conde S, Alarcon-Martinez T, Zhao X, Lamee J, Popp PJO, Calhoun ME, Arribas JI, Schlegel AA, Stasi LLD, Rho JM, Inge L, Otero-Millan J, Treiman DM, Macknik SL (2017) Abnormal capillary Vasodynamics contribute to ictal neurodegeneration in epilepsy. Sci Rep 7:43276
Lee J, Gursoy-Ozdemir Y, Fu B, Boas DA, Dalkara T (2016) Optical coherence tomography imaging of capillary reperfusion after ischemic stroke. Appl Opt 55:9526
Li Q, Puro DG (2001) Adenosine activates ATP-sensitive K(+) currents in pericytes of rat retinal microvessels: role of A1 and A2a receptors. Brain Res 907:93–99
Lindahl P, Johansson BR, Levéen P, Betsholtz C (1997) Pericyte loss and microaneurysm formation in PDGF-B-deficient mice. Science 277(5323):242–245
Lindblom P, Gerhardt H, Liebner S, Abramsson A, Enge M, Hellstrom M, Backstrom G, Fredriksson S, Landegren U, Nystrom HC, Bergström G, Dejana E, Östman A, Lindahl P, Betsholtz C (2003) Endothelial PDGF-B retention is required for proper investment of pericytes in the microvessel wall. Genes Dev 17(15):1835–1840
Liu S, Agalliu D, Yu C, Fisher M (2012) The role of pericytes in blood-brain barrier function and stroke. Curr Pharm Des 18:3653–3662
Mayhan WG (1999) VEGF increases permeability of the blood-brain barrier via a nitric oxide synthase/cGMP-dependent pathway. Am J Phys Cell Phys 276(5):C1148–C1153
Metea MR, Newman EA (2006) Glial cells dilate and constrict blood vessels: a mechanism of neurovascular coupling. J Neurosci 26:2862–2870
Mishra A, Reynolds JP, Chen Y, Gourine AV, Rusakov DA, Attwell D (2016) Astrocytes mediate neurovascular signaling to capillary pericytes but not to arterioles. Nat Neurosci 19:1619–1627
Motiejūnaitė R, Kazlauskas A (2008) Pericytes and ocular diseases. Exp Eye Res 86(2):171–177
Murphy DD, Wagner RC (1994) Differential contractile response of cultured micro-vascular pericytes to vasoactive agents. Microcirculaiton 1:121–128

Nakagomi T, Nakano-Doi A, Kawamura M, Matsuyama T (2015) Do vascular Pericytes contribute to neurovasculogenesis in the central nervous system as multipotent vascular stem cells? Stem Cells Dev 24:1730–1739
Nakahara T, Hoshino M, Hoshino S-I, Mori A, Sakamoto K, Ishii K (2015) Structural and functional changes in retinal vasculature induced by retinal ischemia-reperfusion in rats. Exp Eye Res 135:134–145
Nakaizumi A, Puro DG (2011) Vulnerability of the retinal microvasculature to hypoxia: role of polyamine-regulated K(ATP) channels. Invest Ophthalmol Vis Sci 52:9345–9352
Nakamura K, Kamouchi M, Kitazono T, Kuroda J, Shono Y, Hagiwara N, Ago T, Oo-boshi H, Ibayashi S, Iida M (2009) Amiloride inhibits hydrogen peroxide-induced Ca2+ responses in human CNS pericytes. Microvasc Res 77:327–334
Nakamura K, Arimura K, Nishimura A, Tachibana M, Yoshikawa Y, Makihara N, Wakisaka Y, Kuroda J, Kamouchi M, Ooboshi H, Kitazono T, Ago T (2016) Possible involvement of basic FGF in the upregulation of PDGFRβ in pericytes after ischemic stroke. Brain Res 1630:98–108
Nehls V, Drenckhahn D (1991) Heterogeneity of microvascular pericytes for smooth muscle type alpha-actin. J Cell Biol 113:147–154
Newman EA (2013) Functional hyperemia and mechanisms of neurovascular coupling in the retinal vasculature. J Cereb Blood Flow Metab 33:1685–1695
O'Farrell FM, Mastitskaya S, Hammond-Haley M, Freitas F, Wah WR, Attwell D (2017) Capillary pericytes mediate coronary no-reflow after myocardial ischaemia. eLife 6
Ogura S, Kurata K, Hattori Y, Takase H, Ishiguro-Oonuma T, Hwang Y, Ahn S, Park I, Ikeda W, Kusuhara S, Fukushima Y, Nara H, Sakai H, Fujiwara T, Matsushita J, Ema M, Hirashima M, Minami T, Shibuya M, Takakura N, Kim P, Miyata T, Ogura Y, Uemura A (2017) Sustained inflammation after pericyte depletion induces irreversible blood-retina barrier breakdown. JCI Insight 2(3):e90905
Ozerdem U, Stallcup WB (2004) Pathological angiogenesis is reduced by targeting pericytes via the NG2 proteoglycan. Angiogenesis 7:269–276
Park YS, Kim NH, Jo I (2003) Hypoxia and vascular endothelial growth factor acutely up-regulate angiopoietin-1 and Tie2 mRNA in bovine retinal pericytes. Microvasc Res 65:125–131
Park TS, Bhutto I, Zimmerlin L, Huo JS, Nagaria P, Miller D, Rufaihah AJ, Talbot C, Aguilar J, Grebe R, Merges C, Reijo-Pera R, Feldman RA, Rassool F, Cooke J, Lutty G, Zambidis ET (2014a) Vascular progenitors from cord blood–derived induced pluripotent stem cells possess augmented capacity for regenerating ischemic retinal vasculature. Circulation 129:359–372
Park SW, Yun J-H, Kim JH, Kim K-W, Cho C-H, Kim JH (2014b) Angiopoietin 2 induces pericyte apoptosis via α3β1 integrin signaling in diabetic retinopathy. Diabetes 63:3057–3068
Park DY, Lee J, Kim J, Kim K, Hong S, Han S, Kubota Y, Augustin HG, Ding L, Kim JW, Kim H, He Y, Adams RH, Koh GY (2017) Plastic roles of pericytes in the blood–retinal barrier. Nat Commun 8(1)
Patel JI, Hykin PG, Gregor ZJ, Boulton M, Cree IA (2005) Angiopoietin concentrations in diabetic retinopathy. Br J Ophthalmol 89(4):480–483
Peppiatt CM, Howarth C, Mobbs P, Attwell D (2006) Bidirectional control of CNS capillary diameter by pericytes. Nature 443:700–704
Puro DG (2007) Physiology and pathobiology of the pericyte-containing retinal microvasculature: new developments. Microcirculation 14:1–10
Puro DG (2012) Retinovascular physiology and pathophysiology: new experimental approach/new insights. Prog Retin Eye Res 31:258–270
Riva CE, Titze P, Hero M, Petrig BL (1997) Effect of acute decreases of perfusion pressure on choroidal blood flow in humans. Invest Ophthalmol Vis Sci 38:1752–1760
Romeo G, Liu W-H, Asnaghi V, Kern TS, Lorenzi M (2002) Activation of nuclear factor-kappaB induced by diabetes and high glucose regulates a proapoptotic program in retinal pericytes. Diabetes 51:2241–2248
Rossi DJ, Brady JD, Mohr C (2007) Astrocyte metabolism and signaling during brain ischemia. Nat Neurosci 10:1377–1386

Roy CS, Sherrington CS (1890) On the regulation of the blood-supply of the brain. J Physiol 11:85–158.17

Rungta RL, Chaigneau E, Osmanski B-F, Charpak S (2018) Vascular compartmentalization of functional hyperemia from the synapse to the pia. Neuron 99:362–375.e4

Sakuma R, Kawahara M, Nakano-Doi A, Takahashi A, Tanaka Y, Narita A, Kuwahara-Otani S, Hayakawa T, Yagi H, Matsuyama T, Nakagomi T (2016) Brain pericytes serve as micro-glia-generating multipotent vascular stem cells following ischemic stroke. J Neuroinflammation 13(1):57

Santos GSP, Prazeres PHDM, Mintz A, Birbrair A (2018) Role of pericytes in the retina. Eye 32:483–486

Schallek J, Geng Y, Nguyen H, Williams DR (2013) Morphology and topography of retinal pericytes in the living mouse retina using in vivo adaptive optics imaging and ex vivo characterization. Invest Ophthalmol Vis Sci 54:8237–8250

Schor AM, Allen TD, Canfield AE, Sloan P, Schor SL (1990) Pericytes derived from the retinal microvasculature undergo calcification in vitro. J Cell Sci 97(Pt 3):449–461

Shepro D, Morel NM (1993) Pericyte physiology. FASEB J 7:1031–1038

Srienc AI, Kurth-Nelson ZL, Newman EA (2010) Imaging retinal blood flow with laser speckle flowmetry. Front Neuroenerg 2

Stebbins MJ, Gastfriend BD, Canfield SG, Lee M-S, Richards D, Faubion MG, Li W-J, Daneman R, Palecek SP, Shusta EV (2019) Human pluripotent stem cell–derived brain pericyte–like cells induce blood-brain barrier properties. Sci Adv 5(3):eaau7375

Suzuki T, Matsunami T, Hisa Y, Takata K, Takamatsu T, Oyamada M (2009) Roles of gap junctions in glucose transport from glucose transporter 1-positive to -negative cells in the lateral wall of the rat cochlea. Histochem Cell Biol 131:89–102

Sweeney MD, Ayyadurai S, Zlokovic BV (2016) Pericytes of the neurovascular unit: key functions and signaling pathways. Nat Neurosci 19:771–783

Sweeney MD, Kisler K, Montagne A, Toga AW, Zlokovic BV (2018) The role of brain vasculature in neurodegenerative disorders. Nat Neurosci 21:1318–1331

Takahashi A, Park HK, Melgar MA, Alcocer L, Pinto J, Lenzi T, Diaz FG, Rafols JA (1997) Cerebral cortex blood flow and vascular smooth muscle contractility in a rat model of ischemia: a correlative laser Doppler flowmetric and scanning electron microscopic study. Acta Neuropathol (Berl) 93:354–368

Takihara Y, Inatani M, Eto K, Inoue T, Kreymerman A, Miyake S, Ueno S, Nagaya M, Nakanishi A, Iwao K, Takamura Y, Sakamoto H, Satoh K, Kondo M, Sakamoto T, Goldberg JL, Nabekura J, Tanihara H (2015) In vivo imaging of axonal transport of mitochondria in the diseased and aged mammalian CNS. Proc Natl Acad Sci 112:10515–10520

Taskiran-Sag A, Yemisci M, Gursoy-Ozdemir Y, Erdener SE, Karatas H, Yuce D, Dalkara T (2018) Improving microcirculatory reperfusion reduces parenchymal oxy-gen radical formation and provides neuroprotection. Stroke 49:1267–1275

Teichert M, Milde L, Holm A, Stanicek L, Gengenbacher N, Savant S, Ruckdeschel T, Hasanov Z, Srivastava K, Hu J, Hertel S, Bartol A, Schlereth K, Augustin HG (2017) Pericyte-expressed Tie2 controls angiogenesis and vessel maturation. Nat Commun 8(1)

Toft-Kehler AK, Skytt DM, Svare A, Lefevere E, Van Hove I, Moons L, Waagepetersen HS, Kolko M (2017) Mitochondrial function in Müller cells – does it matter? Mitochondrion 36:43–51

Trost A, Lange S, Schroedl F, Bruckner D, Motloch KA, Bogner B, Kaser-Eichberger A, Strohmaier C, Runge C, Aigner L, Rivera FJ, Reitsamer HA (2016) Brain and retinal pericytes: origin, function and role. Front Cell Neurosci 10

Uemura A, Ogawa M, Hirashima M, Fujiwara T, Koyama S, Takagi H, Honda Y, Wiegand SJ, Yancopoulos GD, Nishikawa S-I (2002) Recombinant angiopoietin-1 restores higher-order architecture of growing blood vessels in mice in the absence of mural cells. J Clin Investig 110:1619–1628

Underly RG, Levy M, Hartmann DA, Grant RI, Watson AN, Shih AY (2016) Pericytes as inducers of rapid, matrix metalloproteinase-9-dependent capillary damage during ischemia. J Neurosci 37(1):129–140

Vanlandewijck M, He L, Mäe MA, Andrae J, Ando K, Del Gaudio F, Nahar K, Lebouvier T, Laviña B, Gouveia L, Sun Y, Raschperger E, Räsänen M, Zarb Y, Mochizuki N, Keller A, Lendahl U, Betsholtz C (2018) A molecular atlas of cell types and zonation in the brain vasculature. Nature 554:475–480

Vidhya S, Ramya R, Coral K, Sulochana K, Bharathidevi S (2018) Free amino acids hydroxyproline, lysine, and glycine promote differentiation of retinal pericytes to adipocytes: a protective role against proliferative diabetic retinopathy. Exp Eye Res 173:179–187

Wallow IH, Burnside B (1980) Actin filaments in retinal pericytes and endothelial cells. Invest Ophthalmol Vis Sci 19:1433–1441

Wang YL, Hui YN, Guo B, Ma JX (2007) Strengthening tight junctions of retinal microvascular endothelial cells by pericytes under normoxia and hypoxia involving angiopoietin-1 signal way. Eye 21:1501–1510

Wasilewa P, Hockwin O, Korte I (1976) Glycogen concentration changes in retina, vitreous body and other eye tissues caused by disturbances of blood circulation. Albrecht Von Graefes Arch Für Klin Exp Ophthalmol Albrecht Von Graefes Arch Clin Exp Ophthalmol 199:115–120

Wilson CA, Berkowitz BA, Funatsu H, Metrikin DC, Harrison DW, Lam MK, Sonkin PL (1995) Blood-retinal barrier breakdown following experimental retinal ischemia and reperfusion. Exp Eye Res 61:547–557

Winkler EA, Sagare AP, Zlokovic BV (2014) The pericyte: a forgotten cell type with important implications for Alzheimers disease? Brain Pathol 24(4):371–386

Wong TY, Sun J, Kawasaki R, Ruamviboonsuk P, Gupta N, Lansingh VC, Maia M, Mathenge W, Moreker S, Muqit MMK, Resnikoff S, Verdaguer J, Zhao P, Ferris F, Aiello LP, Taylor HR (2018) Guidelines on diabetic eye care: the International Council of Ophthalmology Recommendations for screening, follow-up, referral, and treatment based on resource settings. Ophthalmology 125:1608–1622

Wu DM, Kawamura H, Sakagami K, Kobayashi M, Puro DG (2003) Cholinergic regulation of pericyte-containing retinal microvessels. Am J Physiol Heart Circ Physiol 284:H2083–2090

Yamagishi S, Yonekura H, Yamamoto Y, Fujimori H, Sakurai S, Tanaka N, Yamamoto H (1999) Vascular endothelial growth factor acts as a pericyte mitogen under hypoxic conditions. Lab Investig 79(4):501–509

Yamanishi S, Katsumura K, Kobayashi T, Puro DG (2006) Extracellular lactate as a dynamic vasoactive signal in the rat retinal microvasculature. Am J Physiol Heart Circ Physiol 290:H925–H934

Yemisci M, Gursoy-Ozdemir Y, Vural A, Can A, Topalkara K, Dalkara T (2009) Pericyte contraction induced by oxidative-nitrative stress impairs capillary reflow despite successful opening of an occluded cerebral artery. Nat Med 15:1031–1037

Yoshida Y, Kabara M, Kano K, Horiuchi K, Hayasaka T, Tomita Y, Takehara N, Minoshima A, Aonuma T, Maruyama K, Nakagawa N, Azuma N, Hasebe N, Kawabe JI (2019) Capillary-resident EphA7 pericytes are multipotent cells with anti-ischemic effects through capillary formation. Stem Cells Transl Med 9(1):120–130

Yu D-Y, Cringle SJ, Yu PK, Balaratnasingam C, Mehnert A, Sarunic MV, An D, Su E-N (2019a) Retinal capillary perfusion: spatial and temporal heterogeneity. Prog Retin Eye Res 70:23–54

Yu P-K, An D, Balaratnasingam C, Cringle SJ, Yu D-Y (2019b) Topographic distribution of contractile protein in the human macular microvasculature. Invest Ophthalmol Vis Sci 60:4574–4582

Zeisel A, Hochgerner H, Lönnerberg P, Johnsson A, Memic F, van der Zwan J, Häring M, Braun E, Borm LE, La Manno G, Codeluppi S, Furlan A, Lee K, Skene N, Harris KD, Hjerling-Leffler J, Arenas E, Ernfors P, Marklund U, Linnarsson S (2018) Molecular architecture of the mouse nervous system. Cell 174:999–1014

Zhang ZG, Chopp M, Goussev A, Lu D, Morris D, Tsang W, Powers C, Ho KL (1999) Cerebral microvascular obstruction by fibrin is associated with upregulation of PAI-1 acutely after onset of focal embolic ischemia in rats. J Neurosci 19:10898–10907

Zheng L, Gong B, Hatala DA, Kern TS (2007) Retinal ischemia and reperfusion causes capillary degeneration: similarities to diabetes. Investig Opthalmol Visual Sci 48(1):361

Chapter 6
Inflammatory Mediators Released by Brain Pericytes as Sensors and Effectors in Blood-Brain Barrier Dysfunction

Shinya Dohgu, Fuyuko Takata, and Yasufumi Kataoka

Abstract

Introduction

Brain pericytes are located at the cerebral vascular walls where they regulate blood flow in microvessels without smooth muscle cells. *In vitro* and *in vivo* studies have demonstrated that pericytes enhance and maintain blood–brain barrier (BBB) integrity by releasing soluble factors. Another functional feature of pericytes is facilitated through cytokine/chemokine receptors, macrophage markers, innate and adaptive immune system-related receptors and scavenger receptors; these enable pericytes to respond to various inflammatory mediators and pathogens. These features allow pericytes to act as sensors to monitor various pathological stimuli from the periphery and neighboring cells in the central nervous system under the pathological conditions. In this chapter, we summarize pericytic regulation of the neurovascular unit, and then discuss brain pericyte activation and its impacts on neighboring cell types under the pathophysiological conditions. We also discuss phenotypic markers for identifying the activated brain pericytes.

Methods

We surveyed literatures reporting pericytes or pericyte-derived factors in the regulation of BBB or neurovascular unit on PubMed.

Results

Among the cells comprising the neurovascular unit, brain pericytes exhibit the highest sensitivity to TNF-α and thrombin, leading to the release of cytokines, chemokines and matrix metalloproteinase-9 and the enhancement of microglial activation. These properties suggest that pericytes act as effectors to amplify neuroinflammation in the neurovascular unit and consequently induce the impairment of BBB function. In this review, we define the state in which brain pericytes release various inflammatory mediators as "activated brain pericytes". These are operative for

S. Dohgu · F. Takata · Y. Kataoka (✉)
Department of Pharmaceutical Care and Health Sciences, Faculty of Pharmaceutical Sciences, Fukuoka University, Fukuoka, Japan
e-mail: dohgu@fukuoka-u.ac.jp; ftakata@fukuoka-u.ac.jp; ykataoka@fukuoka-u.ac.jp

A. Birbrair (ed.), *Biology of Pericytes – Recent Advances*, Stem Cell Biology and Regenerative Medicine 68, https://doi.org/10.1007/978-3-030-62129-2_6

induction and promotion of the BBB dysfunction. Altered expression levels of pericyte marker proteins under the pathological conditions are potential clues for identifying activated brain pericytes.

Conclusions
Although further studies are needed to clarify the regulation and identification of activated brain pericytes, this activated state of brain pericytes should be considered as a potential therapeutic target for neurodegenerative diseases with BBB dysfunction.

Keywords α-Synuclein · Blood-brain barrier · Heterogeneity · Interleukin-1β · Interleukin-6 · Lipopolysaccharide · Macrophage inflammatory protein -1α · Matrix metalloproteinase-9 · Microglia · Neuroinflammation · Oncostatin M · Pericyte · Platelet-derived growth factor · Thrombin · Tumor necrosis factor-α

6.1 Introduction

The blood–brain barrier (BBB) physically restricts the movement of blood-derived substrates including blood cells, neurotoxins, pathogens and plasma proteins into the brain parenchyma. Impairment of the BBB is observed in patients with various diseases. BBB dysfunction accompanies and/or is responsible for the development and pathogenesis of several diseases (Sweeney et al. 2019). The impaired barrier allows blood cells and neurotoxins to enter the brain, leading to neuroinflammation and neuronal damage. The barrier properties of the BBB are mainly regulated by brain capillary endothelial cells connected by tight junctions; these are composed of occludin, claudin-5, and zonula occludens (ZO)-1, 2 and 3. Brain endothelial barrier integrity is supported by the neighboring cells including brain pericytes and astrocytes. Brain pericytes are located at the capillary wall and share the basement membrane of brain endothelial cells. Pericytes directly interact with brain endothelial cells through peg-and-socket type contacts (Winkler et al. 2011). Pericytic laminin, a major component of the basement membrane, contributes to the maintenance of BBB integrity (Gautam et al. 2016). In addition, brain pericytes release several factors to communicate with brain endothelial cells. This interaction stabilizes the cerebral vasculature, regulates cerebral blood flow, enhances brain endothelial barrier genesis and maintains barrier integrity. Therefore, dysregulated interactions between brain endothelial cells and brain pericytes could induce BBB impairment. In this chapter, we first summarize the regulation of the neurovascular unit by pericyte-derived soluble factors, and then discuss possible molecules involved in brain pericyte activation and the impact of pericyte-derived inflammatory mediators on neighboring cell types, especially brain endothelial cells, under pathophysiological conditions. We also evaluate phenotypic markers of activated brain pericytes that are associated with BBB dysfunction.

6.2 Intercellular Communication Between Brain Pericytes and Other Cell Types in the Neurovascular Unit in Health and Disease

The use of *in vitro* BBB models using brain endothelial cells co-cultured with pericytes has revealed that pericytes tighten the barrier (Nakagawa et al. 2007, 2009) and increase the efflux of transporters in brain endothelial cells (Berezowski et al. 2004). Pericyte-derived transforming growth factor (TGF)-β (Dohgu et al. 2005), angiopoietin-1 (Hori et al. 2004) and plasminogen activator inhibitor (PAI)-1 (Dohgu et al. 2011b) enhance and maintain the integrity of the brain endothelial barrier. In addition, brain pericyte-secreted soluble factors enhance the insulin sensitivity of hypothalamic neurons (Takahashi et al. 2015) and the differentiation of oligodendrocyte progenitor cells to mature oligodendrocytes (De La Fuente et al. 2017). Accumulating evidence using genetically modified mice with pericyte deficiencies during aging (e.g. platelet-derived growth factor receptor (PDGFR)$\beta^{+/-}$, PDGFb$^{ret/ret}$, PDGFR$\beta^{F7/F7}$) has revealed that reduced pericyte coverage of the cerebral capillaries induces BBB impairment, reduces cerebral blood flow, enhances microglial activation and induces neuronal dysfunction (Winkler et al. 2010; Daneman et al. 2010; Bell et al. 2010; Armulik et al. 2010). Dysregulation of the interactions between brain endothelial cells and pericytes can be caused by detachment from and/or loss of the brain pericytes in the cerebral capillary wall. PDGFRβ is a pericyte-specific marker, and brain endothelial cells secrete its ligand PDGF-BB. This signal transmission regulates pericytic survival, proliferation and differentiation (Brandt et al. 2019; Lindahl et al. 1997). Thus, PDGF-BB/PDGFRβ signaling is likely to be crucial for brain endothelial cell-to-pericyte communication to hold pericytes in the capillary wall. Bmal1−/− or cathepsin D−/− mice also exhibit pericyte loss in the cerebral vasculature, which leads to BBB impairment (Okada et al. 2015; Nakazato et al. 2017). These loss-of-function *in vivo* studies using genetically modified mice indicate that brain pericytes are required for maintenance and formation of the BBB. The findings using pericyte-deficient mice are supported in part by studies of Alzheimer's disease (AD) patients. Apolipoprotein E4 carriers with AD exhibit pericyte loss and BBB impairment (Halliday et al. 2016). Patients with AD show lower levels of vascular markers of brain mural cells (α-smooth muscle actin (α-SMA), PDGFRβ and CD13) including brain pericytes (Bourassa et al. 2020). Experimental data from animals show that pericyte degradation or detachment occurs in both brain and peripheral diseases. Traumatic brain injury induces rapid pericyte loss (Zehendner et al. 2015) and pericyte migration from the cerebral vascular walls (Dore-Duffy et al. 2000). Detachment of brain pericytes is closely associated with BBB dysfunction in lipopolysaccharide (LPS)-induced septic mice (Nishioku et al. 2009). Oxidative stress induces pericyte loss in the cerebral microvasculature at 12 weeks in streptozotocin-induced diabetic mice (Price et al. 2012).

Another possible way that brain pericytes induce BBB dysfunction without pericyte deficiency in and/or detachment from cerebral capillary walls is the dysregulation of brain endothelial barrier integrity through pericyte-derived substances. Brain pericytes possess the ability to release various proinflammatory mediators for BBB disruption in response to various stimuli (Gaceb et al. 2018a; Rustenhoven et al. 2017). Indeed, brain pericytes co-cultured with brain endothelial cells induce and/or aggravate brain endothelial barrier dysfunction in response to various disease-associated factors including thrombin (Machida et al. 2017b), oncostatin M (Takata et al. 2019), α-synuclein (Dohgu et al. 2019), advanced glycation endproducts (AGEs) (Shimizu et al. 2013) and LPS (Pieper et al. 2013; Dohgu and Banks 2013). Tumor necrosis factor (TNF)-α stimulates brain pericytes to release various cytokines and chemokines, and subsequently induces microglial activation (Matsumoto et al. 2014). Activated microglia are involved in BBB dysfunction (Nishioku et al. 2010; Shigemoto-Mogami et al. 2018; Sumi et al. 2010). Thus, brain pericyte-released factors impair BBB function directly and indirectly through microglial activation under pathological conditions. Therefore, we propose that brain pericytes could act as (1) sensors to receive inflammatory signals from the periphery or neighboring cells in the neurovascular unit and (2) effectors to amplify neuroinflammation leading to BBB dysfunction. Here, we define a state of brain pericytes in which they release factors associated with BBB dysfunction in response to various pathological stimuli as "activated brain pericytes" (Table 6.1).

Table 6.1 Soluble factors derived from activated brain pericytes

	Activated brain pericytes		
Activators	BBB impairment	Activated brain pericyte-derived soluble factors	References
TNF-α	○	G-CSF, GRO/KC/IL-8, IL-12(p70), IL-17, IL-1α, IL-2, IL-5, IL-6, IP-10, MCP-1, MIP-1α, MMP-9, NO, RANTES	Jansson et al. (2014), Persidsky et al. (2016), Matsumoto et al. (2014), Takata et al. (2011)
Thrombin	○	MMP-9, IL-1β, IL-6, TNF-α	Machida et al. (2017a, b, 2015)
α-Synuclein	○	IL-1β, IL-6, MCP-1, TNF-α, MMP-9	Dohgu et al. (2019)
OSM	○	IL-6	Takata et al. (2019)
LPS	○	CCL2, CCL3, CCL4, CXCL1/2/3, eotaxin, G-CSF, GM-CSF, GRO/KC, IL-10, IL-12(p40), IL-12(p40), IL-13, IL-17, IL-1α, IL-3, IL-6, IL-9, INF-γ, IP-10, MCP-1, NO, TNF-α	Dohgu and Banks (2013), Kovac et al. (2011), Guijarro-Munoz et al. (2014)
PDGF-BB	–	VEGF, IL-6, IL-8, IP-10, BDNF, bFGF, βNGF	Gaceb et al. (2018b)
IL-1β	○	IL-8, IP-10, MCP-1, NO, RANTES	Persidsky et al. (2016), Pieper et al. (2013, 2014)

6.3 Extracellular Molecules as Pericyte Activators

- TNF-α

In the central nervous system (CNS), TNF-α, a proinflammatory cytokine, is produced by glial cells under brain insults such as stroke (Low et al. 2014) or may penetrate from the blood to the brain parenchyma across the BBB after injury (Pan et al. 1999). TNF-α induces the release of matrix metalloproteinase (MMP)-9, a mediator of BBB dysfunction, from the cell types constituting the BBB such as endothelial cells, pericytes and astrocytes. Interestingly, among these cell types brain pericytes are the major source of MMP-9 release at the BBB level (Takata et al. 2011). TNF-α converts brain pericytes from the normal state to the reactive state through activation of MAPKs and the Akt signaling pathway, leading to MMP-9 release. Therefore, activated pericytes are responsible for TNF-α-induced BBB dysfunction (Fig. 6.1). Along with MMP-9, TNF-α-activated pericytes release several cytokines and chemokines such as macrophage inflammatory protein (MIP)-1α, interleukin (IL)-6 and monocyte chemotactic protein (MCP)-1. These proinflammatory molecules are associated with the development of neuroinflammation. Surprisingly, the sensitivity of brain pericytes for induction of this cytokine and chemokine release is much higher than that of microglia, which are well known to be responsible for immune defense in the CNS (Fig. 6.2). Besides releasing cytokines and chemokines, brain pericytes facilitate migration, phagocytic activity, and inducible nitric oxide synthase (iNOS) synthesis due to TNF-α stimulation; these are essential inflammatory reactions (Pieper et al. 2014; Takata et al. 2011). Based on these findings, brain pericytes could be considered to be the cell type that acts as a trigger for the induction and development of TNF-α-related neuroinflammation. In fact, TNF-α-reactive pericytes elicit the activation of microglia; this microglial state is known to mediate CNS inflammation (Matsumoto et al. 2014; Cianciulli et al. 2020). Increased release of IL-6 and other, unknown, soluble factors derived from TNF-α-activated pericytes elevates the iNOS and IL-1β levels in microglia (Matsumoto et al. 2014, 2018). The ability of TNF-α-reactive pericytes to release IL-6 is much higher than that of TNF-α-reactive astrocytes. This may be due to a difference in the intracellular signaling mechanisms underlying IL-6 production between pericytes and astrocytes (Montagne et al. 2018). Brain pericytes possess a unique amplification mechanism characterized by cooperation between STAT3 and NF-kB in TNF-α-reactive pericytes but not in TNF-α-reactive astrocytes. These data suggest that, in response to TNF-α, pericytes rather than astrocytes may act as the predominant cell type in mediating mechanisms of microglial activation during TNF-α-related neuroinflammatory processes such as stroke.

Based on these data, brain pericytes are the most TNF-α-sensitive cell type among the cell types constituting the CNS and act specifically as effectors of TNF-α through the production of several molecules during brain inflammatory processes. Thus, the interactions between activated pericytes and other cell types could be attractive therapeutic targets for brain inflammation.

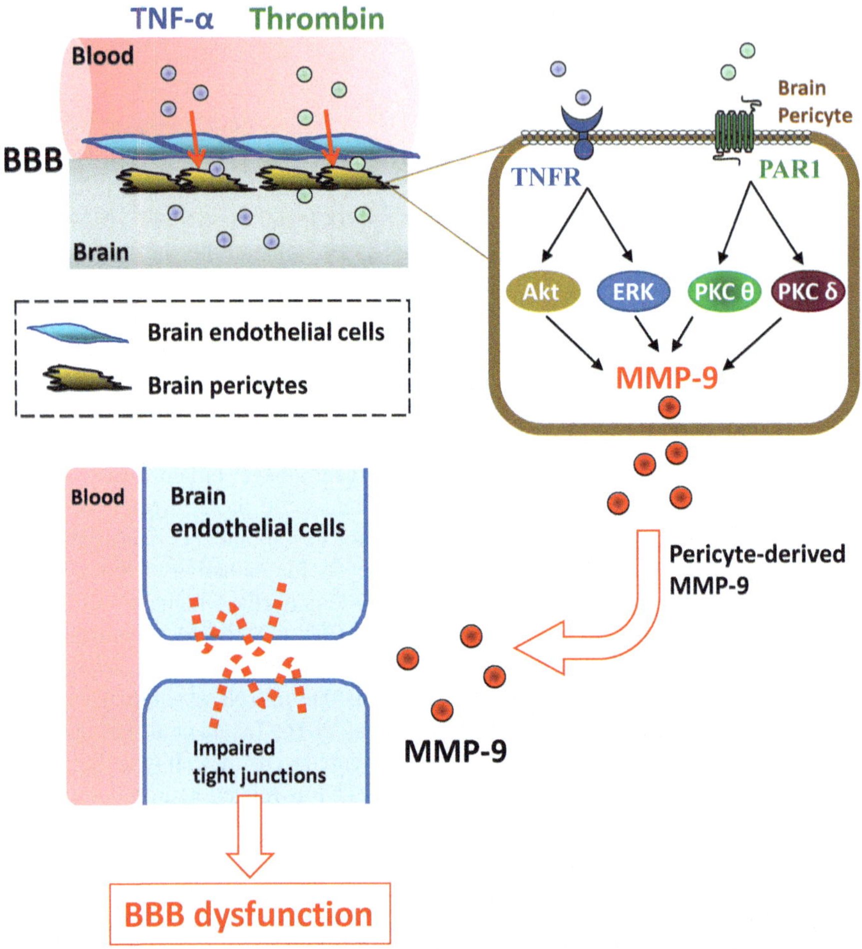

Fig. 6.1 Schematic diagram of the mechanisms by which TNF-α- and thrombin-activated brain pericytes produce BBB dysfunction through MMP-9 release. TNF-α activates MAPK and Akt signaling pathways in brain pericytes to release MMP-9. Thrombin elevates MMP-9 release through PAR1 - PKC θ/δ axis in brain pericytes. MMP-9 release from these activated brain pericytes induces BBB dysfunction through an impairment of tight junctions

- Thrombin

  Thrombin is a component of the coagulation cascade and also directly modulates cellular function through the protease-activated receptors (PARs), which are specific receptors for thrombin. The PARs are expressed on brain pericytes. Pericytes have the highest levels of PAR1 expression among the cell types constituting the BBB such as brain endothelial cells and astrocytes. PAR-2, 3 and 4 expression levels in brain pericytes are lower than those in brain endothelial cells (Machida et al. 2015). Exposure of brain pericytes to thrombin elevates MMP-9 release through the PAR1- PKC θ/δ axis (Machida et al. 2017a), and brain pericytes are the most thrombin-sensitive MMP-9-releasing cell type constituting the BBB (Machida et al.

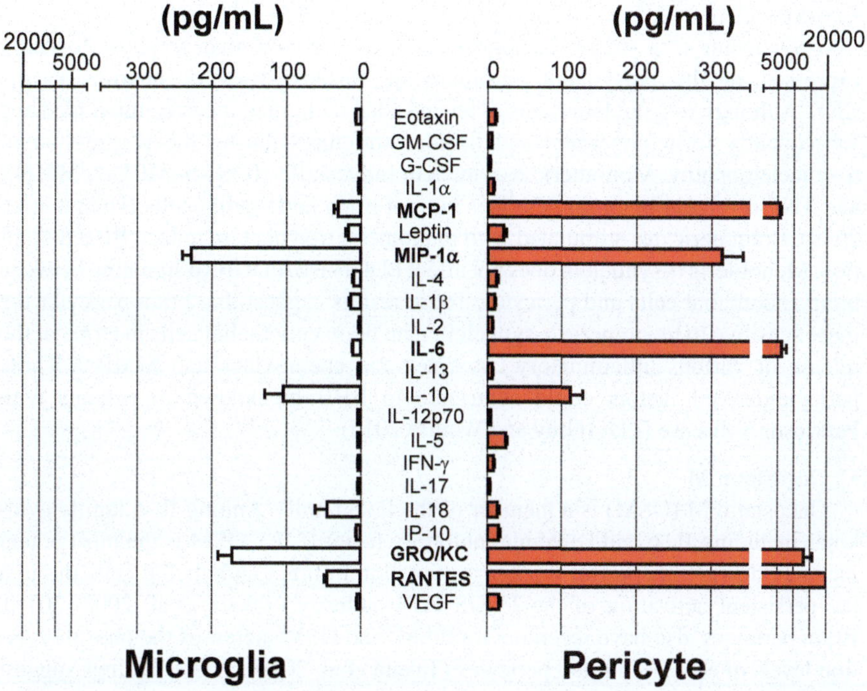

Fig. 6.2 Comparison of the TNF-α-induced cytokine and chemokine release profile between rat brain pericytes (right) and microglia (left) in primary culture. Cells were treated with TNF-α (20 ng/mL) for 24 h. Brain pericytes released higher levels of MCP-1, MIP-1α, IL-6 GRO/KC and RANTES than microglia. Data are means ± SEM (n = 9)

2015). It has been shown that MMP-9 is a critical factor evoking BBB dysfunction (Yang et al. 2007). Indeed, thrombin-stimulated brain pericytes display increased BBB permeability with decreased expression levels of tight junction proteins such as ZO-1 and occludin (Machida et al. 2017b). Besides MMP-9, several inflammatory cytokines including IL-1β, IL-6 and TNF-α are released from thrombin-stimulated brain pericytes (Machida et al. 2017b). Therefore, these mediators released from pericytes could be associated with the development of BBB dysfunction and neuroinflammation in the CNS under pathological conditions. Thrombin is able to penetrate the BBB, suggesting that pericytes may be exposed to thrombin under both physiological and pathological conditions. In a mouse model of obesity-associated diabetes, thrombin levels in blood and brain are elevated and BBB permeability is also increased (Machida et al. 2017b). Considering these findings, thrombin transported into the brain from the circulating blood is likely to produce BBB dysfunction in obesity-associated diabetes through the release of mediators including MMP-9 from brain pericytes (Fig. 6.1). Thrombin-activated pericytes with various mediator release properties should be considered as a possible hallmark for the development and progression of CNS disease accompanied by increased thrombin penetration into the brain.

- α-Synuclein

α-Synuclein is a major component of Lewy bodies that is physiologically expressed in the presynaptic terminals of various types of neurons in the CNS. Although α-synuclein exists as an unfolded monomer, it can be misfolded and subsequently form oligomers and fibrillar structures during the progression of α-synucleinopathy. Monomeric α-synuclein induces IL-1β, IL-6, MCP-1, MMP-9 and TNF-α release by brain pericytes but not brain endothelial cells (Dohgu et al. 2019). Brain pericytes stimulated with monomeric α-synuclein induce BBB disruption. Monomeric α-synuclein does not affect PDGF-BB/PDGFRβ signaling between brain endothelial cells and pericytes. These results suggest that brain pericytes are more sensitive to monomeric α-synuclein than brain endothelial cells in terms of the release of various inflammatory cytokines and chemokines and MMP-9. These pericyte-derived factors could contribute to BBB breakdown in patients with Parkinson's disease (PD) (Gray and Woulfe 2015).

- Oncostatin M

Oncostatin M (OSM) is a member of the IL-6 family. Among IL-6 family cytokines including IL-6 and leukemia inhibitory factor (LIF), OSM is the most potent factor in downregulation of the barrier function in brain endothelial cells through the persistent activation of the JAK/STAT3 pathway (Takata et al. 2008, 2018). Brain pericytes also have receptors for OSM, and OSM increases the phosphorylation levels of STAT3 in brain pericytes (Takata et al. 2019). These findings suggest that brain pericytes have the ability to respond to OSM as well as brain endothelial cells. Indeed, it was reported that brain pericytes aggravate OSM-induced BBB dysfunction through soluble factors in the *in vitro* BBB model. The aggravated BBB dysfunction was abolished by ruxolitinib, a JAK inhibitor (Takata et al. 2019). These findings suggest that soluble factors released from OSM-activated pericytes via JAK signaling contribute to an aggravation of OSM-induced BBB dysfunction (Fig. 6.3). In brain pericytes, IL-6 expression levels were increased by OSM treatment through activation of the JAK/STAT3 pathway. However, the increased IL-6 expression levels in OSM-reactive pericytes were not related to the exacerbated BBB dysfunction caused by OSM-reactive pericytes (Takata et al. 2019). In future experiments, it will be important to identify the pericyte-derived mediators responsible for the OSM-induced BBB dysfunction and also to clarify the role of soluble mediators (including IL-6) from OSM-reactive pericytes in CNS diseases.

- Lipopolysaccharide

Lipopolysaccharide (LPS) is an activator of the innate immune system that is found in the circulating blood of sepsis (Opal et al. 1999) and HIV-1 (Brenchley et al. 2006) patients. LPS evokes BBB dysfunction that facilitates penetration of both immune cells (Pieper et al. 2013) and virus into the brain (Dohgu et al. 2011a). Brain pericytes respond to LPS stimulation, leading to the detachment of brain pericytes from microvessels and BBB dysfunction including enhanced transcytosis of the HIV-1 virus across the BBB (Nishioku et al. 2009; Dohgu et al. 2011a). LPS induces nuclear translocation of the transcription factor NF-kB through Toll-like

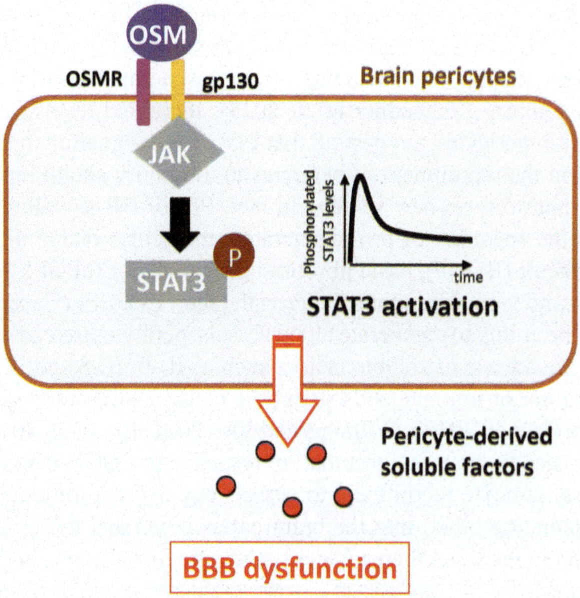

Fig. 6.3 Schematic diagram of the mechanisms by which OSM-activated pericytes induce BBB dysfunction. OSM activates JAK/STAT3 signaling to release soluble factors from pericytes. The released soluble factors from OSM-activated pericytes impair barrier function in the brain endothelial cells

receptor 4 in stimulated pericytes (Guijarro-Munoz et al. 2014). Stimulation of brain pericytes with LPS produces several proinflammatory cytokines and chemokines, including TNF-α and IL-8 (Guijarro-Munoz et al. 2014; Kovac et al. 2011), which are cytokines that downregulate BBB function (Nishioku et al. 2010; Sun et al. 2016). These soluble factors derived from LPS-reactive pericytes may be associated with LPS-induced BBB dysfunction. Besides the release of proinflammatory soluble factors, nitric oxide (NO) is also produced from LPS-reactive pericytes (Kovac et al. 2011; Pieper et al. 2014). NO production in microglia, the immunocompetent cells of the CNS, plays an important role in the death of dopaminergic neurons in PD (Dehmer et al. 2000). Thus, both NO production in LPS-reactive pericytes and NO derived from activated microglia likely participate in CNS pathogenesis. Microglia are activated by several factors including TNF-α, granulocyte-macrophage colony-stimulating factor-deficient (GM-CSF) and interferon (IFN)-γ, and contribute to the progression of CNS disease (Nakamura 2002). Brain pericytes stimulated by LPS release the aforementioned molecules and are likely to induce microglia activation, suggesting that LPS-activated pericytes facilitate the activation of microglia due to LPS.

- PDGF-BB

Brain pericytes express receptors for PDGF-BB, which is known as a marker of pericytes. PDGF-BB signaling is a critical pathway for the recruitment of

PDGFRβ-positive mesenchymal cells to the developing brain capillaries, the stabilization of blood vessels and the normal formation of vessels (Lindahl et al. 1997). Under pathological conditions including cerebral ischemia (Renner et al. 2003) and traumatic brain injury (Zehendner et al. 2015), increased levels of PDGFRβ are observed in brain pericytes, suggesting that PDGF-BB signaling in pericytes could be implicated in the recruitment of pericytes to the injury site to repair the injured blood vessels and to form new vessels. In fact, PDGF-BB signaling in brain pericytes induces the secretion of pro-regenerative molecules including brain-derived neurotrophic factor (BDNF), basic fibroblast growth factor (bFGF), β-nerve growth factor (βNGF) and vascular endothelial growth factor (VEGF) (Gaceb et al. 2018b). In addition to the ability to regenerate blood vessels, pericytes treated with PDGF-BB can facilitate the release of soluble factors such as IL-6, IL-8 and IP-10 associated with the activation of immune cells such as T cells, neutrophils macrophages and microglia (Gaceb et al. 2018b; Williams and Jose 2001; Lee et al. 2017; Matsumoto et al. 2018). These findings suggest that, in response to PDGF-BB, pericytes could induce not only pericyte recruitment to vessels but also the infiltration of immune cells from circulating blood into the brain parenchyma and the migration of activated microglia to the injured area under pathological conditions in the CNS.

- IL-1β

  IL-1β, a proinflammatory cytokine, is a mediator of BBB impairment (Wang et al. 2014). Disrupted BBB integrity is observed in CNS disease such as multiple sclerosis (MS) and during HIV-1 infection (Ortiz et al. 2014; Al-Obaidi and Desa 2018). BBB dysfunction under pathological conditions allows immune cells including neutrophils, and viruses in the blood, to infiltrate the brain parenchyma; these effects are enhanced by the presence of brain pericytes (Dohgu et al. 2011a; Pieper et al. 2013). IL-1β levels are increased in the cerebral spinal fluid of HIV-infected patients (Persidsky et al. 2016) and IL-1β is present in and around MS lesions (Brosnan et al. 1995). These findings raise the possibility that IL-1β is a critical factor in the control of infiltration of immune cells and viruses across the BBB during CNS diseases and that the response of brain pericytes to IL-1β is associated with their infiltration of the BBB. IL-1β stimulation induces the production of several cytokines and chemokines in brain pericytes, including IL-8, MCP-1, regulated on activation, normal T cell expressed and secreted (RANTES) and Interferon gamma-induced protein 10 (IP-10) (Persidsky et al. 2016; Pieper et al. 2013). MCP-1 and IL-8 elevate BBB permeability (Stamatovic et al. 2005; Sun et al. 2016), suggesting that pericyte-derived cytokines and chemokines contribute to downregulation of BBB integrity. Indeed, IL-1β-reactive pericytes support neutrophil transmigration across the BBB via IL-8 production (Pieper et al. 2013). Along with chemokines and cytokines, iNOS expression is increased in IL-1β-reactive pericytes. NO generated by iNOS is a radical gas known to kill invading pathogens.

6.4 Identification of Activated Brain Pericytes Associated with BBB Dysfunction

The expression levels of pericyte marker proteins are likely to be changed in accordance with the functional modification of brain pericytes. Pericytes express PDGFRβ (Winkler et al. 2010; Bondjers et al. 2003), CD13 (aminopeptidase N) (Kunz et al. 1994), NG2 (neuron-glial antigen 2 or chondroitin sulfate proteoglycan 4) (Virgintino et al. 2007; Bondjers et al. 2003), α-SMA (Verbeek et al. 1994), regulator of G protein signaling 5 (RGS-5) (Bondjers et al. 2003), ATP-sensitive potassium channel protein Kir6.1 (Bondjers et al. 2006), desmin (Hellstrom et al. 1999), vimentin (Bandopadhyay et al. 2001), nestin (Alliot et al. 1999) and CD146 (Chen et al. 2017). Expression levels of these proteins in pericytes have been widely used for identifying brain pericytes in *in vitro* and *in vivo* studies. Brain pericytes exhibit different morphologies depending on their cerebral vascular locations (Attwell et al. 2016). At precapillary arterioles and venules, pericytes exhibit mesh-like structures. Thin strand or helical structures are observed in pericytes at capillaries (Hartmann et al. 2015). These "mesh pericytes" and "thin-strand pericytes" are categorized as "capillary pericytes" (Grant et al. 2019); these cells express lower levels of α-SMA than pericytes located at first-branch arterioles ("ensheathing pericytes") and smooth muscle cells distributed at penetrating arterioles. Recent studies using RNA sequencing analyses indicated that pericyte characterizations differ according to their differing vascular zonation (Chasseigneaux et al. 2018; Vanlandewijck et al. 2018). Capillary pericyte-enriched genes include *Pdgfrb*, *Cspg4* (chondroitin sulfate proteoglycan 4 or NG2), *Anpep* (aminopeptidase N or CD13), *Rgs5*, *Abcc9* (SUR2 subunit of K^+-ATP channel), *Kcnj8* (Kir 6.1), *Vtn* (vitronectin) and *Ifitm1* (interferon-induced transmembrane protein 1). Arteriole smooth muscle cell-enriched genes are *Acta2* (α-smooth muscle actin), *Tagln* (transgelin), *Myh11* (myosin heavy chain 11), *Myl9* (myosin light chain 9) and *Mylk* (myosin light chain kinase). Although these RNA sequencing data clearly showed that capillary brain pericytes exhibit strong mRNA expression of PDGFRβ and NG2 but not α-SMA, cultured brain pericytes express detectable protein levels of α-SMA *in vitro* (Smyth et al. 2018; Thanabalasundaram et al. 2011; Dore-Duffy et al. 2006). This discrepancy may be explained by a possible contamination of smooth muscle cells in the isolated capillaries during the primary culture preparation. Cell sorting methods using a combination of cell surface markers for capillary pericytes (PDGFRβ, CD13 and NG2) may be important for the isolation of pure capillary pericytes from brain capillaries. However, it is possible that α-SMA expression in capillary pericytes (α-SMA negative pericytes) is upregulated in culture. Indeed, the culture conditions including culture media, supplements and serum concentration modify their morphology and expression levels of the common pericyte marker proteins CD146, α-SMA, desmin, NG2 and PDGFRβ (Rustenhoven et al. 2018). Importantly, Thanabalasundaram et al. showed that pericytes differentially affect BBB integrity *in vitro* depending on the expression levels of α-SMA (Thanabalasundaram et al. 2011). Pericytic expression of α-SMA is likely affected by bFGF or TGF-β, which could be released by neighboring cells. bFGF-treated pericytes showing lower expression levels of α-SMA stabilize the brain endothelial barrier integrity, while TGF-β-treated pericytes showing enhanced α-SMA expression impair the barrier integrity. These authors also found that expression levels of other pericyte markers, desmin, vimentin and nestin, are lower in TGF-β-treated pericytes (α-SMA-positive pericytes) compared with bFGF-treated pericytes. In addition, TGF-β stimulates brain pericytes to release MMP-9 (Takahashi et al. 2014), suggesting that the increased α-SMA expression in brain pericytes is correlated with MMP-9 release. bFGF increases PDGFRβ expression in brain pericytes (Nakamura et al. 2016). In response to the PDGFRβ ligand PDGF-BB, brain pericytes release VEGF to induce BBB hyperpermeability (Gaceb et al. 2018b). Increased PDGFRβ expression in brain pericytes was observed in models of ischemic brain injury (Arimura et al. 2012) and 6-hydroxydopamine-induced PD (Padel et al. 2016). PDGFRβ-positive pericytes contribute to the localized proteolytic degradation of the BBB through MMP-9 release during cerebral ischemia (Underly et al. 2017). These results suggest that brain pericytes with increased expression levels of α-SMA or PDGFRβ exhibit their activated state, releasing BBB disruption factors. Other studies have suggested that the increased expression of RGS-5 and NG2 in brain pericytes reflects their activation state during vascular remodeling (Berger et al. 2005; Ozerdem et al. 2001). This phenotype of brain pericytes was associated with microvascular changes in the brain of Huntington's disease patients (Padel et al. 2018) and with pathological changes in the lesioned striatum of PD mice (Padel et al. 2016). However, these studies did not examine the impact of activated pericytes on BBB function.

Alterations of marker protein expression in brain pericytes may reflect their immunological responses. LPS induces enhanced α-SMA expression in brain pericytes in parallel with the expression of iNOS, an inflammatory marker protein (Pieper et al. 2014). In response to LPS, brain pericytes release various cytokines and chemokines, including factors that contribute to BBB disruption (Dohgu and Banks 2013; Kovac et al. 2011) and increase expressions of CD11b, CD14 (Gaceb et al. 2018b), ICAM-1 and VCAM-1 (Guijarro-Munoz et al. 2014). IL-1β-induced MCP-1 mRNA expression and phagocytic activity are enhanced in pericytes with decreased expression of the aforementioned pericyte marker proteins (Rustenhoven et al. 2018).

Another classification of pericyte subtypes is type 1 (Nestin-/NG2+) and type 2 (Nestin+/NG2+) pericytes expressing both PDGFRβ and CD146 (Birbrair et al. 2013a). These two pericyte subtypes are functionally different and both are present in the brain. Type 1 pericytes participate in scar formation after brain injury (Birbrair et al. 2013b). Type 2 pericytes are involved in normal and tumoral angiogenesis (Birbrair et al. 2014) and generate neural progenitors (Birbrair et al. 2013a). Birbrair et al. reported that type 2 pericytes were increased in the tumor after the intracranial injection of glioblastoma cells. In contrast, both type 1 and type 2 pericytes were observed in normal brain tissue (Birbrair et al. 2014). These results suggest that tumor-derived factors induce the recruitment of type 2 pericytes during tumoral angiogenesis or the transformation from type 1 to type 2 pericytes. However, their precise roles in the maintenance of BBB integrity under pathological conditions are still unknown.

6.5 Conclusion and Perspective

The studies described here have demonstrated that brain pericytes have at least two separate functional states in terms of the regulation of BBB function, as shown in Fig. 6.4. Brain pericytes enhance and maintain BBB function under normal physiological conditions. During the progression of pathological conditions, brain pericytes are sensitized to various inflammatory signals as sensors of the disease state or severity in the periphery and central nervous system, and then are transformed to the activation state with altered expression of their common marker proteins. Finally, as effectors, the activated brain pericytes exhibit various inflammatory mediator-releasable properties and/or migrate from cerebral vascular walls, leading to BBB dysfunction. Recent advances in transcriptomics analysis show that this approach would be a useful tool for identifying and profiling activated brain pericytes *in vivo* in various mouse disease models with BBB dysfunction. Determination of the key molecule(s) that induce the switch from "normal" to "activated" pericytes could accelerate our understanding of brain pericyte pathophysiology and provide new therapeutic targets for neurodegenerative diseases with BBB dysfunction.

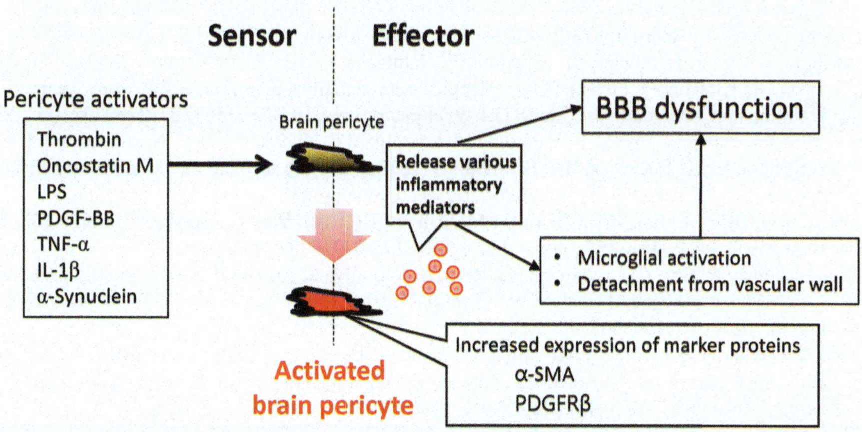

Fig. 6.4 Schematic diagram of the function of activated brain pericytes. Brain pericytes act as sensors of various substances ("Pericyte activators"), the levels of which are elevated under various pathological conditions. In response to pericyte activators, brain pericytes are converted or transformed to the activation state. Activated brain pericytes also exhibit alterations in expression of pericyte marker proteins such as α-SMA and PDGFRβ. These activated brain pericytes release various inflammatory mediators including cytokines, chemokines and matrix metalloproteinases, which directly disrupt BBB function. Activated brain pericyte-derived factors concurrently induce microglial activation and pericyte detachment from the cerebral vascular walls, leading to BBB dysfunction

Acknowledgments Shinya Dohgu and Fuyuko Takata contributed equally. We thank Melony Black, PhD, from Edanz Group (https://en-author-services.edanzgroup.com/) for editing a draft of this manuscript and the members of the Fukuoka BBB Research Association (Fukuoka 811-4313, Japan) for their helpful comments on this review.

Disclosure of Interests All authors declare they have no conflict of interests.

Ethical Approval This article does not contain any studies with animals or human participants performed by any of the authors.

Compliance with Ethical Standards

Funding This work was supported in part by Grants-in-Aid for Scientific Research (KAKENHI) Grant Numbers JP16K08429 (to Y.K.), JP16K08566, JP19K07338 (to S.D.), JP26460113, JP18K06712 (to F.T.) from the Japan Society for the Promotion of Science (https://www.jsps.go.jp/), funds to F.T. (grant number 181044) from the Central Research Institute of Fukuoka University (http://www.suisin.fukuoka-u.ac.jp/home1/) and funds to F.T. from Fukuoka University Program to support the research activities of female researchers.

References

Alliot F, Rutin J, Leenen PJ, Pessac B (1999) Pericytes and periendothelial cells of brain parenchyma vessels co-express aminopeptidase N, aminopeptidase A, and nestin. J Neurosci Res 58(3):367–378

Al-Obaidi MMJ, Desa MNM (2018) Mechanisms of blood brain barrier disruption by different types of bacteria, and bacterial-host interactions facilitate the bacterial pathogen invading the brain. Cell Mol Neurobiol 38(7):1349–1368. https://doi.org/10.1007/s10571-018-0609-2

Arimura K, Ago T, Kamouchi M, Nakamura K, Ishitsuka K, Kuroda J, Sugimori H, Ooboshi H, Sasaki T, Kitazono T (2012) PDGF receptor beta signaling in pericytes following ischemic brain injury. Curr Neurovasc Res 9(1):1–9. https://doi.org/10.2174/156720212799297100

Armulik A, Genove G, Mae M, Nisancioglu MH, Wallgard E, Niaudet C, He L, Norlin J, Lindblom P, Strittmatter K, Johansson BR, Betsholtz C (2010) Pericytes regulate the blood-brain barrier. Nature 468(7323):557–561. https://doi.org/10.1038/nature09522

Attwell D, Mishra A, Hall CN, O'Farrell FM, Dalkara T (2016) What is a pericyte? J Cereb Blood Flow Metab 36(2):451–455. https://doi.org/10.1177/0271678X15610340

Bandopadhyay R, Orte C, Lawrenson JG, Reid AR, De Silva S, Allt G (2001) Contractile proteins in pericytes at the blood-brain and blood-retinal barriers. J Neurocytol 30(1):35–44. https://doi.org/10.1023/a:1011965307612

Bell RD, Winkler EA, Sagare AP, Singh I, LaRue B, Deane R, Zlokovic BV (2010) Pericytes control key neurovascular functions and neuronal phenotype in the adult brain and during brain aging. Neuron 68(3):409–427. https://doi.org/10.1016/j.neuron.2010.09.043

Berezowski V, Landry C, Dehouck MP, Cecchelli R, Fenart L (2004) Contribution of glial cells and pericytes to the mRNA profiles of P-glycoprotein and multidrug resistance-associated proteins in an in vitro model of the blood-brain barrier. Brain Res 1018(1):1–9. https://doi.org/10.1016/j.brainres.2004.05.092

Berger M, Bergers G, Arnold B, Hammerling GJ, Ganss R (2005) Regulator of G-protein signaling-5 induction in pericytes coincides with active vessel remodeling during neovascularization. Blood 105(3):1094–1101. https://doi.org/10.1182/blood-2004-06-2315

Birbrair A, Zhang T, Wang ZM, Messi ML, Enikolopov GN, Mintz A, Delbono O (2013a) Skeletal muscle pericyte subtypes differ in their differentiation potential. Stem Cell Res 10(1):67–84. https://doi.org/10.1016/j.scr.2012.09.003

Birbrair A, Zhang T, Wang ZM, Messi ML, Mintz A, Delbono O (2013b) Type-1 pericytes participate in fibrous tissue deposition in aged skeletal muscle. Am J Physiol Cell Physiol 305(11):C1098–C1113. https://doi.org/10.1152/ajpcell.00171.2013

Birbrair A, Zhang T, Wang ZM, Messi ML, Olson JD, Mintz A, Delbono O (2014) Type-2 pericytes participate in normal and tumoral angiogenesis. Am J Physiol Cell Physiol 307(1):C25–C38. https://doi.org/10.1152/ajpcell.00084.2014

Bondjers C, Kalen M, Hellstrom M, Scheidl SJ, Abramsson A, Renner O, Lindahl P, Cho H, Kehrl J, Betsholtz C (2003) Transcription profiling of platelet-derived growth factor-B-deficient mouse embryos identifies RGS5 as a novel marker for pericytes and vascular smooth muscle cells. Am J Pathol 162(3):721–729. https://doi.org/10.1016/s0002-9440(10)63868-0

Bondjers C, He L, Takemoto M, Norlin J, Asker N, Hellstrom M, Lindahl P, Betsholtz C (2006) Microarray analysis of blood microvessels from PDGF-B and PDGF-Rbeta mutant mice identifies novel markers for brain pericytes. FASEB J 20(10):1703–1705. https://doi.org/10.1096/fj.05-4944fje

Bourassa P, Tremblay C, Schneider JA, Bennett DA, Calon F (2020) Brain mural cell loss in the parietal cortex in Alzheimer's disease correlates with cognitive decline and TDP-43 pathology. Neuropathol Appl Neurobiol. https://doi.org/10.1111/nan.12599

Brandt MM, van Dijk CGM, Maringanti R, Chrifi I, Kramann R, Verhaar MC, Duncker DJ, Mokry M, Cheng C (2019) Transcriptome analysis reveals microvascular endothelial cell-dependent pericyte differentiation. Sci Rep 9(1):15586. https://doi.org/10.1038/s41598-019-51838-x

Brenchley JM, Price DA, Schacker TW, Asher TE, Silvestri G, Rao S, Kazzaz Z, Bornstein E, Lambotte O, Altmann D, Blazar BR, Rodriguez B, Teixeira-Johnson L, Landay A, Martin JN, Hecht FM, Picker LJ, Lederman MM, Deeks SG, Douek DC (2006) Microbial translocation is a cause of systemic immune activation in chronic HIV infection. Nat Med 12(12):1365–1371. https://doi.org/10.1038/nm1511

Brosnan CF, Cannella B, Battistini L, Raine CS (1995) Cytokine localization in multiple sclerosis lesions: correlation with adhesion molecule expression and reactive nitrogen species. Neurology 45(6 Suppl 6):S16–S21. https://doi.org/10.1212/wnl.45.6_suppl_6.s16

Chasseigneaux S, Moraca Y, Cochois-Guegan V, Boulay AC, Gilbert A, Le Crom S, Blugeon C, Firmo C, Cisternino S, Laplanche JL, Curis E, Decleves X, Saubamea B (2018) Isolation and differential transcriptome of vascular smooth muscle cells and mid-capillary pericytes from the rat brain. Sci Rep 8(1):12272. https://doi.org/10.1038/s41598-018-30739-5

Chen J, Luo Y, Hui H, Cai T, Huang H, Yang F, Feng J, Zhang J, Yan X (2017) CD146 coordinates brain endothelial cell-pericyte communication for blood-brain barrier development. Proc Natl Acad Sci USA 114(36):E7622–E7631. https://doi.org/10.1073/pnas.1710848114

Cianciulli A, Porro C, Calvello R, Trotta T, Lofrumento DD, Panaro MA (2020) Microglia mediated Neuroinflammation: focus on PI3K modulation. Biomol Ther 10(1). https://doi.org/10.3390/biom10010137

Daneman R, Zhou L, Kebede AA, Barres BA (2010) Pericytes are required for blood-brain barrier integrity during embryogenesis. Nature 468(7323):562–566. https://doi.org/10.1038/nature09513

De La Fuente AG, Lange S, Silva ME, Gonzalez GA, Tempfer H, van Wijngaarden P, Zhao C, Di Canio L, Trost A, Bieler L, Zaunmair P, Rotheneichner P, O'Sullivan A, Couillard-Despres S, Errea O, Mae MA, Andrae J, He L, Keller A, Batiz LF, Betsholtz C, Aigner L, Franklin RJM, Rivera FJ (2017) Pericytes stimulate oligodendrocyte progenitor cell differentiation during CNS remyelination. Cell Rep 20(8):1755–1764. https://doi.org/10.1016/j.celrep.2017.08.007

Dehmer T, Lindenau J, Haid S, Dichgans J, Schulz JB (2000) Deficiency of inducible nitric oxide synthase protects against MPTP toxicity in vivo. J Neurochem 74(5):2213–2216. https://doi.org/10.1046/j.1471-4159.2000.0742213.x

Dohgu S, Banks WA (2013) Brain pericytes increase the lipopolysaccharide-enhanced transcytosis of HIV-1 free virus across the in vitro blood-brain barrier: evidence for cytokine-mediated pericyte-endothelial cell crosstalk. Fluids Barriers CNS 10(1):23. https://doi.org/10.1186/2045-8118-10-23

Dohgu S, Takata F, Yamauchi A, Nakagawa S, Egawa T, Naito M, Tsuruo T, Sawada Y, Niwa M, Kataoka Y (2005) Brain pericytes contribute to the induction and up-regulation of blood-brain barrier functions through transforming growth factor-beta production. Brain Res 1038(2):208–215. https://doi.org/10.1016/j.brainres.2005.01.027

Dohgu S, Fleegal-DeMotta MA, Banks WA (2011a) Lipopolysaccharide-enhanced transcellular transport of HIV-1 across the blood-brain barrier is mediated by luminal microvessel IL-6 and GM-CSF. J Neuroinflammation 8:167. https://doi.org/10.1186/1742-2094-8-167

Dohgu S, Takata F, Matsumoto J, Oda M, Harada E, Watanabe T, Nishioku T, Shuto H, Yamauchi A, Kataoka Y (2011b) Autocrine and paracrine up-regulation of blood-brain barrier function by plasminogen activator inhibitor-1. Microvasc Res 81(1):103–107. https://doi.org/10.1016/j.mvr.2010.10.004

Dohgu S, Takata F, Matsumoto J, Kimura I, Yamauchi A, Kataoka Y (2019) Monomeric alpha-synuclein induces blood-brain barrier dysfunction through activated brain pericytes releasing inflammatory mediators in vitro. Microvasc Res 124:61–66. https://doi.org/10.1016/j.mvr.2019.03.005

Dore-Duffy P, Owen C, Balabanov R, Murphy S, Beaumont T, Rafols JA (2000) Pericyte migration from the vascular wall in response to traumatic brain injury. Microvasc Res 60(1):55–69. https://doi.org/10.1006/mvre.2000.2244

Dore-Duffy P, Katychev A, Wang X, Van Buren E (2006) CNS microvascular pericytes exhibit multipotential stem cell activity. J Cereb Blood Flow Metab 26(5):613–624. https://doi.org/10.1038/sj.jcbfm.9600272

Gaceb A, Barbariga M, Ozen I, Paul G (2018a) The pericyte secretome: potential impact on regeneration. Biochimie 155:16–25. https://doi.org/10.1016/j.biochi.2018.04.015

Gaceb A, Ozen I, Padel T, Barbariga M, Paul G (2018b) Pericytes secrete pro-regenerative molecules in response to platelet-derived growth factor-BB. J Cereb Blood Flow Metab 38(1):45–57. https://doi.org/10.1177/0271678X17719645

Gautam J, Zhang X, Yao Y (2016) The role of pericytic laminin in blood brain barrier integrity maintenance. Sci Rep 6:36450. https://doi.org/10.1038/srep36450

Grant RI, Hartmann DA, Underly RG, Berthiaume AA, Bhat NR, Shih AY (2019) Organizational hierarchy and structural diversity of microvascular pericytes in adult mouse cortex. J Cereb Blood Flow Metab 39(3):411–425. https://doi.org/10.1177/0271678X17732229

Gray MT, Woulfe JM (2015) Striatal blood-brain barrier permeability in Parkinson's disease. J Cereb Blood Flow Metab 35(5):747–750. https://doi.org/10.1038/jcbfm.2015.32

Guijarro-Munoz I, Compte M, Alvarez-Cienfuegos A, Alvarez-Vallina L, Sanz L (2014) Lipopolysaccharide activates toll-like receptor 4 (TLR4)-mediated NF-kappaB signaling pathway and proinflammatory response in human pericytes. J Biol Chem 289(4):2457–2468. https://doi.org/10.1074/jbc.M113.521161

Halliday MR, Rege SV, Ma Q, Zhao Z, Miller CA, Winkler EA, Zlokovic BV (2016) Accelerated pericyte degeneration and blood-brain barrier breakdown in apolipoprotein E4 carriers with Alzheimer's disease. J Cereb Blood Flow Metab 36(1):216–227. https://doi.org/10.1038/jcbfm.2015.44

Hartmann DA, Underly RG, Grant RI, Watson AN, Lindner V, Shih AY (2015) Pericyte structure and distribution in the cerebral cortex revealed by high-resolution imaging of transgenic mice. Neurophotonics 2(4):041402. https://doi.org/10.1117/1.NPh.2.4.041402

Hellstrom M, Kalen M, Lindahl P, Abramsson A, Betsholtz C (1999) Role of PDGF-B and PDGFR-beta in recruitment of vascular smooth muscle cells and pericytes during embryonic blood vessel formation in the mouse. Development 126(14):3047–3055

Hori S, Ohtsuki S, Hosoya K, Nakashima E, Terasaki T (2004) A pericyte-derived angiopoietin-1 multimeric complex induces occludin gene expression in brain capillary endothelial cells through Tie-2 activation in vitro. J Neurochem 89(2):503–513. https://doi.org/10.1111/j.1471-4159.2004.02343.x

Jansson D, Rustenhoven J, Feng S, Hurley D, Oldfield RL, Bergin PS, Mee EW, Faull RL, Dragunow M (2014) A role for human brain pericytes in neuroinflammation. J Neuroinflammation 11:104. https://doi.org/10.1186/1742-2094-11-104

Kovac A, Erickson MA, Banks WA (2011) Brain microvascular pericytes are immunoactive in culture: cytokine, chemokine, nitric oxide, and LRP-1 expression in response to lipopolysaccharide. J Neuroinflammation 8:139. https://doi.org/10.1186/1742-2094-8-139

Kunz J, Krause D, Kremer M, Dermietzel R (1994) The 140-kDa protein of blood-brain barrier-associated pericytes is identical to aminopeptidase N. J Neurochem 62(6):2375–2386. https://doi.org/10.1046/j.1471-4159.1994.62062375.x

Lee JH, Kim B, Jin WJ, Kim HH, Ha H, Lee ZH (2017) Pathogenic roles of CXCL10 signaling through CXCR3 and TLR4 in macrophages and T cells: relevance for arthritis. Arthritis Res Ther 19(1):163. https://doi.org/10.1186/s13075-017-1353-6

Lindahl P, Johansson BR, Leveen P, Betsholtz C (1997) Pericyte loss and microaneurysm formation in PDGF-B-deficient mice. Science 277(5323):242–245. https://doi.org/10.1126/science.277.5323.242

Low PC, Manzanero S, Mohannak N, Narayana VK, Nguyen TH, Kvaskoff D, Brennan FH, Ruitenberg MJ, Gelderblom M, Magnus T, Kim HA, Broughton BR, Sobey CG, Vanhaesebroeck B, Stow JL, Arumugam TV, Meunier FA (2014) PI3Kdelta inhibition reduces TNF secretion and neuroinflammation in a mouse cerebral stroke model. Nat Commun 5:3450. https://doi.org/10.1038/ncomms4450

Machida T, Takata F, Matsumoto J, Takenoshita H, Kimura I, Yamauchi A, Dohgu S, Kataoka Y (2015) Brain pericytes are the most thrombin-sensitive matrix metalloproteinase-9-releasing cell type constituting the blood-brain barrier in vitro. Neurosci Lett 599:109–114. https://doi.org/10.1016/j.neulet.2015.05.028

Machida T, Dohgu S, Takata F, Matsumoto J, Kimura I, Koga M, Nakamoto K, Yamauchi A, Kataoka Y (2017a) Role of thrombin-PAR1-PKCθ/δ axis in brain pericytes in thrombin-induced MMP-9 production and blood-brain barrier dysfunction in vitro. Neuroscience 350:146–157. https://doi.org/10.1016/j.neuroscience.2017.03.026

Machida T, Takata F, Matsumoto J, Miyamura T, Hirata R, Kimura I, Kataoka Y, Dohgu S, Yamauchi A (2017b) Contribution of thrombin-reactive brain pericytes to blood-brain barrier dysfunction in an in vivo mouse model of obesity-associated diabetes and an in vitro rat model. PLoS One 12(5):e0177447. https://doi.org/10.1371/journal.pone.0177447

Matsumoto J, Takata F, Machida T, Takahashi H, Soejima Y, Funakoshi M, Futagami K, Yamauchi A, Dohgu S, Kataoka Y (2014) Tumor necrosis factor-alpha-stimulated brain pericytes possess a unique cytokine and chemokine release profile and enhance microglial activation. Neurosci Lett 578:133–138. https://doi.org/10.1016/j.neulet.2014.06.052

Matsumoto J, Dohgu S, Takata F, Machida T, Bölükbaşi Hatip FF, Hatip-Al-Khatib I, Yamauchi A, Kataoka Y (2018) TNF-α-sensitive brain pericytes activate microglia by releasing IL-6 through cooperation between IκB-NFκB and JAK-STAT3 pathways. Brain Res 1692:34–44. https://doi.org/10.1016/j.brainres.2018.04.023

Montagne A, Nikolakopoulou AM, Zhao Z, Sagare AP, Si G, Lazic D, Barnes SR, Daianu M, Ramanathan A, Go A, Lawson EJ, Wang Y, Mack WJ, Thompson PM, Schneider JA, Varkey J, Langen R, Mullins E, Jacobs RE, Zlokovic BV (2018) Pericyte degeneration causes white matter dysfunction in the mouse central nervous system. Nat Med 24(3):326–337. https://doi.org/10.1038/nm.4482

Nakagawa S, Deli MA, Nakao S, Honda M, Hayashi K, Nakaoke R, Kataoka Y, Niwa M (2007) Pericytes from brain microvessels strengthen the barrier integrity in primary cultures of rat brain endothelial cells. Cell Mol Neurobiol 27(6):687–694. https://doi.org/10.1007/s10571-007-9195-4

Nakagawa S, Deli MA, Kawaguchi H, Shimizudani T, Shimono T, Kittel A, Tanaka K, Niwa M (2009) A new blood-brain barrier model using primary rat brain endothelial cells, pericytes and astrocytes. Neurochem Int 54(3–4):253–263. https://doi.org/10.1016/j.neuint.2008.12.002

Nakamura Y (2002) Regulating factors for microglial activation. Biol Pharm Bull 25(8):945–953. https://doi.org/10.1248/bpb.25.945

Nakamura K, Arimura K, Nishimura A, Tachibana M, Yoshikawa Y, Makihara N, Wakisaka Y, Kuroda J, Kamouchi M, Ooboshi H, Kitazono T, Ago T (2016) Possible involvement of basic FGF in the upregulation of PDGFRbeta in pericytes after ischemic stroke. Brain Res 1630:98–108. https://doi.org/10.1016/j.brainres.2015.11.003

Nakazato R, Kawabe K, Yamada D, Ikeno S, Mieda M, Shimba S, Hinoi E, Yoneda Y, Takarada T (2017) Disruption of Bmal1 impairs blood-brain barrier integrity via pericyte dysfunction. J Neurosci 37(42):10052–10062. https://doi.org/10.1523/JNEUROSCI.3639-16.2017

Nishioku T, Dohgu S, Takata F, Eto T, Ishikawa N, Kodama KB, Nakagawa S, Yamauchi A, Kataoka Y (2009) Detachment of brain pericytes from the basal lamina is involved in disruption of the blood-brain barrier caused by lipopolysaccharide-induced sepsis in mice. Cell Mol Neurobiol 29(3):309–316. https://doi.org/10.1007/s10571-008-9322-x

Nishioku T, Matsumoto J, Dohgu S, Sumi N, Miyao K, Takata F, Shuto H, Yamauchi A, Kataoka Y (2010) Tumor necrosis factor-α mediates the blood–brain barrier dysfunction induced by activated microglia in mouse brain microvascular endothelial cells. J Pharmacol Sci 112(2):251–254. https://doi.org/10.1254/jphs.09292SC

Okada R, Wu Z, Zhu A, Ni J, Zhang J, Yoshimine Y, Peters C, Saftig P, Nakanishi H (2015) Cathepsin D deficiency induces oxidative damage in brain pericytes and impairs the blood-brain barrier. Mol Cell Neurosci 64:51–60. https://doi.org/10.1016/j.mcn.2014.12.002

Opal SM, Scannon PJ, Vincent JL, White M, Carroll SF, Palardy JE, Parejo NA, Pribble JP, Lemke JH (1999) Relationship between plasma levels of lipopolysaccharide (LPS) and LPS-binding protein in patients with severe sepsis and septic shock. J Infect Dis 180(5):1584–1589. https://doi.org/10.1086/315093

Ortiz GG, Pacheco-Moisés FP, Macías-Islas M, Flores-Alvarado LJ, Mireles-Ramírez MA, González-Renovato ED, Hernández-Navarro VE, Sánchez-López AL, Alatorre-Jiménez MA (2014) Role of the blood-brain barrier in multiple sclerosis. Arch Med Res 45(8):687–697. https://doi.org/10.1016/j.arcmed.2014.11.013

Ozerdem U, Grako KA, Dahlin-Huppe K, Monosov E, Stallcup WB (2001) NG2 proteoglycan is expressed exclusively by mural cells during vascular morphogenesis. Dev Dyn 222(2):218–227. https://doi.org/10.1002/dvdy.1200

Padel T, Ozen I, Boix J, Barbariga M, Gaceb A, Roth M, Paul G (2016) Platelet-derived growth factor-BB has neurorestorative effects and modulates the pericyte response in a partial 6-hydroxydopamine lesion mouse model of Parkinson's disease. Neurobiol Dis 94:95–105. https://doi.org/10.1016/j.nbd.2016.06.002

Padel T, Roth M, Gaceb A, Li JY, Bjorkqvist M, Paul G (2018) Brain pericyte activation occurs early in Huntington's disease. Exp Neurol 305:139–150. https://doi.org/10.1016/j.expneurol.2018.03.015

Pan W, Kastin AJ, Bell RL, Olson RD (1999) Upregulation of tumor necrosis factor alpha transport across the blood-brain barrier after acute compressive spinal cord injury. J Neurosci 19(9):3649–3655. https://doi.org/10.1523/jneurosci.19-09-03649.1999

Persidsky Y, Hill J, Zhang M, Dykstra H, Winfield M, Reichenbach NL, Potula R, Mukherjee A, Ramirez SH, Rom S (2016) Dysfunction of brain pericytes in chronic neuroinflammation. J Cereb Blood Flow Metab 36(4):794–807. https://doi.org/10.1177/0271678X15606149

Pieper C, Pieloch P, Galla HJ (2013) Pericytes support neutrophil transmigration via interleukin-8 across a porcine co-culture model of the blood-brain barrier. Brain Res 1524:1–11. https://doi.org/10.1016/j.brainres.2013.05.047

Pieper C, Marek JJ, Unterberg M, Schwerdtle T, Galla HJ (2014) Brain capillary pericytes contribute to the immune defense in response to cytokines or LPS in vitro. Brain Res 1550:1–8. https://doi.org/10.1016/j.brainres.2014.01.004

Price TO, Eranki V, Banks WA, Ercal N, Shah GN (2012) Topiramate treatment protects blood-brain barrier pericytes from hyperglycemia-induced oxidative damage in diabetic mice. Endocrinology 153(1):362–372. https://doi.org/10.1210/en.2011-1638

Renner O, Tsimpas A, Kostin S, Valable S, Petit E, Schaper W, Marti HH (2003) Time- and cell type-specific induction of platelet-derived growth factor receptor-beta during cerebral ischemia. Brain Res Mol Brain Res 113(1-2):44–51. https://doi.org/10.1016/s0169-328x(03)00085-8

Rustenhoven J, Jansson D, Smyth LC, Dragunow M (2017) Brain Pericytes as mediators of neuroinflammation. Trends Pharmacol Sci 38(3):291–304. https://doi.org/10.1016/j.tips.2016.12.001

Rustenhoven J, Smyth LC, Jansson D, Schweder P, Aalderink M, Scotter EL, Mee EW, Faull RLM, Park TI, Dragunow M (2018) Modelling physiological and pathological conditions to study pericyte biology in brain function and dysfunction. BMC Neurosci 19(1):6. https://doi.org/10.1186/s12868-018-0405-4

Shigemoto-Mogami Y, Hoshikawa K, Sato K (2018) Activated microglia disrupt the blood-brain barrier and induce chemokines and cytokines in a rat in vitro model. Front Cell Neurosci 12:494. https://doi.org/10.3389/fncel.2018.00494

Shimizu F, Sano Y, Tominaga O, Maeda T, Abe MA, Kanda T (2013) Advanced glycation end-products disrupt the blood-brain barrier by stimulating the release of transforming growth factor-beta by pericytes and vascular endothelial growth factor and matrix metalloproteinase-2 by endothelial cells in vitro. Neurobiol Aging 34(7):1902–1912. https://doi.org/10.1016/j.neurobiolaging.2013.01.012

Smyth LCD, Rustenhoven J, Scotter EL, Schweder P, Faull RLM, Park TIH, Dragunow M (2018) Markers for human brain pericytes and smooth muscle cells. J Chem Neuroanat 92:48–60. https://doi.org/10.1016/j.jchemneu.2018.06.001

Stamatovic SM, Shakui P, Keep RF, Moore BB, Kunkel SL, Van Rooijen N, Andjelkovic AV (2005) Monocyte chemoattractant protein-1 regulation of blood-brain barrier permeability. J Cereb Blood Flow Metab 25(5):593–606. https://doi.org/10.1038/sj.jcbfm.9600055

Sumi N, Nishioku T, Takata F, Matsumoto J, Watanabe T, Shuto H, Yamauchi A, Dohgu S, Kataoka Y (2010) Lipopolysaccharide-activated microglia induce dysfunction of the blood-brain barrier in rat microvascular endothelial cells co-cultured with microglia. Cell Mol Neurobiol 30(2):247–253. https://doi.org/10.1007/s10571-009-9446-7

Sun Y, Li N, Zhang J, Liu H, Liu J, Xia X, Sun C, Feng X, Gu J, Du C, Han W, Lei L (2016) Enolase of streptococcus Suis serotype 2 enhances blood-brain barrier permeability by inducing IL-8 release. Inflammation 39(2):718–726. https://doi.org/10.1007/s10753-015-0298-7

Sweeney MD, Zhao Z, Montagne A, Nelson AR, Zlokovic BV (2019) Blood-brain barrier: from physiology to disease and Back. Physiol Rev 99(1):21–78. https://doi.org/10.1152/physrev.00050.2017

Takahashi Y, Maki T, Liang AC, Itoh K, Lok J, Osumi N, Arai K (2014) p38 MAP kinase mediates transforming-growth factor-beta1-induced upregulation of matrix metalloproteinase-9 but not −2 in human brain pericytes. Brain Res 1593:1–8. https://doi.org/10.1016/j.brainres.2014.10.029

Takahashi H, Takata F, Matsumoto J, Machida T, Yamauchi A, Dohgu S, Kataoka Y (2015) Brain pericyte-derived soluble factors enhance insulin sensitivity in GT1-7 hypothalamic neurons. Biochem Biophys Res Commun 457(4):532–537. https://doi.org/10.1016/j.bbrc.2015.01.016

Takata F, Sumi N, Nishioku T, Harada E, Wakigawa T, Shuto H, Yamauchi A, Kataoka Y (2008) Oncostatin M induces functional and structural impairment of blood-brain barriers comprised of rat brain capillary endothelial cells. Neurosci Lett 441(2):163–166. https://doi.org/10.1016/j.neulet.2008.06.030

Takata F, Dohgu S, Matsumoto J, Takahashi H, Machida T, Wakigawa T, Harada E, Miyaji H, Koga M, Nishioku T, Yamauchi A, Kataoka Y (2011) Brain pericytes among cells constituting the blood-brain barrier are highly sensitive to tumor necrosis factor-alpha, releasing matrix metalloproteinase-9 and migrating in vitro. J Neuroinflammation 8:106. https://doi.org/10.1186/1742-2094-8-106

Takata F, Dohgu S, Matsumoto J, Machida T, Sakaguchi S, Kimura I, Yamauchi A, Kataoka Y (2018) Oncostatin M-induced blood-brain barrier impairment is due to prolonged activation of STAT3 signaling in vitro. J Cell Biochem 119(11):9055–9063. https://doi.org/10.1002/jcb.27162

Takata F, Dohgu S, Sakaguchi S, Sakai K, Yamanaka G, Iwao T, Matsumoto J, Kimura I, Sezaki Y, Tanaka Y, Yamauchi A, Kataoka Y (2019) Oncostatin-M-reactive pericytes aggravate blood-brain barrier dysfunction by activating JAK/STAT3 signaling in vitro. Neuroscience 422:12–20. https://doi.org/10.1016/j.neuroscience.2019.10.014

Thanabalasundaram G, Schneidewind J, Pieper C, Galla HJ (2011) The impact of pericytes on the blood-brain barrier integrity depends critically on the pericyte differentiation stage. Int J Biochem Cell Biol 43(9):1284–1293. https://doi.org/10.1016/j.biocel.2011.05.002

Underly RG, Levy M, Hartmann DA, Grant RI, Watson AN, Shih AY (2017) Pericytes as inducers of rapid, matrix metalloproteinase-9-dependent capillary damage during ischemia. J Neurosci 37(1):129–140. https://doi.org/10.1523/jneurosci.2891-16.2016

Vanlandewijck M, He L, Mae MA, Andrae J, Ando K, Del Gaudio F, Nahar K, Lebouvier T, Lavina B, Gouveia L, Sun Y, Raschperger E, Rasanen M, Zarb Y, Mochizuki N, Keller A, Lendahl U, Betsholtz C (2018) A molecular atlas of cell types and zonation in the brain vasculature. Nature 554(7693):475–480. https://doi.org/10.1038/nature25739

Verbeek MM, Otte-Holler I, Wesseling P, Ruiter DJ, de Waal RM (1994) Induction of alpha-smooth muscle actin expression in cultured human brain pericytes by transforming growth factor-beta 1. Am J Pathol 144(2):372–382

Virgintino D, Girolamo F, Errede M, Capobianco C, Robertson D, Stallcup WB, Perris R, Roncali L (2007) An intimate interplay between precocious, migrating pericytes and endothelial cells governs human fetal brain angiogenesis. Angiogenesis 10(1):35–45. https://doi.org/10.1007/s10456-006-9061-x

Wang Y, Jin S, Sonobe Y, Cheng Y, Horiuchi H, Parajuli B, Kawanokuchi J, Mizuno T, Takeuchi H, Suzumura A (2014) Interleukin-1β induces blood-brain barrier disruption by downregulating sonic hedgehog in astrocytes. PLoS One 9(10):e110024. https://doi.org/10.1371/journal.pone.0110024

Williams TJ, Jose PJ (2001) Neutrophils in chronic obstructive pulmonary disease. Novartis Found Symp 234:136–141.; discussion 141-138. https://doi.org/10.1002/0470868678.ch9

Winkler EA, Bell RD, Zlokovic BV (2010) Pericyte-specific expression of PDGF beta receptor in mouse models with normal and deficient PDGF beta receptor signaling. Mol Neurodegener 5:32. https://doi.org/10.1186/1750-1326-5-32

Winkler EA, Bell RD, Zlokovic BV (2011) Central nervous system pericytes in health and disease. Nat Neurosci 14(11):1398–1405. https://doi.org/10.1038/nn.2946

Yang Y, Estrada EY, Thompson JF, Liu W, Rosenberg GA (2007) Matrix metalloproteinase-mediated disruption of tight junction proteins in cerebral vessels is reversed by synthetic matrix metalloproteinase inhibitor in focal ischemia in rat. J Cereb Blood Flow Metab 27(4):697–709. https://doi.org/10.1038/sj.jcbfm.9600375

Zehendner CM, Sebastiani A, Hugonnet A, Bischoff F, Luhmann HJ, Thal SC (2015) Traumatic brain injury results in rapid pericyte loss followed by reactive pericytosis in the cerebral cortex. Sci Rep 5:13497. https://doi.org/10.1038/srep13497

Chapter 7
TLR-4 Signaling in Pericytes

Alessandra Stasi, Rossana Franzin, Giuseppe Stefano Netti, Elena Ranieri, Loreto Gesualdo, Giovanni Stallone, and Giuseppe Castellano

Abstract

Introduction

Pericytes are cells with intriguing properties that have only recently attracted the attention of numerous researchers. In the past years, their function was mainly associated with microvascular homeostasis, angiogenesis and maintenance of blood-tissue barriers. Recently, several studies suggest that pericytes play a predominant role in acute as well as chronic diseases contributing to damage or tissue repair in an organ specific manner. Given their role in inflammatory and fibrotic context, the activation of pericyte Toll like Receptor-4 (TLR-4) signaling was suggested as a powerful molecular mechanism of pericytes activation and dysfunction. Here we review emerging works regarding the involvement of the TLR-4 signaling in pericytes activation both in healthy and pathological conditions.

Methods

For the selection of literature cited we used MEDLINE/Pubmed database and in particular the MESH vocabulary. The keywords used in the MEDLINE research were: pericytes; TLR-4 signaling; MyD88 pathway, mesenchymal stem cells; pericyte to myofibroblast transition (PMT); endotoxin; LPS and LPS-Binding protein (LBP).

Results

Brain and lung pericytes are sensitive to microbial or non-microbial injury by TLR-4 signaling and exhibit non-professional Antigen-Presenting Cell (APC) characteristics suggesting a prominent role in immunosurveillance. Moreover, renal

A. Stasi · R. Franzin · L. Gesualdo
Nephrology, Dialysis and Transplantation Unit, Department of Emergency and Organ Transplantation, University of Bari, Bari, Italy

G. S. Netti · G. Stallone · G. Castellano (✉)
Nephrology, Dialysis and Transplantation Unit, Department of Medical and Surgical Sciences, University of Foggia, Foggia, Italy
e-mail: giuseppe.castellano@unifg.it

E. Ranieri
Clinical Pathology, Center of Molecular Medicine, Department of Medical and Surgical Sciences, University of Foggia, Foggia, Italy

A. Birbrair (ed.), *Biology of Pericytes – Recent Advances*, Stem Cell Biology and Regenerative Medicine 68, https://doi.org/10.1007/978-3-030-62129-2_7

pericytes activate fibrogenesis via the TLR-4 and MyD88-dependent mechanism. TLR-4 signaling has been reported to be pivotal in various pathological conditions, including cardiovascular diseases, neuronal degeneration and correlated disorders, sepsis, lung and renal diseases.

Conclusions
There is an urgent need for exploring TLR-4 signaling in different pericyte populations in order to develop new therapeutic approaches trying to modulate their behavior in pathological conditions.

Keywords Pericytes · Heterogeneity · Plasticity · Non-professional antigen-presenting cell · Mesenchymal stem cells · Inflammation · Fibrosis · Myofibroblasts · Sepsis · Pericyte to myofibroblast transition (PMT) · Endotoxin · LPS-binding protein (LBP) · TLR-4 signaling · MyD88 pathway · Innate immune response

7.1 Introduction

As bridge cells, the peculiar characteristic of pericytes is the ability to integrate information from systemic circulation to organ parenchymal cells and vice versa (Harrell et al. 2018). Several studies demonstrated that pericytes are responsible for sensing microorganisms and endogenous molecules by Pattern Recognition Receptors (PRR), in particular by the Toll like Receptor-4 (TLR-4), likewise sentinels in innate immunity (Danese et al. 2007; Pober and Sessa 2015; Guijarro-Muñoz et al. 2014). As we will discuss, these effects are particularly evident in the brain and lung after endotoxemic insult (Edelman et al. 2006; Krueger and Bechmann 2010; Pieper et al. 2014).

In addition, pericytes are not mere gatekeepers in the inflammatory response, but they directly contribute to the onset of innate immune responses. Indeed, pericytes are able to control leukocytes activation and trafficking to damaged or infected tissue (Stark et al. 2018) and, under certain conditions, they can modulate T cell activation and proliferation (Fabry et al. 1993; Balabanov et al. 1999). In pioneering studies, pericytes were compared to macrophage-like, non-professional Antigen-Presenting Cells (APC) (Balabanov et al. 1996; Shepro and Morel 1993; Navarro et al. 2016). Herein, we will review available evidences on pericyte role in different organs and the TLR-4 signaling involved in pericytes functioning. We will also comment on potential strategies to modulate TLR-4 activity and the possible consequences in pericytes behavior in different pathological conditions.

7.1.1 The Heterogeneity of Pericytes

Pericytes represent a morphologically heterogeneous cell population ubiquitously found in different organs (Dias Moura Prazeres et al. 2017). Their heterogeneity is correlated to their embryonic origin between tissues (Dias Moura Prazeres et al.

2017). Several studies showed that pericytes in brain and thymus originate from the neuroectodermis (Foster et al. 2008); those localized in gut, liver, lungs and heart arise from the mesothelium, while in the other organs such as kidneys, liver and pancreas they derive from the mesoderm (Armulik et al. 2011). Recently, Yamazaki et al. showed that a subpopulation of pericytes within the embryonic skin and brain developed from the hematopoietic lineage by transforming growth factor-β1 (TGF-β1) signaling (Yamazaki et al. 2017). Therefore, also in the same tissue, pericytes may be heterogeneous in their origin.

Their diversity influences their behavior and functions in normal and pathological conditions; indeed, emerging evidences demonstrated that pericytes with immunomodulatory characteristics arise from mesoderm and are able to suppress immune response and inflammation as perivascular mesenchymal stem cells and multipotent, self-renewable cells (Harrell et al. 2018).

Moreover, pericytes development and functions are regulated by several signals coming from parenchymal and circulating immune cells as platelet-derived growth factor-β (PDGF-β), TGFβ, heparin-binding epidermal growth factor (HB-EGF), stromal-derived factor 1-a (SDF-1α), Sonic hedgehog (Shh), Jagged-1 (Jag-1) Ephrin and macrophages molecules (Levéen et al. 1994; Gaengel et al. 2009; Stratman et al. 2010; Song et al. 2009; Nielsen and Dymecki 2010; Duque and Descoteaux 2014). In several pathological settings, such as fibrosis, ischemic organ failure, tumor growth and metastasis, infections, cardiovascular and neurological diseases, pericytes are also influenced by other molecular mechanisms (Pober and Sessa 2015).

Reconstructing the origin of pericytes in several organs is crucial to better understand the molecular signaling responsible for their different functions in physiological, as well as in pathological conditions. This understanding may bring new approaches in clinical practice in which pericyte-target therapy will be efficiently applied.

7.1.2 The Main Sensor of LPS: The TLR-4

The establishment of an efficient immune response against pathogenic microorganisms such as bacteria, fungi, viruses or protozoan parasites depends on the recognition of highly conserved microbial products called Pathogen-Associated Molecular Patterns (PAMPs) by PRRs. Interestingly, these host sensors, also called Primitive Recognition Receptors evolve before other parts of adaptive immunity and play a vital role in the function of the innate immune system (Akira et al. 2006; Mogensen 2009).

The PRR are expressed, mainly, by cells of the innate immune system, such as dendritic cells, macrophages, monocytes, neutrophils but also on epithelial, endothelial cells and pericytes (Navarro et al. 2016). The most representative PRR are the Toll-like Receptors (TLRs) (Medzhitov 2001). Beside circulating myeloid cells, TLR expression has also been observed in different kind of resident cells of brain, heart, kidney, lung, liver, small intestine, spinal cord, spleen, and reproductive organs (Nishimura and Naito 2005). TLRs can recognize several PAMPs including lipid-based bacterial cell wall components such as lipopolysaccharide (LPS) and lipopeptides, microbial protein components (i.e. flagellin), nucleic acids such as single-stranded or double-stranded RNA and CpG DNA (Yang et al. 2016). TLRs recognize and respond also to Damage-Associated Molecular Patterns (DAMPs), danger signals, or alarmins which are released by infected, damaged or necrotic cells. PRRs also mediate the initiation of antigen-specific adaptive immune response and release of inflammatory cytokines (Yang et al. 2016).

TLRs are a family of PRRs initially identified as Toll proteins in 1984 for their roles in the early embryogenesis of the fruit fly Drosophila melanogaster (Yang et al. 2016). In the mammalian TLRs family, all the ten human members are type I transmembrane proteins characterized by an extracellular leucine-rich ligand binding domain (LRR) and a signaling, intracellular tail domain called Toll/Interleukin-1 (IL-1) receptor homology (TIR) (Yang et al. 2016). Some TLRs, TLR-1, 2, 4, 5, 6, and 10, localize to the cell surface, while TLR-3, 7, 8, 9, 11, and 13 are located within cytoplasmic compartments (Chaturvedi and Pierce 2009). From a molecular point of view, the binding of TLR ligand to the extracellular LRR region causes TLR dimerization and conformational changes of the intracellular TIR domains, with recruitment and activation of the adaptor molecules MyD88, and/or TRIF, with signal transduction (Chaturvedi and Pierce 2009). Among TLRs, TLR-4 is considered the major sensor of LPS, one of the outer membrane molecular components of Gram-negative (i.e. *Staphylococcus aureus, Escherichia coli*) also defined as endotoxin (Munford 2008). LPS contains a lipid A moiety and a sugar moiety of a core polysaccharide and an *O*-polysaccharide of variable length (Munford 2008). In the LPS structure, the lipid A is extremely toxic to humans, therefore plays an essential role in evoking the inflammatory response and is responsible for systemic vasodilatation and hypotension occurring in septic shock.

Normally, the host defense response to LPS includes expression of a variety of pro-inflammatory cytokines, such as tumor necrosis factor-α (TNF-α) and interferon-β (IFN-β), and also other pro-inflammatory enzyme as iNOS (inducible NO synthase) (Terrell et al. 2006). Nonetheless, in addition to LPS, TLR-4 also recognizes lipoteichoic acid (LTA), fibronectin (Okamura et al. 2001), the fusion protein of respiratory syncytial virus (RSV) and taxol, a plant diterpene structurally unrelated to LPS but exhibiting LPS mimetic effects on murine cells (Kawasaki et al. 2000).

In the TLR-4 pathway, the activation is triggered by interaction between LPS and LPS binding protein (LBP). Like other acute-phase protein, LBP is produced in the liver, released in the bloodstream where it recognizes and forms a high-affinity complex with the lipid A moiety of LPS, as free molecules, fragments, or still bound to the outer membrane of intact bacteria(Schumann et al. 1990). Recently, the molecule High mobility group box 1 protein (HMGB1) is emerged as capable to interact with LPS and to transfer LPS to CD14 for enhancing LPS-mediated inflammation (Youn et al. 2008). Interestingly, HMGB1 shows structural similarity to LBP although HMGB1 has less amino acid (215 aa) (Beamer et al. 1997; Lotze and

Tracey 2005). Ju Ho Youn et al., demonstrated that HMGB1 triggers TLR-4 signaling in human PBMCs, increasing TNF-α synthesis and amplifying inflammatory response (Youn et al. 2008).

After the binding of LPS, LBP transfers a single molecule of LPS from aggregates in solution to the adapter CD14, that in turns shuttles LPS to the TLR-4/MD-2 complex, in which MD2 is an accessory protein. CD14 is found in two forms (Youn et al. 2008); the first is soluble CD14 (sCD14), which occurs in plasma where it helps to transduce LPS signaling in cells lacking membrane-bound CD14, such as endothelial and epithelial cells. The second, more extensively studied form of CD14, is membrane bound (mCD14), attached to the surface of myeloid cells via a glycosylphosphatidylinositol tail, enabling CD14 to be membrane proximal despite lacking a transmembrane domain (Youn et al. 2008). The formation of the LPS/MD-2/TLR-4 complex on plasma membrane leads to MyD88 or to TRIF recruiting. The early MyD88-dependent signaling pathway, which is dependent by MyD88-like adapter (Mal), recruits IRAKs and activates TRAF6 and the IkB kinase (IKK), leading to NF-kB activation (Yang et al. 2016). The later response, the MyD88-independent signaling, uses the TIR-domain-containing adapter-inducing interferon-β (TRIF) and TRIF-related adapter molecule (TRAM) leading to the late activation of both IRF3 and NF-kB leading to interferon production, induction of cytokines, chemokines, and other transcription factors (Yang et al. 2016). Importantly, TLR-4 is the only TLR that engages both MyD88 and TRIF for intracellular signaling.

Regardless its pivotal role in infections, TLR-4 can recognize also a wide spectrum of DAMPs molecules in noninfectious diseases such as in ischemia/reperfusion injury (IRI), autoimmune diseases and neuroinflammation (Williams et al. 2006). Physiologically, the pro-inflammatory TLR-4 signaling leads to tissue integrity restoration and functional recovery after the acute phase response to an infection, with protective and beneficial effects. However, in prolonged and excessive TLR-4 stimulation, the initial inflammation can evolve in severe forms, as observed during sepsis (Guijarro-Muñoz et al. 2014; Kuzmich et al. 2017; Edelman et al. 2007a).

7.1.3 The Molecular Mechanisms of Pericytes in Sepsis: TLR-4 Immunomodulation

Sepsis is defined as life-threatening organ dysfunction induced by an overwhelmed immune response to infection (Fani et al. 2018). Despite the development of new therapeutically approaches the mortality rate remains very high in the intensive care units (Chirico et al. 2007). During sepsis, the increased levels of endotoxin, proinflammatory cytokines and the activation of coagulation system induce a systemic endothelial dysfunction via TLR signaling pathway (Keynan et al. 2011). From a cellular point of view, endothelial cells are activated by microbial factors or endogenous inflammatory agonists via TLR signaling and participate in early and delayed inflammatory responses (Danese et al. 2007; Mai et al. 2013). In particular the activation of endothelial TLR-2, TLR-4 and TLR-9 modulates the coagulation pathway, the innate immune pathway and organ fibrosis (Khakpour et al. 2015). However other cells can take part in vascular dysfunction as pericytes (Navarro et al. 2016). Because of their proximity to endothelial cells, pericytes contribute to modify blood flow, vascular permeability, coagulation activity and fibrosis (Castellano et al. 2019a) during sepsis.

Several studies suggested that pericytes were not passive mural cells in the interface between circulating blood and surrounding tissues but they activated the inflammatory response via PRR, in particular TLR-4 (Edelman et al. 2006). The engagement of pericytes innate immune receptors, as TLR-4, with microbial and host-derived agonists, up-regulated the expression of specific cytokines and chemokines, and adhesion molecules such as ICAM-1 and VCAM-1 (Guijarro-Muñoz et al. 2014). The activation of innate immune pathways by TLR agonists have been reported to potentiate inflammatory response and to control leucocytes trafficking and activation. This effect was particularly evident in the brain, where the vascular pericytes expressed TLR-4; interestingly, the TLR-4 protein levels increased within cytoplasm after LPS stimulation (Guijarro-Muñoz et al. 2014).

Maintaining pericyte/endothelial cell interactions is crucial to avoid vascular hyper-permeability and excessive leukocyte recruitment in sepsis (Rudziak et al. 2019). Recent studies showed that LPS treatment induced increased angiogenin 2 (Ang-2), and decreased Ang-1 and tyrosine-protein kinase receptor (Tie-2) expression in endotoxemic mouse model. Ang-2 is an antagonist of Ang-1 and Tie-2 and contributed to the pathophysiology of sepsis-induced organ dysfunction resulting in an overall loss of pericyte coverage and increased hyper-permeability (Zeng et al. 2016). Moreover, the decrease in pericyte density augmented neutrophil/macrophage infiltration in both the heart and lung after LPS infusion. Interestingly, the overexpression of Sirt3 reduced Ang-2 expression and attenuated LPS-induced pericyte loss and vascular leakage (Zeng et al. 2016).

Therefore, the available data implicate pericytes TLR-dependent pathways in the inflammatory response during sepsis and organ damage (Rudziak et al. 2019). For instance, therapies could target pericytes intracellular pathways through specific inhibitors to control systemic inflammatory response. Indeed, pericytes also maintain physiologic homeostasis and contribute to the inflammatory response in term of protection against infections; as consequences, pericytes-targeted therapies could modify physiologic responses causing more detrimental effects. Therefore, a better understanding of the mechanisms by which TLR activation induces harmful response in sepsis, is necessary to find a specific target to prevent dysfunctional pericytes activation and organ failure.

7.1.4 Pericytes and TLR-4 Signaling in Cardiac Diseases: From Inflammation to Fibrosis?

Cardiac dysfunction is strongly integrated with inflammatory processes. Thus, excessive inflammatory response in several cardiac diseases with microbial or non-microbial etiology can be harmful for the organism, inducing a severe damage to heart tissue (Yang et al. 2016) and other organs as kidney (Scrascia et al. 2017; Sarnak 2014). During myocardial inflammation, host response is initiated mainly by TLR-4 (Yang et al. 2016). High levels of TLR-4 expression in myocardial tissue are associated with a several progression of cardiovascular diseases including myocarditis (Fairweather et al. 2003), injury (Ding et al. 2013), heart failure (Liu et al. 2015), atherosclerosis (Bagheri et al. 2014) and hypertension (Satoh et al. 2016).

TLR-4 is constitutively expressed in cardiac myocytes (Frantz et al. 1999) and immune cells and is up-regulated in both cell types after injury (Fallach et al. 2010). Moreover, endothelial cells also express TLR-4 under normal conditions, and increase its expression after stimulation with pro-inflammatory cytokines (Faure et al. 2001). To date, it is still not defined which cell type is actually responsible for the myocardial dysfunction. Tavener et al. (2004) demonstrated that, during myocardial dysfunction, the increased expression of TLR-4 in immune cells had more detrimental effects than in cardiac myocytes. However, they showed that activation of TLR-4 in myocytes caused a significant mitochondrial dysfunction that could amplify cardiac injury (Tavener et al. 2004). Moreover in atherosclerosis disease, TLR-4 expression was largely confined to endothelial cells and macrophages, suggesting that these two cell types were the primary responders to injury in the plaque (Edfeldt et al. 2002).

Despite the increased knowledge about the role of myocytes, endothelial and immune cells, it is not yet clear whether cardiac pericytes are implicated in the development of disease states (Alex and Frangogiannis 2019). Several studies elucidated the origin of cardiac pericytes. These cells could derive from the epicardium, a single layer of epithelial cells that surrounds the outer layer of the myocardium (Mikawa and Gourdie 1996). Through the epithelial to mesenchymal transition (EMT) process, epicardial cells acquired mesenchymal phenotype and gave rise to cardiac fibroblasts, pericytes, and coronary vascular smooth muscle cells.

Pericytes can effectively mediate multiple functions in the homeostasis of cardiac muscle, including regulation of microvascular compartment and angiogenesis. During angiogenesis, endothelial cells are activated by vascular endothelial growth factor (VEGF) and start to form sprouting vessel tips. In this context, they produce and release PDGF-BB, which recruits pericytes that stabilize the vessel wall via direct contact or paracrine signaling. Pericytes secrete proteins and nucleic acids, in a free form or encapsulated into macrovesicles, inducing endothelial cell growth and migration for new vessel formation (Caporali et al. 2017). Several studies suggested the involvement of pericytes in regulation of capillary permeability after myocardial infarction. It has been shown that pro-nerve growth factor (NGF) and angiopoietin-2 secreted by cardiomyocytes and endothelial cells respectively, mediated their activation and detachment from blood vessels increasing microvascular permeability (Fu et al. 2018; Siao et al. 2012).

In addition, cardiac pericytes may secrete a plethora of chemokines and cytokines in response to post-infarction inflammatory stimuli. Similarly to brain and lung pericytes, these cells could activate inflammatory signaling through TLR, in particular TLR-4 (Edelman et al. 2007a; Nyúl-Tóth et al. 2017; Stark et al. 2013), thereby amplifying the inflammatory damage. Moreover, cardiac pericytes could also participate in leucocytes recruitment, acquiring an inflammatory phenotype with an increased expression of adhesion molecules and chemo attractant mediators (Alex and Frangogiannis 2019). Recently, O'Farrell et al. demonstrated in a rat model of myocardial IRI that the activation of pericytes and subsequent constriction contributed to the 'no-reflow phenomenon' causing thrombi formation and microvascular obstruction (O'Farrel et al. 2017; O'Farrell and Attwell 2014).

Multiple cardiovascular diseases are associated with cardiac fibrosis that is characterized by the proliferation of fibroblasts and their differentiation into myofibroblasts that contribute to the accumulation of extracellular matrix proteins (ECM). Several evidences demonstrated that also endothelial cells are an important source for myofibroblasts in cardiac diseases. Despite recent studies showed that pericytes contributed to organ fibrosis (Leaf et al. 2017; Sava et al. 2017; Sundberg et al. 1996), there are few clear evidences of their involvement in cardiac fibrosis. The absence of findings that demonstrated the direct involvement of pericytes in cardiac fibrosis could be probably due to the lack of canonical pericytes markers during the acquisition of myofibroblast phenotype. However, the fate tracing analysis performed by Kramann et al. revealed new insights in the cellular origin of myofibroblast during cardiac fibrosis. The authors of the study, identified for the first time Gli1+ cells, a minority of the mesenchymal PDGFR-β + cells that formed a niche of perivascular progenitor cells with characteristics of mesenchymal stem cells. The major finding was that upon injury of liver, lung, kidney or heart, these Gli1+ cells differentiated into myofibroblasts and genetic ablation of these Gli1+ cells ameliorated fibrosis and organ function. In particular, after angiotensin-2-induced myocardial fibrosis and ascending aortic constriction, the ablation of Gli1+ myofibroblast progenitor population reduced fibrosis severity, cardiac hypertrophy and rescued systolic left ventricular function following injury (Murray et al. 2017; Kramann et al. 2015). These papers provide the proof of concept that perivascular pericytes, despite their mesenchymal cell's properties, are not beneficial and do not necessarily exert regenerative effects; on the contrary they can be committed to the myofibroblast lineage during organ injury. Therefore, it is important to characterize cardiac pericyte populations, in order to identify the subsets with distinct functional properties and explore the role of pericyte-mediated actions in cardiac injury, repair and remodeling (Alex and Frangogiannis 2019).

7.1.5 *Pericytes in Lung Diseases: From Sentinel of Immune Response to Mediators of Fibrosis*

The lung is the principal organ in constant interaction with external environment and presents several immune and non-immune cells that play an important role in the first line defense during infections (Hung et al. 2017a). These cells express PRR which bind PAMPs and or DAMPs from injured tissue inducing an inflammatory response which could exacerbate lung injury (Suresh and Mosser 2013). Immune regulation of this response is controlled by functional interstitial immune sentinel cells that are identified as pericyte-like cells (Hung et al. 2013, 2017b). These cells express functional TLR and may act as immunosurveillance cells capable to control infection and initiate repair of damaged tissue.

In several studies, Edelman et al. demonstrated the expression and the activation of TLR-4 in rat lung pericytes that increased vessel permeability and production of IL-1, underlying their involvement in the inflammatory cascade (Edelman et al. 2006, 2007a, b). After TLR activation, these cells up-regulated the leukocyte adhesion molecules, greatly enhancing leucocyte extravasation to lung parenchyma (Hung et al. 2017a). Therefore, the presence of pericytes in the lung interstitial is pivotal for early detection of PAMPs and DAMPs from damaged tissue and to enhance immune cells to scan the local microenvironment (Hung et al. 2017a).

It has been extensively demonstrated that a sustained inflammatory response to PAMPs or DAMPs results in progressive fibrosis (Rowley and Johnson 2014; Wynn 2008). Lung fibrotic diseases are correlated to different environmental exposure (asbestos, antigens and viruses) or chronic inflammatory diseases (rheumatoid arthritis) or, more commonly, are idiopathic (Moore et al. 2013); particularly, idiopathic lung fibrosis has a poor prognosis with a high mortality rate (Wynn 2011).

The mechanism that underlies the interaction of the inflammatory process and fibrosis remains unclear; several studies suggested that the overexpression of TLR-4 in lung tissue was accompanied by the aggravation of fibrosis (He et al. 2009a, b). The activation of TLR-4 in lung fibroblasts induced the acquisition of myofibroblasts phenotype (He et al. 2009a; Wang et al. 2017), which are the major contributors of this process by synthesizing ECM components that contribute to progressive interstitial fibrosis (Greenhalgh et al. 2013).

The origin of myofibroblasts, particularly *in vivo*, remains under debate. These cells could derive from resident fibroblasts, bone marrow-derived MSCs, epithelial cells and endothelial cells (Greenhalgh et al. 2013); numerous studies demonstrated that epithelial and endothelial cells acquired a mesenchymal phenotype and contributed to fibrosis through a process known as EMT or endothelial-to-mesenchymal transition (EndMT) (Greenhalgh et al. 2015).

Recent lineage tracing studies overall supported the assertion that also pericytes participated to the myofibroblast population in lung fibrosis (Hung et al. 2013). It is clear that other cells, such as fibroblasts are also responsible for collagen deposition. Further studies are certainly warranted to investigate whether targeting the pericyte-to myofibroblast transition (PMT) is a viable treatment strategy for lung fibrosis.

In conclusion, since dysregulated TLR-4 signaling is implicated in lung inflammatory process, fibroblasts activation and progressive fibrosis, we can speculate that this pathway could be involved in PMT and strategies to selectively inhibit TLR-4 signaling could also control pericytes behavior in lung diseases. As emerged by the literature, these cells not only support the vasculature and regulate immune response, but are also involved in fibrogenic processes. However, the mechanisms of the PMT and migration to fibrotic foci need to be addressed.

7.1.6 Pericytes in Neurovascular Diseases and the Impact of TLR-4 Signaling

The blood–brain barrier (BBB) has a unique structure organized by specialized capillary endothelial cells that interacts with pericytes, astrocyte end-feet, microglia and neurons (Rustenhoven et al. 2017). The function of BBB is essential to prevent the exposure of neurons to dangerous molecules, that can modify brain parenchyma (Rustenhoven et al. 2017; Winkler et al. 2011).

The BBB can be affected by inflammation arising in the brain or the periphery such as in sepsis. As well known, inflammation is an important process to resolve infections, but this process could arise detrimental effects in brain, for its limited regenerative capacity (Horner and Gage 2002). Indeed, the loss of neurons as results of an inflammatory response cannot be fully recovered (Horner and Gage 2002). It is clear that the loss of BBB integrity and the inflammation are the principal hallmarks in neurological diseases such as Alzheimer's disease (Heppner et al. 2015), epilepsy (Milesi et al. 2014), stroke (Sakuma et al. 2016), and multiple sclerosis (Lévesque et al. 2016).

In neuro-inflammatory diseases, the principal cells investigated are microglia and astrocytes. As previously underlined, also blood vessels are another important site of cerebral inflammation. In particular, endothelial cells that localize at the interface of the bloodstream and the surrounding brain parenchyma, contribute to amplify inflammatory process, allowing entry of peripheral immune cells and inflammatory molecules causing neurological injury (Smyth et al. 2018).

Recently, brain pericytes are defined as gatekeepers of inflammation tightly regulating leukocytes infiltration, blood flow as well as the entry of circulating mediators and metabolites (Stark et al. 2018). Once leukocytes adhere to the endothelium and start to transmigrate, they interact with pericytes, which usually form a barrier to avoid infiltration into the brain parenchyma (Smyth et al. 2018). In the setting of inflammation, pericytes alter their morphology and up-regulate the expression of the adhesion molecule, ICAM-1 and release chemokines and chemoattractant factors as MIF. These encourage the recruitment and infiltration of peripheral immune cells through the damaged BBB. Thus, pericytes drive migration of monocytes by

MIF and CCL2 facilitating neutrophil extravasation by MIF and CXCL8 (Rustenhoven et al. 2017; Proebstl et al. 2012). These interactions are crucial for the efficient navigation of immune cells to the loci of inflammation to execute their effector functions.

Moreover, pericytes assume a behavior similar to macrophages, phagocytosing other cells and presenting antigens to immune cells (Rustenhoven et al. 2017). Herland et al. showed that the exposure of pericytes to cytokines induced the release of inflammatory molecules and matrix metalloprotease 9 (MMP9), that contributed to BBB breakdown (Herland et al. 2016). Therefore, pericytes attracted and interacted with myeloid leukocytes 'instructing'' them with PRR and motility programs (Brown et al. 2019).

In addition, the responses to IL-1β, TNFα, and LPS/TLR-4 through NF-κB signaling is conserved between endothelial cells and pericytes, as well as the response to IFN-γ through STAT1, and to TGFβ1 through the SMAD2/3 pathway (Smyth et al. 2018). Smyth et al. demonstrated that endothelial cells and pericytes shared the majority of secreted factors but differed only for the levels of secretion (Smyth et al. 2018). Interestingly, pericytes secreted high levels of IGFBP-2 and IGFBP-3 that have a key role in endothelial viability, angiogenesis, inflammation and pericyte coverage (Smyth et al. 2018).

Dohgu S et al., showed that LPS/TLR-4 signaling stimulated a crosstalk between brain endothelial cells and pericytes, altering BBB and enhancing the infiltration of HIV to brain parenchyma. Another study from *Jansson* et al. confirmed that human brain pericytes are sensitive to LPS, demonstrating that pericytes play an important role in communicating inflammatory signals and may be involved in blood brain barrier disruption (Jansson et al. 2014). During meningitis, TLR signaling represents a key pathway for the dysregulated inflammatory cascade. Then, interfering with TLR activation may represent a strategic approach to prevent neurological diseases.

However, the inhibition of MyD88-TLR-2/4 signaling showed detrimental effects in mouse model of meningitis as a consequence of impaired bacterial eradication. Thus, the choice of a specific therapy is necessary to control immune response without inducing immunosuppression (Jansson et al. 2014).

Several studies investigated the role of astrocytes in scar formation in the central nervous system (CNS). In a recent study, Göritz et al. evaluated the behavior of a subpopulation of pericytes lining blood vessels in the spinal cord parenchyma before and after injury (Göritz et al. 2011). They demonstrated that these cells acquired a myofibroblast phenotype and started to secrete fibrotic factors and increased connective tissue in the scar (Göritz et al. 2011). The impact of TLR engagement at the BBB remains to be determined but is expected to be an attractive therapeutic target to reduce pathogen and immune cells infiltration across the BBB. The regulation of TLR signaling is considered a therapeutic strategy, but at the same time it could be associated with a considerable risk. The use of TLR agonists to induce the adaptive immune responses may be efficient for vaccine design but may also trigger and amplify inflammatory disorders. Therefore, additional studies addressing these questions within the CNS microenvironment are warranted (Hanke and Kielian 2011).

7.1.7 Pericytes in Renal Diseases: From Inflammation to Fibrosis

Kidney diseases are often primarily characterized by damage to the vasculature. Renal pericytes are mural cells of the microcirculation, which have been shown to play a key role in regulating blood flow, angiogenesis, hematopoiesis, progenitor cell functions, immunomodulation and trophic effects following injury (Shaw et al. 2018). Several studies identified different populations of pericytes such as the glioma associated oncogene Gli1-expressing pericytes which contribute to renal fibrosis (Kramann et al. 2015), pericytes of the juxtaglomerular arterioles which produce renin and regulate blood pressure, and mesangial cells which support the glomeruli architecture and functions (Stefanska et al. 2016).

In renal parenchyma, pericytes are closely associated to endothelial cells and the crosstalk between these two populations plays a key role in basement membrane formation, maintenance and remodeling (Shaw et al. 2018). The number of pericytes varies considerably in relation to different regions of tissues and to the age of organs. Stefanska et al. showed that the number of pericytes and the relative endothelial coverage decreased in the cortex and medulla of aged mice (Stefanska et al. 2015). Moreover the loss of pericytes is a peculiar hallmark of acute kidney injury (AKI) and it is associated to a declining of kidney function (Shaw et al. 2018).

Besides their known involvement in angiogenesis and vascular homeostasis, these cells have an active role in the amplification of inflammatory response through the activation of TLR-4 and MyD88 pathway (Leaf et al. 2017). Pericytes sense renal damage through TLR-4 and activate MyD88 and IRAK4 inducing the synthesis and the release of inflammatory mediators as IL-1β. Through autocrine mechanism pericytes respond to IL-1β and amplify inflammatory response (Leaf et al. 2017). Thus, pericytes promote the activation and the recruitment of the innate immune cells, critical for eradicating microbial infection and endogenous danger signal. Chemoattractants facilitate leukocyte extravasation from the bloodstream to the renal parenchyma and instruct leucocytes to exert their effector functions (Leaf et al. 2017).

Moreover, TLR-4/MyD88/IRAK4 axis mediates pro-fibrotic pathways, contributing to fibroblast accumulation and activation in damaged renal parenchyma (Castellano et al. 2019a; Leaf et al. 2017). Several studies demonstrated that TLR-4-deficient mice were protected from renal fibrosis with a strong decrease of extracellular matrix deposition at tubulointerstitial level (Souza et al. 2015). Recently, Leaf I.A. et al. demonstrated that both profibrotic TGF-β receptors and TLR-4/MyD88 are required for PMT process (Leaf et al. 2017). These pathways cooperated in triggering PMT process and TLR-4 pathway transduced profibrotic signal to TGF-β-R

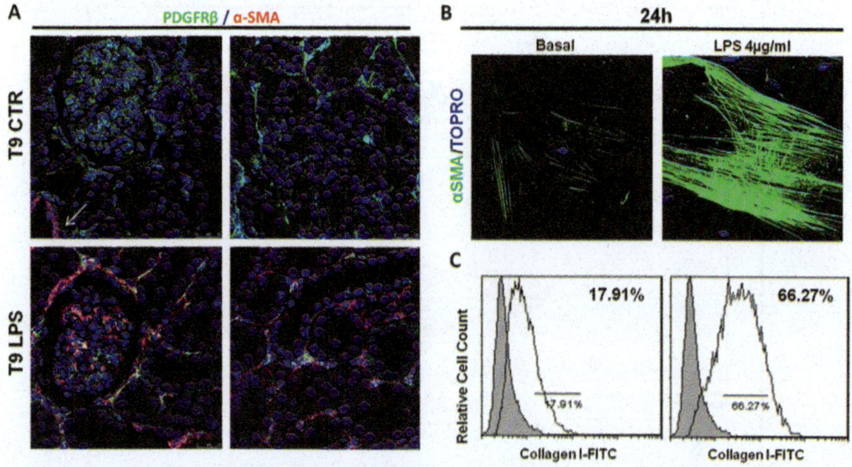

Fig. 7.1 LPS profibrotic effects. (**a**) Dysfunctional activation of renal pericytes in a swine model of LPS-induced AKI. Pericytes were double-stained for PDGFRβ (green) and α-SMA(red). In healthy pigs (T9 CTR) PDGFRβ⁺ pericytes were weakly positive for α-SMA. After 9 h from LPS infusion pericytes acquired myofibroblast phenotype, as shown by the colocalization of the two markers in the interstitium and at glomerular level. (**b–c**) *In vitro* effects of LPS. Pericytes were stimulated with LPS for 24 h. These cells showed high level of α-SMA expression localized in stress fibers and increased collagen I synthesis. Modified from Fig. 1 and Fig. 2 in Int. J. Mol. Sci. 2019, 20, 3682; doi:https://doi.org/10.3390/ijms20153682 (Castellano et al. 2019a). License: (http://creativecommons.org/licenses/by/4.0/)

pathway (Leaf et al. 2017). Accordingly, in our recent study we showed the profibrotic effects of LPS on pericytes (Fig. 7.1b–c) and the molecular mechanism of MyD88 and TGFβRI/II transactivation (Castellano et al. 2019a). We showed that after TGFβ-Receptor-blocking, LPS induced the SMAD2/3 phosphorylation in a TGF-β independent manner and pericytes underwent PMT. In addition, LPS induced the synthesis and the secretion of TGF-β in pericytes, promoting an autocrine self-sustaining positive feedback loop that amplified renal fibrosis progression (Fig. 7.2) (Castellano et al. 2019a). Moreover, in a swine model of LPS-induced AKI, we demonstrated the occurrence of PMT. After 9 h from LPS infusion (T9 LPS), renal pericytes increased the expression of α-SMA marker and acquired a myofibroblast phenotype (Castellano et al. 2019a). The co-localization of their constitutive marker PDGFβR and α-SMA was more evident in arterioles, peritubular capillaries and mesangial cells (Fig. 7.1a). Finally, we also underlined the key role of LBP in triggering inflammatory and fibrotic response also at low concentration of LPS (Castellano et al. 2019a).

Our findings are in line with other studies that described LBP as a “biological taxi service” for maximizing cellular response to endotoxin (Castellano et al. 2014, 2019a; Stasi et al. 2016). As well known, LBP is an acute phase protein synthesized by hepatocytes after microbial or non-microbial insult. In our study, we demonstrated a local renal synthesis by pericytes after TLR-4 activation, underlying their

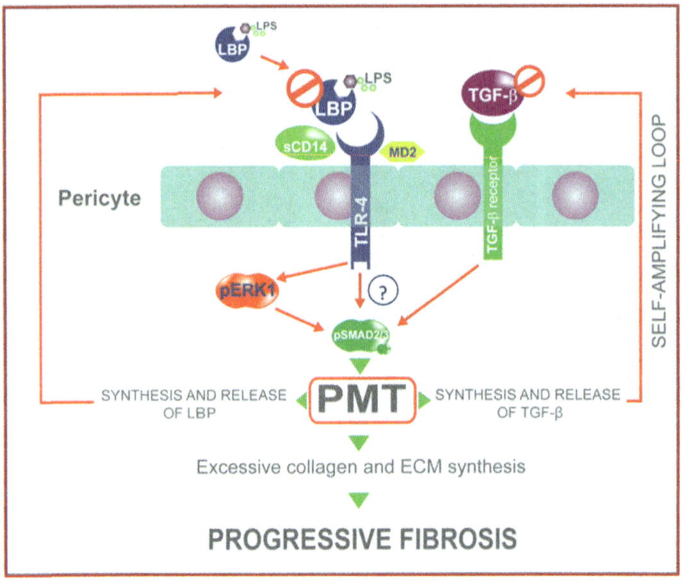

Fig. 7.2 Fibrotic mechanism driven by TLR-4 on renal pericytes. After endotoxemic injury, there is a relevant increase of LBP in blood circulation. LBP engages endotoxin monomers and facilitates the binding to TLR-4 on renal pericytes. The presence of LBP amplifies LPS signaling and increases pericytes response. LPS/TLR-4 axis induces the phosphorylation of SMAD2/3 complex in a TGFβ independent manner. After activation, pericytes acquire a myofibroblast phenotype and synthesize and secrete collagen and ECM components. Interestingly, pericytes contribute to further accumulation of LBP and TGFβ developing a self-sustaining feed-forward loop that increases renal fibrosis. The removal of LBP or other factors as TGFβ leads to a significant decrease of TLR-4 activation preserving pericytes phenotype and function. Reprinted from (Castellano et al. 2019a). License: (http://creativecommons.org/licenses/by/4.0/)

powerful role in renal diseases and the necessity to find several approach to prevent their dysfunctional activation (Fig. 7.2) (Castellano et al. 2019a).

Therefore, pericytes are involved both in inflammation and fibrogenesis via TLR-4/MyD88/IRAK4 axis, and also contribute to the generation of components of the complement pathway that in turn amplify immune response and prime fibrotic process. Specifically, *Xavier S.* et al. demonstrated a local synthesis and secretion of C1q by pericytes in two different animal model of CKD. (Xavier et al. 2017). In this study, the increased synthesis of C1q in pericytes triggered renal inflammation through the release of cytokines such as IL-6, MCP-1, and MIP1-α that also contributed to fibrosis (Xavier et al. 2017). Moreover C1q promotes complement activation, increasing the release of C3 fragments in interstitial cells, which induce extracellular matrix accumulation and fibrosis (Xavier et al. 2017).

Therefore, next to LPS and TGFβ, several components of complement pathway (Xavier et al. 2017) may lead to extracellular matrix deposition with the acquirement of myofibroblast phenotype via PMT. Accordingly, we recently demonstrated that complement activation mediates PMT and early tubulointerstitial fibrosis in a

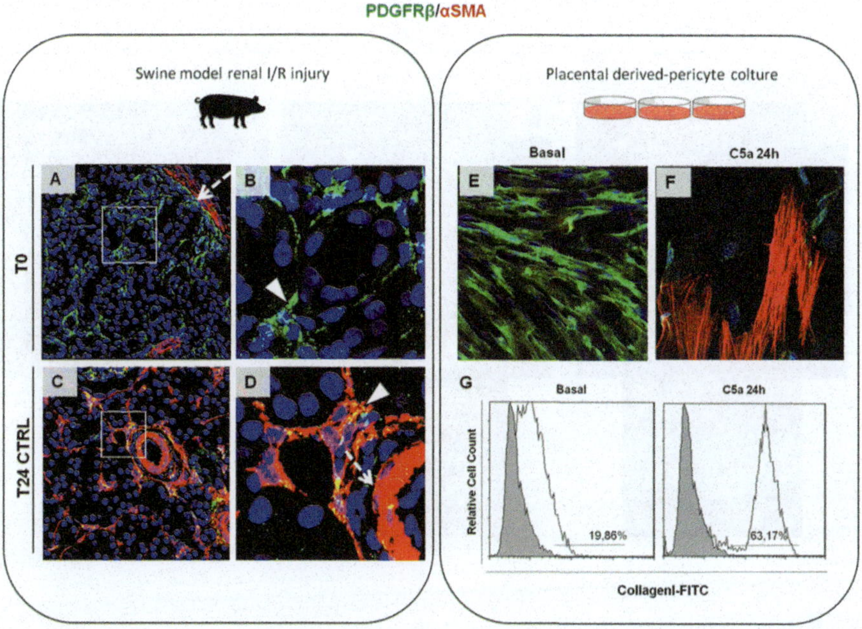

Fig. 7.3 Complement activation induces PMT. (**a–d**) PMT in a swine model of renal IRI. Pericytes were double stained for PDGFRβ (green) and α-SMA(red). Under normal condition (T0, before IRI) pericytes were detected in interstitial peritubular capillaries and did not express myofibroblast marker, α-SMA. After 24 hours from IRI, peritubular pericytes expressed both PDGFRβ (green) and αSMA, demonstrating the occurrence of PMT. (**e–g**) *In vitro* effects of C5a. After 24 h from C5a stimulation, pericytes reduced PDGFRβ expression and acquired a contractile phenotype with a strong increase of αSMA-stress fibers. (**g**) FACS analysis showed increased collagen I expression in C5a-stimulated pericytes. Modified from Fig. 4 and Fig. 7 in Front. Immunol. 9:1002. doi: https://doi.org/10.3389/fimmu.2018.01002 (Castellano et al. 2018). License: (http://creativecommons.org/licenses/by/4.0/)

swine model of renal IRI (Fig. 7.3 a–d) (Castellano et al. 2018; Simone et al. 2014). We also described the in vitro effects of the C5a in promoting the acquirement of myofibroblast phenotype by pericytes (Fig. 7.3 e–f) assessing the occurrence of PMT also in a mouse model of renal IRI (Fig. 7.4a–c) (Castellano et al. 2018). Interestingly, we observed that C5aR1-deficient mice were protected from renal IRI and showed significantly lower number of PDGFRβ$^+$/αSMA$^+$ cells compared with the Wild type mice (Fig. 7.4a–c) (Castellano et al. 2018). Thus, in accordance with Peng, Qi et al. study, C5a/C5aR1 signaling is strongly involved in renal tubulointerstitial fibrosis (Peng et al. 2019); in fact, anti- inflammatory strategies targeting TLR pathways, cytokines and complement activation might also exert beneficial effects on renal pericytes (Xavier et al. 2017).

Renal interstitial fibrosis is the principal key event in the progression of CKD (Shaw et al. 2018; Castellano et al. 2019b; Fiorentino et al. 2018). The principal mechanisms in the cascade of events leading to CKD include the activation of

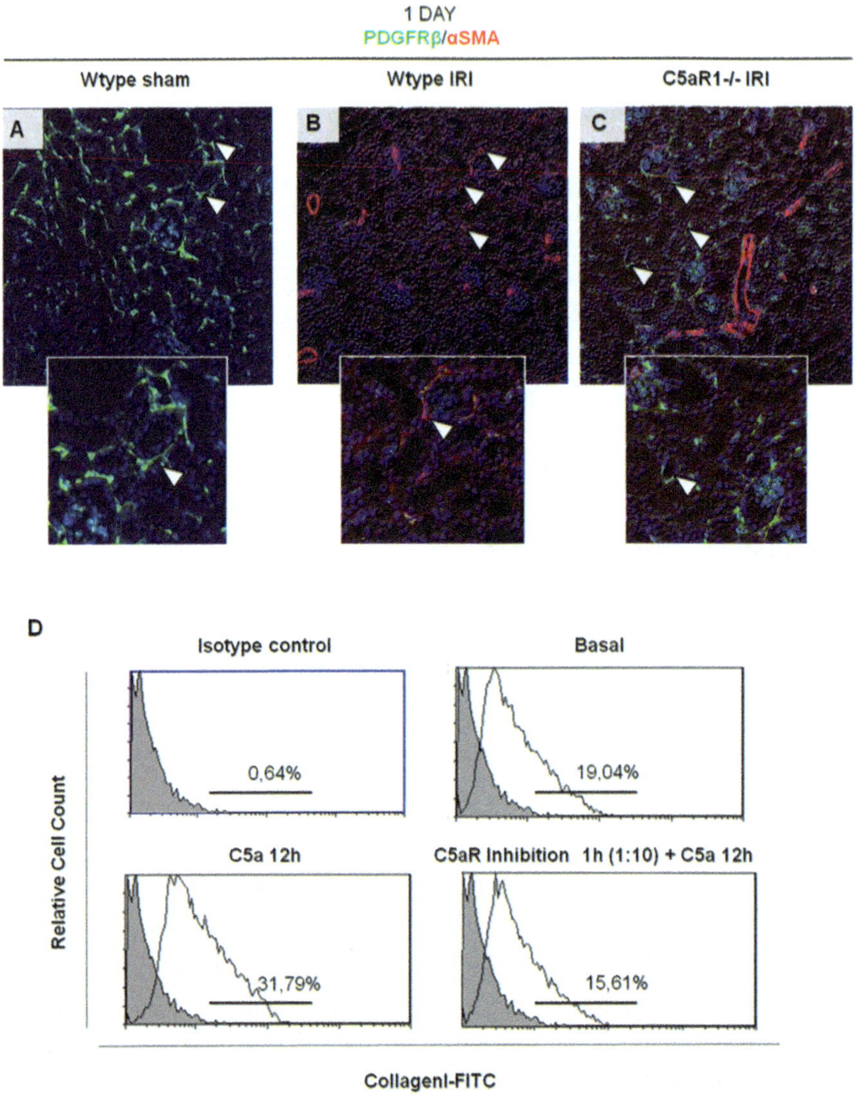

Fig. 7.4 C5a/C5aR1 pathway activation induces PMT. (**a**–**c**) Renal IRI was performed for 24 h in a mouse model of IRI. Renal biopsies were double stained for PDGFRβ (green) and α-SMA(red). Wild type sham mice showed PDGFRβ+ pericytes weakly positive for αSMA marker. After 24 h of IRI, the number of PDGFRβ+/αSMA+ cells strongly increased at peritubular level. In C5aR1−/− mice, PDGFRβ+/αSMA+ perivascular cells were barely detectable, demonstrating the key role of C5a/C5aR1 pathway in PMT process. (D) *In vitro* C5aR inhibition assay; FACS analysis showed that upon C5aR1 blocking, C5a-stimulated pericytes did not increase Collagen I synthesis. Modified from Fig. 5 and Fig. 8 in Front. Immunol. 9:1002. doi: https://doi.org/10.3389/fimmu.2018.01002 (Castellano et al. 2018). License: (http://creativecommons.org/licenses/by/4.0/)

inflammatory cells and the local priming of resident fibroblasts. Recently, pericytes have been identified as an important source for inflammatory process and renal fibrosis under pathological conditions (Shaw et al. 2018). In addition to pericytes, podocytes express TLR-4 and are essential cells involved in inflammation and in the progression of AKI toward CKD (Netti et al. 2019). These cells share more similar morphological characteristics of pericytes, and are identified as specialized pericyte-like cells.

In addition to the well-known classical TGF-β signaling pathways, the TLR-4 pathway is involved in dysfunctional pericyte activation and renal fibrosis (Castellano et al. 2019a; Leaf et al. 2017). Therefore, the importance of pericytes in both inflammation and fibrogenesis cannot be underestimated in renal diseases. There is a growing interest to explore novel strategies to reduce pericytes activation and to improve renal function avoiding the development of CKD.

7.2 Conclusions

Pericytes play a key role in angiogenesis, maintenance of vascular compartment and have multiple functions and differentiation capacities in several pathological settings. The ubiquitous presence of pericytes in several compartments of human body is in line with the accumulating evidence of their participation in various diseases. The recent knowledge illustrates the molecular mechanisms responsible for pericytes in mediating regulation in angiogenesis and blood flow; however, the principal signaling involved in immunomodulation and fibrogenic process are still not completely revealed. According to recent findings, activation of TLR and in particular TLR-4 in pericytes is associated with the development of exacerbated inflammatory response and fibrosis in several organs and diseases. The most severe disease deriving from TLR-4 excessive activation by PAMPs is sepsis. TLR-4 has been suggested as a promising therapeutic target but TLR-4 antagonists have failed to block acute sepsis and organs damage (Kuzmich et al. 2017). However, new promises results derive from the use of TLR-4 antagonists in chronic inflammatory diseases (Kuzmich et al. 2017). In accordance, beside sepsis, the use of new approaches to modulate TLR-4 activation may influence pericytes behavior and ameliorate organ functions in several pathological conditions, including cardiovascular diseases, neuronal degeneration, lung and renal diseases.

Acknowledgements The authors thank Chiara Divella for excellent technical assistance.

Compliance with Ethical Standards

Funding
Supported by University of Bari 'Aldo Moro', University of Foggia, the Italian Ministry of Health (Giovani Ricercatori 2011–2012, GR-2011-02351027, granted to G.C.) and Fondo Sociale Europeo, Azione I.2 "Attrazione e Mobilità Internazionale dei Ricercatori" (AIM-1810057-activity 2 granted to A.S.).

Disclosure of Interests All authors declare they have no conflict of interest.

Ethical Approval The animal studies described in this chapter were approved by Ethical Committee of the Italian Ministry of Health, Prot. N823/2016-PR and Ministero della Salute (Ricerca Finalizzata 2009, GR-2009-1,608662 as indicated in the references (Castellano et al. 2014, 2018, 2019a).

Informed Consent Informed consent for participation and publication was obtained from all individual participants included in the study. All authors read and approved the final manuscript.

References

Akira S, Uematsu S, Takeuchi O (2006) Pathogen recognition and innate immunity. Cell 124(4):783–801

Alex L, Frangogiannis NG (2019) Pericytes in the infarcted heart. Vasc Biol. https://doi.org/10.1530/vb-19-0007

Armulik A, Genové G, Betsholtz C (2011) Pericytes: developmental, physiological, and pathological perspectives, problems, and promises. Dev Cell 21(2):193–215

Bagheri B, Sohrabi B, Akbar Movassaghpour A et al (2014) Hydrocortisone reduces toll-like receptor 4 expression on peripheral CD14+monocytes in patients undergoing percutanoues coronary intervention. Iran Biomed J. https://doi.org/10.6091/ibj.1275.2013

Balabanov R, Washington R, Wagnerova J, Dore-Duffy P (1996) CNS microvascular pericytes express macrophage-like function, cell surface integrin αM, and macrophage marker ED-2. Microvasc Res. https://doi.org/10.1006/mvre.1996.0049

Balabanov R, Beaumont T, Dore-Duffy P (1999) Role of central nervous system microvascular pericytes in activation of antigen-primed splenic T-lymphocytes. J Neurosci Res. https://doi.org/10.1002/(SICI)1097-4547(19990301)55:5<578::AID-JNR5>3.0.CO;2-E

Beamer LJ, Carroll SF, Eisenberg D (1997) Crystal structure of human BPI and two bound phospholipids at 2.4 angstrom resolution. Science 80. https://doi.org/10.1126/science.276.5320.1857

Brown LS, Foster CG, Courtney JM et al (2019) Pericytes and neurovascular function in the healthy and diseased brain. Front Cell Neurosci

Caporali A, Martello A, Miscianinov V et al (2017) Contribution of pericyte paracrine regulation of the endothelium to angiogenesis. Pharmacol Ther 171:56–64

Castellano G, Stasi A, Intini A et al (2014) Endothelial dysfunction and renal fibrosis in endotoxemia-induced oliguric kidney injury: possible role of LPS-binding protein. Crit Care 18:520. https://doi.org/10.1186/s13054-014-0520-2

Castellano G, Franzin R, Stasi A et al (2018) Complement activation during ischemia/reperfusion injury induces pericyte-to-myofibroblast transdifferentiation regulating peritubular capillary lumen reduction through pERK signaling. Front Immunol 9:1002. https://doi.org/10.3389/fimmu.2018.01002

Castellano G, Stasi A, Franzin R et al (2019a) LPS-binding protein modulates acute renal fibrosis by inducing pericyte-to-myofibroblast trans-differentiation through TLR-4 signaling. Int J Mol Sci. https://doi.org/10.3390/ijms20153682

Castellano G, Franzin R, Sallustio F et al (2019b) Complement component C5a induces aberrant epigenetic modifications in renal tubular epithelial cells accelerating senescence by Wnt4/beta-catenin signaling after ischemia/reperfusion injury. Aging (Albany NY) 11:4382–4406. https://doi.org/10.18632/aging.102059

Chaturvedi A, Pierce SK (2009) How location governs toll-like receptor signaling. Traffic 10(6):621–628

Chirico G, Cortinovis S, Fonte C, Giudici G (2007) Bacterial sepsis. J Chemother

Danese S, Dejana E, Fiocchi C (2007) Immune regulation by microvascular endothelial cells: directing innate and adaptive immunity, coagulation, and inflammation. J Immunol. https://doi.org/10.4049/jimmunol.178.10.6017

Dias Moura Prazeres PH, Sena IFG, Borges I da T et al (2017) Pericytes are heterogeneous in their origin within the same tissue. Dev Biol 427(1):6–11

Ding HS, Yang J, Chen P et al (2013) The HMGB1-TLR4 axis contributes to myocardial ischemia/reperfusion injury via regulation of cardiomyocyte apoptosis. Gene. https://doi.org/10.1016/j.gene.2013.05.041

Duque GA, Descoteaux A (2014) Macrophage cytokines: involvement in immunity and infectious diseases. Front Immunol 5:491

Edelman DA, Jiang Y, Tyburski J et al (2006) Toll-like receptor-4 message is up-regulated in lipopolysaccharide-exposed rat lung pericytes. J Surg Res. https://doi.org/10.1016/j.jss.2006.03.007

Edelman DA, Jiang Y, Tyburski JG et al (2007a) Lipopolysaccharide activation of pericyte's toll-like receptor-4 regulates co-culture permeability. Am J Surg. https://doi.org/10.1016/j.amjsurg.2006.08.086

Edelman DA, Jiang Y, Tyburski JG et al (2007b) Cytokine production in lipopolysaccharide-exposed rat lung pericytes. J Trauma Inj Infect Crit Care. https://doi.org/10.1097/TA.0b013e31802dd712

Edfeldt K, Swedenborg J, Hansson GK, Yan Z (2002) Expression of toll-like receptors in human atherosclerotic lesions. Circulation. https://doi.org/10.1161/circ.105.10.1158

Fabry Z, Sandor M, Gajewski TF et al (1993) Differential activation of Th1 and Th2 CD4+ cells by murine brain microvessel endothelial cells and smooth muscle/pericytes. J Immunol 151(1):38–47

Fairweather D, Yusung S, Frisancho S et al (2003) IL-12 receptor β1 and toll-like receptor 4 increase IL-1β- and IL-18-associated myocarditis and coxsackievirus replication. J Immunol. https://doi.org/10.4049/jimmunol.170.9.4731

Fallach R, Shainberg A, Avlas O et al (2010) Cardiomyocyte toll-like receptor 4 is involved in heart dysfunction following septic shock or myocardial ischemia. J Mol Cell Cardiol. https://doi.org/10.1016/j.yjmcc.2010.02.020

Fani F, Regolisti G, Delsante M et al (2018) Recent advances in the pathogenetic mechanisms of sepsis-associated acute kidney injury. J Nephrol 31(3):351–359

Faure E, Thomas L, Xu H et al (2001) Bacterial lipopolysaccharide and IFN-γ induce toll-like receptor 2 and toll-like receptor 4 expression in human endothelial cells: role of NF-κB activation. J Immunol. https://doi.org/10.4049/jimmunol.166.3.2018

Fiorentino M, Grandaliano G, Gesualdo L, Castellano G (2018) Acute kidney injury to chronic kidney disease transition. Contrib Nephrol 193:45–54. https://doi.org/10.1159/000484962

Foster K, Sheridan J, Veiga-Fernandes H et al (2008) Contribution of neural crest-derived cells in the embryonic and adult Thymus. J Immunol. https://doi.org/10.4049/jimmunol.180.5.3183

Frantz S, Kobzik L, Kim YD et al (1999) Toll4 (TLR4) expression in cardiac myocytes in normal and failing myocardium. J Clin Invest. https://doi.org/10.1172/JCI6709

Fu X, Khalil H, Kanisicak O et al (2018) Specialized fibroblast differentiated states underlie scar formation in the infarcted mouse heart. J Clin Invest. https://doi.org/10.1172/JCI98215

Gaengel K, Genové G, Armulik A, Betsholtz C (2009) Endothelial-mural cell signaling in vascular development and angiogenesis. Arterioscler Thromb Vasc Biol 29(5):630–638

Göritz C, Dias DO, Tomilin N et al (2011) A pericyte origin of spinal cord scar tissue. Science 80. https://doi.org/10.1126/science.1203165

Greenhalgh SN, Iredale JP, Henderson NC (2013) Origins of fibrosis: Pericytes take centre stage. F1000Prime Rep. https://doi.org/10.12703/P5-37

Greenhalgh SN, Conroy KP, Henderson N (2015) Healing scars: targeting pericytes to treat fibrosis. QJM 108(1):3–7

Guijarro-Muñoz I, Compte M, Álvarez-Cienfuegos A et al (2014) Lipopolysaccharide activates toll-like receptor 4 (TLR4)-mediated NF-κB signaling pathway and proinflammatory response in human pericytes. J Biol Chem. https://doi.org/10.1074/jbc.M113.521161

Hanke ML, Kielian T (2011) Toll-like receptors in health and disease in the brain: mechanisms and therapeutic potential. Clin Sci 121(9):367–387
Harrell CR, Simovic Markovic B, Fellabaum C et al (2018) Molecular mechanisms underlying therapeutic potential of pericytes. J Biomed Sci 25(1):21
He Z, Zhu Y, Jiang H (2009a) Toll-like receptor 4 mediates lipopolysaccharide-induced collagen secretion by phosphoinositide3-kinase-akt pathway in fibroblasts during acute lung injury. J Recept Signal Transduct. https://doi.org/10.1080/10799890902845690
He ZY, Sen ZY, Jiang H (2009b) Inhibiting toll-like receptor 4 signaling ameliorates pulmonary fibrosis during acute lung injury induced by lipopolysaccharide: an experimental study. Respir Res. https://doi.org/10.1186/1465-9921-10-126
Heppner FL, Ransohoff RM, Becher B (2015) Immune attack: the role of inflammation in Alzheimer disease. Nat Rev Neurosci 16(6):358–372
Herland A, Van Der Meer AD, FitzGerald EA et al (2016) Distinct contributions of astrocytes and pericytes to neuroinflammation identified in a 3D human blood-brain barrier on a chip. PLoS One. https://doi.org/10.1371/journal.pone.0150360
Horner PJ, Gage FH (2002) Regeneration in the adult and aging brain. Arch Neurol
Hung C, Linn G, Chow YH et al (2013) Role of lung pericytes and resident fibroblasts in the pathogenesis of pulmonary fibrosis. Am J Respir Crit Care Med. https://doi.org/10.1164/rccm.201212-2297OC
Hung CF, Mittelsteadt KL, Brauer R et al (2017a) Lung pericyte-like cells are functional interstitial immune sentinel cells. Am J Physiol Lung Cell Mol Physiol. https://doi.org/10.1152/ajplung.00349.2016
Hung CF, Chow YH, Liles WC et al (2017b) Ablation of pericyte-like cells in lungs by oropharyngeal aspiration of diphtheria toxin. Am J Respir Cell Mol Biol. https://doi.org/10.1165/rcmb.2016-0083MA
Jansson D, Rustenhoven J, Feng S et al (2014) A role for human brain pericytes in neuroinflammation. J Neuroinflammation. https://doi.org/10.1186/1742-2094-11-104
Kawasaki K, Akashi S, Shimazu R et al (2000) Mouse toll-like receptor 4·MD-2 complex mediates lipopolysaccharide- mimetic signal transduction by Taxol. J Biol Chem. https://doi.org/10.1074/jbc.275.4.2251
Keynan Y, Fowke KR, Ball TB, Meyers AFA (2011) Toll-like receptors dysregulation after influenza virus infection: insights into pathogenesis of subsequent bacterial pneumonia. ISRN Pulmonol. https://doi.org/10.5402/2011/142518
Khakpour S, Wilhelmsen K, Hellman J (2015) Vascular endothelial cell toll-like receptor pathways in sepsis. Innate Immun 21(8):827–846
Kramann R, Schneider RK, DiRocco DP et al (2015) Perivascular Gli1+ progenitors are key contributors to injury-induced organ fibrosis. Cell Stem Cell 16:51–66. https://doi.org/10.1016/j.stem.2014.11.004
Krueger M, Bechmann I (2010) CNS pericytes: concepts, misconceptions, and a way out. Glia 58(1):1–10
Kuzmich NN, Sivak KV, Chubarev VN et al (2017) TLR4 signaling pathway modulators as potential therapeutics in inflammation and sepsis. Vaccine 5(4):34
Leaf IA, Nakagawa S, Johnson BG et al (2017) Pericyte MyD88 and IRAK4 control inflammatory and fibrotic responses to tissue injury. J Clin Invest. https://doi.org/10.1172/JCI87532
Levéen P, Pekny M, Gebre-Medhin S et al (1994) Mice deficient for PDGF B show renal, cardiovascular, and hematological abnormalities. Genes Dev. https://doi.org/10.1101/gad.8.16.1875
Lévesque SA, Paré A, Mailhot B et al (2016) Myeloid cell transmigration across the CNS vasculature triggers IL-1ß-driven neuroinflammation during autoimmune encephalomyelitis in mice. J Exp Med. https://doi.org/10.1084/jem.20151437
Liu L, Wang Y, Cao ZY et al (2015) Up-regulated TLR4 in cardiomyocytes exacerbates heart failure after long-term myocardial infarction. J Cell Mol Med. https://doi.org/10.1111/jcmm.12659
Lotze MT, Tracey KJ (2005) High-mobility group box 1 protein (HMGB1): nuclear weapon in the immune arsenal. Nat Rev Immunol

Mai J, Virtue A, Shen J et al (2013) An evolving new paradigm: endothelial cells – conditional innate immune cells. J Hematol Oncol

Medzhitov R (2001) Toll-like receptors and innate immunity. Nat Rev Immunol 1:135–145

Mikawa T, Gourdie RG (1996) Pericardial mesoderm generates a population of coronary smooth muscle cells migrating into the heart along with ingrowth of the epicardial organ. Dev Biol. https://doi.org/10.1006/dbio.1996.0068

Milesi S, Boussadia B, Plaud C et al (2014) Redistribution of PDGFRβ cells and NG2DsRed pericytes at the cerebrovasculature after status epilepticus. Neurobiol Dis. https://doi.org/10.1016/j.nbd.2014.07.010

Mogensen TH (2009) Pathogen recognition and inflammatory signaling in innate immune defenses. Clin Microbiol Rev 22(2):240–273

Moore BB, Lawson WE, Oury TD et al (2013) Animal models of fibrotic lung disease. Am J Respir Cell Mol Biol 49(2):167–179

Munford RS (2008) Sensing gram-negative bacterial lipopolysaccharides: a human disease determinant? Infect Immun

Murray IR, Baily JE, Chen WCW et al (2017) Skeletal and cardiac muscle pericytes: functions and therapeutic potential. Pharmacol Ther 171:65–74

Navarro R, Compte M, Álvarez-Vallina L, Sanz L (2016) Immune regulation by pericytes: modulating innate and adaptive immunity. Front Immunol 7:480

Netti GS, Sangregorio F, Spadaccino F et al (2019) LPS removal reduces CD80-mediated albuminuria in critically ill patients with gram-negative sepsis. Am J Physiol Ren Physiol. https://doi.org/10.1152/ajprenal.00491.2018

Nielsen CM, Dymecki SM (2010) Sonic hedgehog is required for vascular outgrowth in the hindbrain choroid plexus. Dev Biol. https://doi.org/10.1016/j.ydbio.2010.01.032

Nishimura M, Naito S (2005) Tissue-specific mRNA expression profiles of human toll-like receptors and related genes. Biol Pharm Bull. https://doi.org/10.1248/bpb.28.886

Nyúl-Tóth Á, Kozma M, Nagyőszi P et al (2017) Expression of pattern recognition receptors and activation of the non-canonical inflammasome pathway in brain pericytes. Brain Behav Immun. https://doi.org/10.1016/j.bbi.2017.04.010

O'Farrel FM, Mastitskaya S, Hammond-Haley M et al (2017) Capillary pericytes mediate coronary no-reflow after myocardial ischaemia. elife. https://doi.org/10.7554/eLife.29280

O'Farrell FM, Attwell D (2014) A role for pericytes in coronary no-reflow. Nat Rev Cardiol 11(7):427–432

Okamura Y, Watari M, Jerud ES et al (2001) The extra domain a of fibronectin activates toll-like receptor 4. J Biol Chem 276:10229–10233. https://doi.org/10.1074/jbc.M100099200. M100099200 [pii]

Peng Q, Wu W, Wu KY et al (2019) The C5a/C5aR1 axis promotes progression of renal tubulointerstitial fibrosis in a mouse model of renal ischemia/reperfusion injury. Kidney Int. https://doi.org/10.1016/j.kint.2019.01.039

Pieper C, Marek JJ, Unterberg M et al (2014) Brain capillary pericytes contribute to the immune defense in response to cytokines or LPS in vitro. Brain Res. https://doi.org/10.1016/j.brainres.2014.01.004

Pober JS, Sessa WC (2015) Inflammation and the blood microvascular system. Cold Spring Harb Perspect Biol. https://doi.org/10.1101/cshperspect.a016345

Proebstl D, Voisin MB, Woodfin A et al (2012) Pericytes support neutrophil subendothelial cell crawling and breaching of venular walls in vivo. J Exp Med. https://doi.org/10.1084/jem.20111622

Rowley JE, Johnson JR (2014) Pericytes in chronic lung disease. Int Arch Allergy Immunol 164(3):178–188

Rudziak P, Ellis CG, Kowalewska PM (2019) Role and molecular mechanisms of pericytes in regulation of leukocyte diapedesis in inflamed tissues. Mediat Inflamm

Rustenhoven J, Jansson D, Smyth LC, Dragunow M (2017) Brain pericytes as mediators of neuroinflammation. Trends Pharmacol Sci 38(3):291–304

Sakuma R, Kawahara M, Nakano-Doi A et al (2016) Brain pericytes serve as microglia-generating multipotent vascular stem cells following ischemic stroke. J Neuroinflammation. https://doi.org/10.1186/s12974-016-0523-9

Sarnak MJ (2014) A patient with heart failure and worsening kidney function. Clin J Am Soc Nephrol. https://doi.org/10.2215/CJN.11601113

Satoh S, Yada R, Inoue H et al (2016) Toll-like receptor-4 is upregulated in plaque debris of patients with acute coronary syndrome more than toll-like receptor-2. Heart Vessel. https://doi.org/10.1007/s00380-014-0565-9

Sava P, Ramanathan A, Dobronyi A et al (2017) Human pericytes adopt myofibroblast properties in the microenvironment of the IPF lung. JCI Insight. https://doi.org/10.1172/jci.insight.96352

Schumann RR, Leong SR, Flaggs GW et al (1990) Structure and function of lipopolysaccharide binding protein. Science 80. https://doi.org/10.1126/science.2402637

Scrascia G, Rotunno C, Simone S et al (2017) Acute kidney injury in high-risk cardiac surgery patients: roles of inflammation and coagulation. J Cardiovasc Med. https://doi.org/10.2459/JCM.0000000000000343

Shaw I, Rider S, Mullins J et al (2018) Pericytes in the renal vasculature: roles in health and disease. Nat Rev Nephrol 14(8):521–534

Shepro D, Morel NML (1993) Pericyte physiology. FASEB J 7(11):1031–1038

Siao CJ, Lorentz CU, Kermani P et al (2012) ProNGF, a cytokine induced after myocardial infarction in humans, targets pericytes to promote microvascular damage and activation. J Exp Med. https://doi.org/10.1084/jem.20111749

Simone S, Rascio F, Castellano G et al (2014) Complement-dependent NADPH oxidase enzyme activation in renal ischemia/reperfusion injury. Free Radic Biol Med 74:263–273. https://doi.org/10.1016/j.freeradbiomed.2014.07.003

Smyth LCD, Rustenhoven J, Park TIH et al (2018) Unique and shared inflammatory profiles of human brain endothelia and pericytes. J Neuroinflammation. https://doi.org/10.1186/s12974-018-1167-8

Song N, Huang Y, Shi H et al (2009) Overexpression of platelet-derived growth factor-BB increases tumor pericyte content via stromal-derived factor-1α/CXCR4 axis. Cancer Res. https://doi.org/10.1158/0008-5472.CAN-08-2007

Souza ACP, Tsuji T, Baranova IN et al (2015) TLR4 mutant mice are protected from renal fibrosis and chronic kidney disease progression. Physiol Rep 3:e12558. https://doi.org/10.14814/phy2.12558

Stark K, Eckart A, Haidari S et al (2013) Capillary and arteriolar pericytes attract innate leukocytes exiting through venules and "instruct" them with pattern-recognition and motility programs. Nat Immunol. https://doi.org/10.1038/ni.2477

Stark K, Pekayvaz K, Massberg S (2018) Role of pericytes in vascular immunosurveillance. Front Biosci Landmark. https://doi.org/10.2741/4615

Stasi A, Intini A, Divella C, et al (2016) Emerging role of Lipopolysaccharide binding protein in sepsis-induced acute kidney injury. Nephrol Dial Transplant gfw250. https://doi.org/10.1093/ndt/gfw250

Stefanska A, Eng D, Kaverina N et al (2015) Interstitial pericytes decrease in aged mouse kidneys. Aging (Albany NY). https://doi.org/10.18632/aging.100756

Stefanska A, Kenyon C, Christian HC et al (2016) Human kidney pericytes produce renin. Kidney Int. https://doi.org/10.1016/j.kint.2016.07.035

Stratman AN, Schwindt AE, Malotte KM, Davis GE (2010) Endothelial-derived PDGF-BB and HB-EGF coordinately regulate pericyte recruitment during vasculogenic tube assembly and stabilization. Blood. https://doi.org/10.1182/blood-2010-05-286872

Sundberg C, Ivarsson M, Gerdin B, Rubin K (1996) Pericytes as collagen-producing cells in excessive dermal scarring. Lab Investig 74(2):452–466

Suresh R, Mosser DM (2013) Pattern recognition receptors in innate immunity, host defense, and immunopathology. Am J Physiol Adv Physiol Educ. https://doi.org/10.1152/advan.00058.2013

Tavener SA, Long EM, Robbins SM et al (2004) Immune cell toll-like receptor 4 is required for cardiac myocyte impairment during endotoxemia. Circ Res. https://doi.org/10.1161/01.RES.0000144175.70140.8c

Terrell AM, Crisostomo PR, Wairiuko GM et al (2006) JAK/STAT/SOCS signaling circuits and associated cytokine-mediated inflammation and hypertrophy in the heart. Shock 26(3):226–234

Wang Q, Wang J, Wang J et al (2017) HMGB1 induces lung fbroblast to myofbroblast differentiation through NF-κB-mediated TGF-β1 release. Mol Med Rep. https://doi.org/10.3892/mmr.2017.6364

Williams DL, Ozment-Skelton T, Li C (2006) Modulation of the phosphoinositide 3-kinase signaling pathway alters host response to sepsis, inflammation, and ischemia/reperfusion injury. Shock 25(5):432–439

Winkler EA, Bell RD, Zlokovic BV (2011) Central nervous system pericytes in health and disease. Nat Neurosci 14(11):1398–1405

Wynn TA (2008) Cellular and molecular mechanisms of fibrosis. J Pathol 214(2):199–210

Wynn TA (2011) Integrating mechanisms of pulmonary fibrosis. J Exp Med. https://doi.org/10.1084/jem.20110551

Xavier S, Sahu RK, Landes SG et al (2017) Pericytes and immune cells contribute to complement activation in tubulointerstitial fibrosis. Am J Physiol Physiol 312:F516–F532. https://doi.org/10.1152/ajprenal.00604.2016

Yamazaki T, Nalbandian A, Uchida Y et al (2017) Tissue myeloid progenitors differentiate into pericytes through TGF-β signaling in developing skin vasculature. Cell Rep. https://doi.org/10.1016/j.celrep.2017.02.069

Yang Y, Lv J, Jiang S et al (2016) The emerging role of toll-like receptor 4 in myocardial inflammation. Cell Death Dis 7(5):e2234

Youn JH, Oh YJ, Kim ES et al (2008) High mobility group box 1 protein binding to lipopolysaccharide facilitates transfer of lipopolysaccharide to CD14 and enhances lipopolysaccharide-mediated TNF-α production in human monocytes. J Immunol. https://doi.org/10.4049/jimmunol.180.7.5067

Zeng H, He X, Tuo QH et al (2016) LPS causes pericyte loss and microvascular dysfunction via disruption of Sirt3/angiopoietins/Tie-2 and HIF-2α/Notch3 pathways. Sci Rep. https://doi.org/10.1038/srep20931

Chapter 8
EphA7+ Multipotent Pericytes and Their Roles in Multicellular Organisms

Jun-Ichi Kawabe

Abstract Existence of multipotent pericytes (MPCs) implies that microvasculature plays a role not only as ducts for blood, but also as a reservoir for stem cells that contributes to tissue maintenance and regeneration. Nerve network is closely linked to the distribution of microvasculature, namely the 'nerve and vessel wiring'. Thus, microvasculature may function to support the fundamental systems for the maintenance of multicellular organisms, *i.e.* blood circulating-, cell supplementing- and information processing- systems. Although this research field is gaining much attention for their potential importance in biological science and clinical application, the lack of an appropriate marker for MPCs impedes our understanding of their pathophysiological roles. Using the new marker, EphA7, capillary stem cells (CapSCs) can be isolated from crude PC fractions as a cell population with high regenerative potency. This chapter describes the role of MPCs, especially a new subpopulation of MPCs, CapSCs, in the microvascular functions to maintain multicellular organisms.

Keywords Pericytes · Endothelial cells · Mesenchymal stem cells · Neural stem cells · Satellite cells · Schwann cells · Capillary · Skeletal muscle · Angiogenesis · Neurogenesis · Myogenesis · EPH receptor · Ischemia · Muscular dystrophy

Multicellular organisms maintain their total numbers of cells by replenishing the cells that are continually losing during their lifespan, through a process referred to as "dynamic equilibrium." The main cells facilitating this cell replenishment are somatic stem cells and are, therefore, indispensable for multicellular organisms. Among the various types of somatic stem cells, mesenchymal stem cells (MSCs), which replenish mesenchymal cell components were first identified as cells that

J.-I. Kawabe (✉)
Department of Biochemistry, Division of Integrated Life Science, Asahikawa Medical University, Asahikawa, Japan
e-mail: kawabeju@asahikawa-med.ac.jp

A. Birbrair (ed.), *Biology of Pericytes – Recent Advances*, Stem Cell Biology and Regenerative Medicine 68, https://doi.org/10.1007/978-3-030-62129-2_8

dissociated from tissues in *in vitro* cell cultures. However, the precise locations of MSCs within tissues and their *in vivo* roles remain unclear.

Pericytes (PCs) are mural cells that are embedded within the basement membrane of microvessels, *i.e.*, capillaries, postcapillary venules, and terminal arterioles capillaries (Diaz-Flores et al. 2009). Accumulating evidence indicates that multipotent PCs (MPCs) are present in crude populations of PCs. These findings have impacted on somatic stem cell research. Based on the wide distribution of capillaries, MPCs are likely intermixed with MSCs and are considered to be fundamental cells among somatic stem cells (Caplan 2017). In addition, capillaries not only function as ducts for circulating blood but also play a role as stem cell reservoirs. Microvessels act as a reservoir, also termed the "vascular niche," for not only MPCs but also other kinds of stem cells, such as hematopoietic stem cells (HSCs) (Gao et al. 2018). Thus, MPCs may play an important role as the cells that maintain multicellular organisms, and they are gaining much attention for their potential importance in biological science and clinical applications. In this chapter, we provide an overview of recent research on MPCs, in particular the new MPCs that we recently discovered.

8.1 Mutent Cells Within the Microvasculature

8.1.1 Are MPCs the Major MSCs?

Mesoangioblasts (MABs) are perivascular stem cells, originally identified in the dorsal aorta of murine embryos, that have the ability to differentiate into mesenchymal cells (Minasi et al. 2002). MABs are thought to be the ancestors of postnatal somatic stem cells. Cells that propagate from the aorta and skeletal muscles of adult humans and other animals have characteristics similar to those of MABs, *i.e.*, multipotency like MSCs and/or neural stem cells (NSCs), and the expression of PC-specific markers (Sampaolesi et al. 2006; Dellavalle et al. 2007; Berry et al. 2007; Wang et al. 2012). In addition to MAB-like cells, other MPCs have been found in various other tissues, including the brain and adipose tissue (Farrington-Rock et al. 2004; Dore-Duffy et al. 2006; Crisan et al. 2008).

Previous studies have also identified populations of endogenous perivascular cells with MSC-like differentiation potential by lineage-tracing approaches using PDGFRβ- and NG2-Cre mouse models (Tang et al. 2008; Feng et al. 2011; Hosaka et al. 2016), although a recent study using lineage tracing of *Tbx18*-expressing cells have brought into question the current view of endogenous PCs as tissue-resident stem cells (Guimaraes-Camboa et al. 2017). Since MPCs are structural cells of the microvasculature and are widely distributed throughout tissues and organs, it is suggested that MSCs consist of a subset of perivascular PCs and MPCs contribute to MSC's multipotency (Caplan 2017).

It has been proposed that MPCs are defined by the expression of certain markers, such as melanoma cell adhesion molecule (CD146) and alkaline phosphatase (ALP), as well as the absence of the hematopoietic and endothelial markers CD45 and CD31 and the skeletal satellite cell marker CD56 (Dellavalle et al. 2007; Crisan et al. 2008; Corselli et al. 2010). However, most of these markers are not appropriate for distinguishing between MPCs and other PC populations (Armulik et al. 2011; Kelly-Goss et al. 2014; Cathery et al. 2018). Recently, Birbrair et al. reported nestin as a marker that can distinguish two PC sub-populations, *i.e.*, type 1 (nestin$^-$ PCs) and type 2 (nestin$^+$ PCs), in PCs prepared from nestin-promoter driven green fluorescent protein (GFP) transgenic mice (Birbrair et al. 2013a). Type 2 PCs have myogenic potency and participate in the regeneration of injured skeletal muscles. In contrast, type 1 PCs are fibrogenic and contribute to fibrous and adipose deposition in elderly mice (Birbrair et al. 2013b, c). These findings imply that tissue regeneration involves a balance of certain PC sub-populations and preparation of the appropriate PC subpopulation is crucial for clinical applications, such as regenerative medicine. However, nestin is located in an intracellular compartment and cannot be used as a marker for the isolation of targeted living cells from human and animal tissues. Thus far, there is no known molecular marker that can be used to identify or isolate living MPCs from heterogenic PC populations.

8.1.2 *Identification of Novel Multipotent PCs Using an Immortalized PC Line Library*

In general, it is difficult to analyze the characteristics of somatic stem cells as only one or a few cells are localized within certain tissue sites. Even if the targeted cells are isolated, subculture and propagation are usually required for further analysis, and these manipulations may alter their characteristics or induce cellular senescence. Although it is now possible to genetically analyze a single cell, cell functions such as differentiation potential cannot be investigated.

Transgenic mice harboring temperature-sensitive SV40 T-antigen (TmSV40T mice) have been utilized to establish immortalized cell lines to clarify the characteristics of cells that are difficult to prepare from native tissues (Obinata 1997). When the isolated cells were incubated at a lower temperature (33 °C), the cells were easily immortalized by activation of SV40T. The immortalized cells exhibited phenotypes similar to those of the original cells (Kondo et al. 2003; Shimizu et al. 2008), although the possibility of unexpected alterations in their characteristics during immortalization should be considered.

Recent studies have demonstrated the existence of stem/progenitor cells, including vasa vasorum-associated MPCs, within the adventitia of blood vessels (Bautch 2011; Majesky et al. 2011; Kawabe and Hasebe 2014). We established a collagen-coated tube (CCT) method to observe the growing microvessels around the adventitia of injured mouse femoral arteries (Asanome et al. 2014). The CCT method can also be used to prepare microvessel-rich samples in pathological settings. By using this method, we successfully established ten immortalized PC lines from a single microvascular cell isolated from a CCT sample of an SV40T mouse (Kabara et al. 2014).

Several of the obtained clonal PC lines have multipotencies that are similar to those of MABs, *i.e.*, differentiation into mesenchymal and neuronal cells and differentiation into vascular cells, including endothelial cells (ECs), to form capillary-like structures. Their genetic profiles, including the expression of known PC and MSC marker genes, and their differentiation potential were maintained during long-term subculture (Kabara et al. 2014). We performed a comprehensive comparison of the gene expression profiles of three PC lines, which showed different degrees of multipotency, to extract the genes related to multipotency. Among the candidate MPC marker genes, EPH receptor A7 (EphA7), a member of the ephrin receptor subfamily of protein tyrosine kinases, was selected (Yoshida et al. 2020). EphA7-expressed $NG2^+$ cells were observed in the microvessels of peripheral tissues, including subcutaneous adipose tissue and skeletal muscle. $EphA7^+$ PCs were successfully separated from crude PCs derived from these tissues by fluorescence-activated cell sorting using an anti-EphA7 antibody. $EphA7^+$ PCs have multipotency similar to that of immortalized MPCs and form sphere from a single cell (Kabara et al. 2014), whereas the $EphA7^-$ control PCs (ctPCs) show no multipotency. Importantly, there are no differences in any known MSCs, HSCs and PC markers, including CD90, CD105, CD34, and CD146, between $EphA7^+$ PCs and ctPCs (Yoshida et al. 2020).

8.2 Role of $EphA7^+$ PCs in the Multicellular Organisms

8.2.1 Angiogenesis: Do $EphA7^+$ PCs Function as Capillary Progenitor Cells?

In multicellular organisms, the growth of blood vessels, induced via vasculogenesis or angiogenesis, is essential during embryonic development, postnatal growth, and tissue repair. Regardless of the angiogenic mechanism, ECs are indispensable structural cells in blood vessels. Various endothelial progenitor cells (EPCs) have been identified in several tissues, including bone marrow and the endothelial layer of blood vessels (Takahashi et al. 1999; Naito et al. 2012). ECs have the potential to form tube-like structures (EC tubes), which are structurally unstable, immature vessels. PCs are then recruited to cover the EC tubes in the formation of mature, functional blood vessels (Jain 2003).

$EphA7^+$ PCs express PC-specific genes but do not express EC- or bone marrow-derived cell-specific genes (Yoshida et al. 2020). They have characteristics similar to EPCs, *i.e.*, differentiation into ECs in the presence of vascular endothelial growth factor. In three-dimensional (3D) gel culture, clonal EphA7$^+$ PCs form EC tubes, which are simultaneously covered by the remaining cells and finally form capillary-like structures (Kabara et al. 2014; Yoshida et al. 2020). Transplantation of EphA7$^+$ PCs into mouse ischemic limbs enhances angiogenesis and improves recovery from ischemia during reperfusion. Transplanted cells differentiate into blood vessel cells, including ECs and PCs, to form functional microvessels and remain in the tissue for a long time (Yoshida et al. 2020). Therefore, EphA7$^+$ PCs function not only as EPCs but also as so-called PCs to form functional capillary-like structures, which are the smallest units of blood vessels *in vivo*. Accordingly, we termed these cells capillary stem cells (CapSCs) (Fig. 8.1). Several questions remain to be answered: How do CapSCs form capillary-like organs composed of different cell types, *i.e.*, EC tubes surrounded by PCs? Are transplanted CapSCs maintained as multipotent CapSCs *in vivo* when they are located at perivascular sites?

8.2.2 Neurogenesis: Do EphA7$^+$ PCs Play a Central Role in Nerve Vessel Wiring?

The peripheral nervous system (PNS), which is distributed throughout the body, is involved in processing information between the central nervous system (CNS) and the body. Similar to the blood vessel system, which circulates hormones/cytokines and immune cells through the entire body, the PNS is an indispensable information system that allows multicellular components to function as a single organism. To elucidate the mechanisms by which the peripheral nerve network is constructed and maintained, many researchers have taken hints from the parallel branching patterns of blood vessels and nerves, namely, the "nerve and vessel wiring" that was documented by anatomists over five centuries ago (Carmeliet and Tessier-Lavigne 2005). It is also well documented that angiogenesis and neurogenesis are intimately linked under postnatal pathophysiological conditions and during embryonic development (Eichmann and Thomas 2013). Angiogenesis is critical for neural regeneration in the CNS and diseases of the PNS (Muramatsu et al. 2012; Cattin et al. 2015; Tomita et al. 2019). Conversely, the redistribution of nerve axons beside microvessels is crucial for the maturation of neovessels in the forming functional vessels (Asanome et al. 2014; Li et al. 2013).

The microvessels formed at the site of injured nerves are required to provide a scaffold for migrating Schwann cells (SchCs) and are necessary for the regeneration of peripheral nerves (Cattin et al. 2015). SchCs are crucial peripheral glial cells that provide trophic support for the extension and maintenance of peripheral nerve axons. Although it is well documented that SchCs originate from de-differentiated pre-existing SchCs in response to peripheral nerve injury (Menorca et al. 2013), the origin of the peripheral SchC lineage is still uncertain. A certain population of PCs, not only in the CNS but also in the peripheral tissues, functions as NSCs

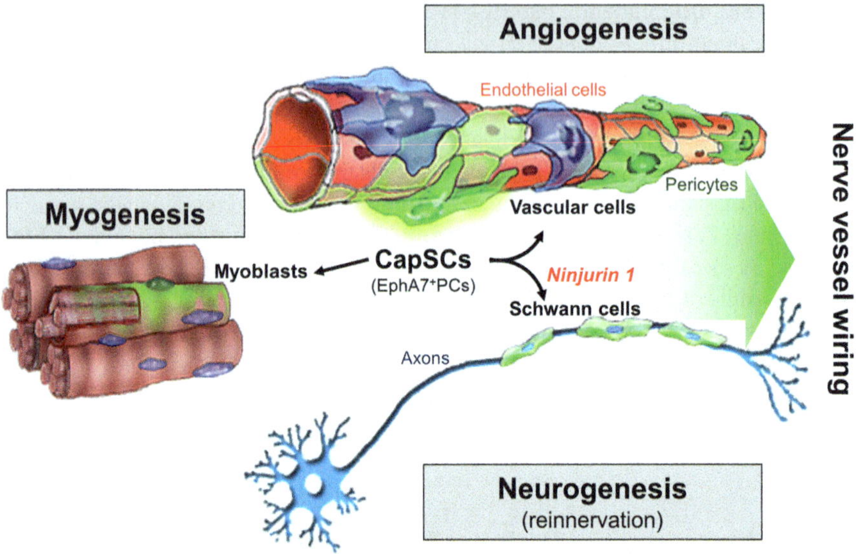

Fig. 8.1 Role of multipotent PCs in nerve and vessel wiring and muscular regeneration
Peripheral nerves are distributed throughout the tissues and are either unmyelinated or myelinated by Schwann cells. Multipotent PCs, such as capillary stem cells (CapSCs), can differentiate into Schwann cells and vascular cells to form the microvasculature. Thus, it was hypothesized that the wiring of the nerve axon along the vessels is mediated by perivascular CapSCs, which are the precursors of Schwann cells. Ninjurin 1 is a key molecule for neurogenesis of CapSCs, contributes to peripheral reinnervation via CapSC's ability to differentiate to Schwann cells. CapSCs have potent muscular regeneration effect via direct myogenic differentiation and microvascular and nerve network formations

(Dore-Duffy et al. 2006; Yoshida et al. 2020; Nakagomi et al. 2015; Birbrair et al. 2013d). CapSCs, which exist in the microvessels of peripheral tissues, have a high ability to differentiate into both vascular cells and neural cells, including SchCs (Yoshida et al. 2020; Tomita et al. 2019). CapSCs are highly nestin-expressing PCs and may be of a similar subpopulation as the nestin$^+$ PCs (type 2 PCs) as reported by Birbrair et al. (2013a). Nestin is a marker of cells originating in the neural crest, which can transdifferentiate across germ layers, *i.e.*, both the mesoderm and ectoderm (Morikawa et al. 2009). To explain why CapSCs located at the perivascular sites of the peripheral tissues have a high ability to differentiate into SchCs, we proposed a hypothesis for the role of CapSCs in the formation of nerve vessel wiring (Fig. 8.1).

Ninjurin1 (Ninj1) was discovered as an adhesion molecule that was induced in the peripheral nerve tissue in response to injury (Araki and Milbrandt 1996). However, the role of Ninj1 in the PNS, especially in neural regeneration, remains unclear. For example, there is no difference in the regeneration of injured nerves in classical Ninj1-knockout mice when compared to that in control mice (*personal communications with Dr Araki*). Recently, we reported that Ninj1 is expressed in PCs and is involved in microvessel formation via the interaction between ECs and

PCs (Matsuki et al. 2015; Minoshima et al. 2018). Ninj1 is also expressed in CapSCs and strongly contributes to their neurogenesis, differentiation into SchCs, and their myelination (Tomita et al. 2019). CapSCs are abundant in the growing microvessels of injured sciatic nerves (Tomita et al. 2019). When Ninj1 gene was deleted in the PCs of NG2 promoter-driven Cre transgenic mice before nerve injury, peripheral nerve regeneration was significantly reduced and the myelination of the regenerated nerves was attenuated. This shows that Ninj1 plays a critical role in neuroregeneration and that CapSCs contribute to peripheral reinnervation through their ability to differentiate into SchCs during the formation of the microvasculature.

8.2.3 *Myogenesis: Do EphA7+ PCs Have Self-Regenerative Potency?*

Skeletal muscle is essential for mobility and metabolic homeostasis and is maintained by constitutive muscular regeneration throughout the life of an organism. Multinucleated myofibers are generated from the iterative fusion of mononucleated myoblasts. Satellite cells (SatCs) are well-known muscle resident myogenic stem cells that support the high regenerative ability of skeletal muscle (Yin et al. 2013). Recent studies using an inducible deletion of SatCs showed that these stem cells are not required to maintain muscle mass throughout the lifespan of an organism (Fry et al. 2015; Keefe et al. 2015). These data imply a new concept, that the regenerative potential of the skeletal muscle is maintained by a heterogenous population of stem/progenitor cells (Yin et al. 2013; Klimczak et al. 2018). Microvascular PCs, including MABs, are important candidate non-SatC myogenic stem cells (Sampaolesi et al. 2006; Dellavalle et al. 2007; Birbrair et al. 2013a). A certain population of PCs contributes to the regeneration of skeletal myofibers, whereas other populations are related to abnormal muscle remodeling, such as fibrotic and fatty metamorphoses (Birbrair et al. 2013c). Thus, it is important to clarify the characteristics of the PC subpopulations and their role in muscular regeneration.

In addition to their multipotency in both microvasculature reconstruction and nerve regeneration, CapSCs also have high muscular regeneration ability. CapSCs transplanted in cardiotoxin-injured muscles are observed in the regenerated myotubes and microvasculature. Importantly, EphA7$^-$ ctPCs induce fibrotic deposition in damaged muscles (Kano et al. 2020). The regenerative effects of CapSCs are also apparent in dystrophin- and utrophin-deficient double knockout (mdx/utrn$^{-/-}$) mice, a phenotypic model for Duchenne muscular dystrophy (Hayashiji et al. 2015) in which endogenous muscle stem cells are exhausted. Compared to the effects of known MSCs, GFP-expressing CapSCs, when engrafted into the muscle of mdx/utrn$^{-/-}$ mice, form GFP-positive muscle fibers that express dystrophin, resulting in improved muscular mass and performance (Kano et al. 2020).

An ideal regenerative therapy for inheritable muscle dystrophy requires at least two conditions: increased regeneration of non-fragile myocytes and maintenance of the pool of myogenic stem cells possessing intact genes within skeletal muscles. In addition to their direct contribution to muscular regeneration, CapSCs have another specific function, differentiation into vascular cells to form microvessels in regenerated tissue (Yoshida et al. 2020; Kano et al. 2020) (Fig. 8.1). The microvasculature is a specific niche for the maintenance of stem cells (Mendelson and Frenette 2014; Morrison and Scadden 2014; Khan et al. 2016). Multipotent CapSCs are originally present in perivascular sites as mural cells (Yoshida et al. 2020; Tomita et al. 2019). Therefore, it is postulated that transplanted CapSCs, which are maintained as myogenic stem cells at perivascular sites, contribute to self-myogenic regeneration for their remaining life span. Whether transplanted CapSCs within microvessels function as stem cells for myogenic regeneration over a relatively long period should be investigated.

8.3 Future Perspectives on MPCs Research

In summary, the following three systems are indispensable for the maintenance of a multicellular organism: (1) a vascular system that delivers nutrients and oxygen to every cell of the entire body; (2) a cell supplementation system that replenishes cells that are continually lost throughout life, which likely involves somatic stem cells and the microvasculature that acts as niche for these stem cells; and (3) the information-linkage and -processing systems that make it possible for the cells to function together as a single living organism. These systems include the vascular system, which circulates hormones, cytokines, and blood cells in the entire body; the neural network, which is key for the relatively quick communication system present in advanced organisms; and the microvasculature, which also functions as a guide and helps maintain the nerve network. Accordingly, the microvasculature may function to support these fundamental systems. CapSCs play an important role in these microvascular functions and, thus, may be fundamental for the maintenance of multicellular organisms (Fig. 8.2).

Although MPCs have been isolated from several human tissues and show promise as therapeutic tools and targets for clinical applications (Perin et al. 2015), some basic scientific issues, *i.e.*, the characteristics of MPCs and their roles in multicellular organisms, still remain to be clarified. In addition, a standardized isolation protocol is needed that is less invasive for the patient and yields a well-characterized, highly pure cell population. Using the new marker, EphA7, CapSCs can be isolated from crude PC fractions as a cell population with high regenerative potency. Importantly, this isolation method can eliminate the PC subpopulation mediating unfavorable effects, such as fibrotic and adipose depositions. CapSCs can be prepared from subcutaneous adipose tissues, which is a less invasive tissue than the skeletal muscle, bone marrow, or the aorta.

Further investigations focused on the characteristics and pathophysiological roles of exogenous and endogenous CapSCs are required to fully realize their

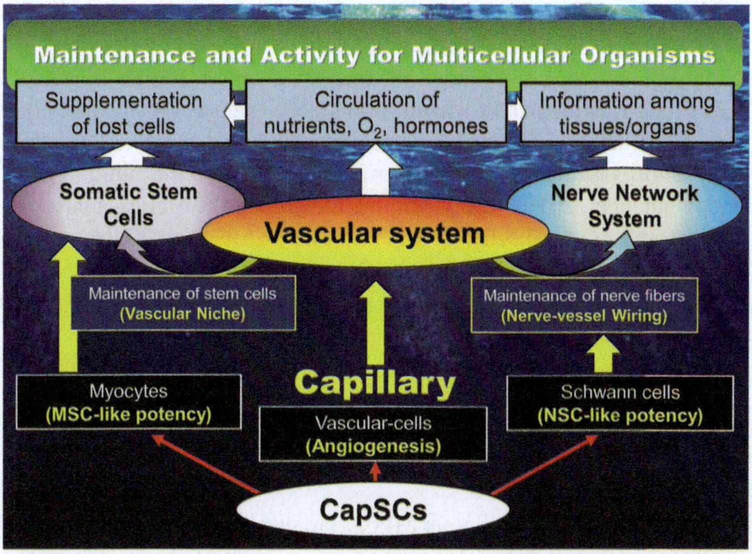

Fig. 8.2 Role of the microvasculature and CapSCs in multicellular organisms
Because most of the life history of all species was spent in the sea, a vascular system is indispensable for maintaining each cell of a multicellular organism in a "sea-like" environment, similar to what was once their home. A second condition that is required for a multicellular organism is the existence of somatic stem cells. Microvasculature functions as major niche for these cells. A third condition is that various information processing networks, such as the nervous system, are required for the cells to act as a single living body. A vascular system is crucial to construct and maintain this network. CapSCs play an important role in these microvascular systems and, thus, may be fundamental for the function and maintenance of multicellular organisms

therapeutic potential. There are several points to consider when clarifying the characteristics of MPCs.

8.3.1 Specificities of PC-Markers

PCs are mural cells that are embedded within the basement membrane of microvessels, *i.e.*, capillaries, postcapillary venules, and terminal arterioles (Diaz-Flores et al. 2009). PCs are often defined by the expression of specific markers, such as platelet-derived growth factor receptor β (PDGFRβ) and neuron-glia 2 (NG2) (Armulik et al. 2011; Cathery et al. 2018; van Dijk et al. 2015). It is necessary to consider the specificity of PCs, as the PC-specific gene profile may change, depending on the patho-physiological conditions and the kinds of tissue or organ in which the microvasculature is located (Vanlandewijck et al. 2018). Alternatively, known PC marker genes may be also expressed in other cell types, although their expression levels are different. For instance, NG2 is expressed not only in PCs but also in other cells, including vascular smooth muscle cells (VSMCs), macrophages, and non-PC neural stem cells (Stallcup 2018).

Owing to the absence of a unique PC marker, a combination of several markers should be used to distinguish PCs from similar cells, such as VSMCs and fibroblasts. A recent study using single-cell analysis provided molecular definitions for the major types of vascular and vessel-associated cells in the adult mouse brain (Vanlandewijck et al. 2018). Transcriptome hierarchical clustering analysis showed two major groups of mural cells, arterial/arteriole VSMCs and PCs/venous VSMCs. Accordingly, several genes, such as *Abcc9* and *Kcnj1*, have been proposed as brain PC-specific markers. Specific antibodies and reporter mice for these candidate marker genes would be useful for studying PCs.

8.3.2 Regulators of MPCs Function

In addition to the use of exogenous MPCs, targeting endogenous MPCs is a less-invasive, more universal option for clinical approaches. However, the mechanisms that regulate the function of MPCs remain to be clarified before the development of clinically valuable drugs. EphA7 and Ninj1 are critical molecules that mediate the angiogenic and neurogenic potencies of CapSCs (Tomita et al. 2019). Eph-ephrin signaling regulates cell assembly and mediates angiogenesis and stem cell function during development and in adulthood (Hellstrom et al. 2007; Pasquale 2008; Genander and Frisen 2010). Ninj1 also functions as an adhesion molecule mediating cell-to-cell interactions and pathophysiological functions (Matsuki et al. 2015; Araki et al. 1997). Thus, further studies will need to investigate the partner cells that interact with CapSCs and the intra- and inter-cellular signaling pathways that involve these molecules.

8.3.3 Transgenic Animal Models Targeting MPCs

It should be noted that the cellular characteristics of exogenous MPCs are modified by the *in vitro* culture conditions. To examine the role of MPCs under pathophysiological conditions, targeted cell-specific gain/loss of function and lineage-tracing experiments using genetically modified animals would be extremely useful. For example, cell marker gene promoter-driven *Cre-loxP* mice have been frequently used to target certain cells. However, as discussed above, no single marker gene can identify PCs, even MPCs. Indeed, CapSCs are identified by the expression of at least two markers, *i.e.*, EphA7 and a PC marker such as NG2. Recently, dual recombinase systems, such as *Nigri-nox/Cre-loxP*, have been used to target certain cells with two markers (Liu et al. 2018; Li et al. 2018). In the future, new genetically modified mouse models will be required to analyze the functions of MPCs.

Acknowledgement I would like to thank the labo members, especially M. Kabara, Y. Yoshida, K. Kano, Y. Tomita, K. Horiuchi, A. Minoshima, T. Matsuki and also Dr. N. Hasebe and T. Araki for useful comments. This work was supported by grants JSPS KAKENHI (17H04170, 17 K19368, 18K16379, 19K16969, 19K16905), and in part by Daiichi Sankyo Co. Ltd., and Asbio Pharmer Co. Ltd. And OideCapiSEA, Inc.

References

Araki T, Milbrandt J (1996) Ninjurin, a novel adhesion molecule, is induced by nerve injury and promotes axonal growth. Neuron 17(2):353–361

Araki T, Zimonjic DB, Popescu NC, Milbrandt J (1997) Mechanism of homophilic binding mediated by ninjurin, a novel widely expressed adhesion molecule. J Biol Chem 272(34):21373–21380

Armulik A, Genove G, Betsholtz C (2011) Pericytes: developmental, physiological, and pathological perspectives, problems, and promises. Dev Cell 21(2):193–215

Asanome A et al (2014) Nerve growth factor stimulates regeneration of perivascular nerve, and induces the maturation of microvessels around the injured artery. Biochem Biophys Res Commun 443(1):150–155

Bautch VL (2011) Stem cells and the vasculature. Nat Med 17(11):1437–1443

Berry SE, Liu J, Chaney EJ, Kaufman SJ (2007) Multipotential mesoangioblast stem cell therapy in the mdx/utrn−/− mouse model for Duchenne muscular dystrophy. Regen Med 2(3):275–288

Birbrair A et al (2013a) Skeletal muscle pericyte subtypes differ in their differentiation potential. Stem Cell Res 10(1):67–84

Birbrair A et al (2013b) Role of pericytes in skeletal muscle regeneration and fat accumulation. Stem Cells Dev 22(16):2298–2314

Birbrair A et al (2013c) Type-1 pericytes participate in fibrous tissue deposition in aged skeletal muscle. Am J Physiol Cell Physiol 305(11):C1098–C1113

Birbrair A et al (2013d) Skeletal muscle neural progenitor cells exhibit properties of NG2-glia. Exp Cell Res 319(1):45–63

Caplan AI (2017) New MSC: MSCs as pericytes are sentinels and gatekeepers. J Orthop Res 35(6):1151–1159

Carmeliet P, Tessier-Lavigne M (2005) Common mechanisms of nerve and blood vessel wiring. Nature 436(7048):193–200

Cathery W, Faulkner A, Maselli D, Madeddu P (2018) Concise review: the regenerative journey of pericytes toward clinical translation. Stem Cells 36(9):1295–1310

Cattin AL et al (2015) Macrophage-induced blood vessels guide Schwann cell-mediated regeneration of peripheral nerves. Cell 162(5):1127–1139

Corselli M, Chen CW, Crisan M, Lazzari L, Peault B (2010) Perivascular ancestors of adult multipotent stem cells. Arterioscler Thromb Vasc Biol 30(6):1104–1109

Crisan M et al (2008) A perivascular origin for mesenchymal stem cells in multiple human organs. Cell Stem Cell 3(3):301–313

Dellavalle A et al (2007) Pericytes of human skeletal muscle are myogenic precursors distinct from satellite cells. Nat Cell Biol 9(3):255–267

Diaz-Flores L et al (2009) Pericytes. Morphofunction, interactions and pathology in a quiescent and activated mesenchymal cell niche. Histol Histopathol 24(7):909–969

Dore-Duffy P, Katychev A, Wang X, Van Buren E (2006) CNS microvascular pericytes exhibit multipotential stem cell activity. J Cereb Blood Flow Metab 26(5):613–624

Eichmann A, Thomas JL (2013) Molecular parallels between neural and vascular development. Cold Spring Harb Perspect Med 3(1):a006551

Farrington-Rock C et al (2004) Chondrogenic and adipogenic potential of microvascular pericytes. Circulation 110(15):2226–2232

Feng J, Mantesso A, De Bari C, Nishiyama A, Sharpe PT (2011) Dual origin of mesenchymal stem cells contributing to organ growth and repair. Proc Natl Acad Sci USA 108(16):6503–6508
Fry CS et al (2015) Inducible depletion of satellite cells in adult, sedentary mice impairs muscle regenerative capacity without affecting sarcopenia. Nat Med 21(1):76–80
Gao X, Xu C, Asada N, Frenette PS (2018) The hematopoietic stem cell niche: from embryo to adult. Development 145(2)
Genander M, Frisen J (2010) Ephrins and Eph receptors in stem cells and cancer. Curr Opin Cell Biol 22(5):611–616
Guimaraes-Camboa N et al (2017) Pericytes of multiple organs do not behave as mesenchymal stem cells in vivo. Cell Stem Cell 20(3):345–359. e345
Hayashiji N et al (2015) G-CSF supports long-term muscle regeneration in mouse models of muscular dystrophy. Nat Commun 6:6745
Hellstrom M et al (2007) Dll4 signalling through Notch1 regulates formation of tip cells during angiogenesis. Nature 445(7129):776–780
Hosaka K et al (2016) Pericyte-fibroblast transition promotes tumor growth and metastasis. Proc Natl Acad Sci USA 113(38):E5618–E5627
Jain RK (2003) Molecular regulation of vessel maturation. Nat Med 9(6):685–693
Kabara M et al (2014) Immortalized multipotent pericytes derived from the vasa vasorum in the injured vasculature. A cellular tool for studies of vascular remodeling and regeneration. Lab Investig 94(12):1340–1354
Kano et al (2020) EphA7(+) perivascular cells as myogenic and angiogenic precursors improving skeletal muscle regeneration in a muscular dystrophic mouse model. Stem Cell Res 47:101914
Kawabe J, Hasebe N (2014) Role of the vasa vasorum and vascular resident stem cells in atherosclerosis. Biomed Res Int 2014:701571
Keefe AC et al (2015) Muscle stem cells contribute to myofibres in sedentary adult mice. Nat Commun 6:7087
Kelly-Goss MR, Sweat RS, Stapor PC, Peirce SM, Murfee WL (2014) Targeting pericytes for angiogenic therapies. Microcirculation 21(4):345–357
Khan JA et al (2016) Fetal liver hematopoietic stem cell niches associate with portal vessels. Science 351(6269):176–180
Klimczak A, Kozlowska U, Kurpisz M (2018) Muscle stem/progenitor cells and mesenchymal stem cells of bone marrow origin for skeletal muscle regeneration in muscular dystrophies. Arch Immunol Ther Exp 66(5):341–354
Kondo T et al (2003) Establishment of conditionally immortalized rat retinal pericyte cell lines (TR-rPCT) and their application in a co-culture system using retinal capillary endothelial cell line (TR-iBRB2). Cell Struct Funct 28(3):145–153
Li W et al (2013) Peripheral nerve-derived CXCL12 and VEGF-A regulate the patterning of arterial vessel branching in developing limb skin. Dev Cell 24(4):359–371
Li Y et al (2018) Genetic lineage tracing of nonmyocyte population by dual recombinases. Circulation 138(8):793–805
Liu K et al (2018) A dual genetic tracing system identifies diverse and dynamic origins of cardiac valve mesenchyme. Development 145(18)
Majesky MW, Dong XR, Hoglund V, Mahoney WM Jr, Daum G (2011) The adventitia: a dynamic interface containing resident progenitor cells. Arterioscler Thromb Vasc Biol 31(7):1530–1539
Matsuki M et al (2015) Ninjurin1 is a novel factor to regulate angiogenesis through the function of pericytes. Circ J 79(6):1363–1371
Mendelson A, Frenette PS (2014) Hematopoietic stem cell niche maintenance during homeostasis and regeneration. Nat Med 20(8):833–846
Menorca RM, Fussell TS, Elfar JC (2013) Nerve physiology: mechanisms of injury and recovery. Hand Clin 29(3):317–330
Minasi MG et al (2002) The meso-angioblast: a multipotent, self-renewing cell that originates from the dorsal aorta and differentiates into most mesodermal tissues. Development 129(11):2773–2783

Minoshima A et al (2018) Pericyte-specific Ninjurin1 deletion attenuates vessel maturation and blood flow recovery in hind limb ischemia. Arterioscler Thromb Vasc Biol 38(10):2358–2370
Morikawa S et al (2009) Development of mesenchymal stem cells partially originate from the neural crest. Biochem Biophys Res Commun 379(4):1114–1119
Morrison SJ, Scadden DT (2014) The bone marrow niche for haematopoietic stem cells. Nature 505(7483):327–334
Muramatsu R et al (2012) Angiogenesis induced by CNS inflammation promotes neuronal remodeling through vessel-derived prostacyclin. Nat Med 18(11):1658–1664
Naito H, Kidoya H, Sakimoto S, Wakabayashi T, Takakura N (2012) Identification and characterization of a resident vascular stem/progenitor cell population in preexisting blood vessels. EMBO J 31(4):842–855
Nakagomi T et al (2015) Brain vascular pericytes following ischemia have multipotential stem cell activity to differentiate into neural and vascular lineage cells. Stem Cells 33(6):1962–1974
Obinata M (1997) Conditionally immortalized cell lines with differentiated functions established from temperature-sensitive T-antigen transgenic mice. Genes Cells 2(4):235–244
Pasquale EB (2008) Eph-ephrin bidirectional signaling in physiology and disease. Cell 133(1):38–52
Perin EC et al (2015) A phase II dose-escalation study of allogeneic mesenchymal precursor cells in patients with ischemic or nonischemic heart failure. Circ Res 117(6):576–584
Sampaolesi M et al (2006) Mesoangioblast stem cells ameliorate muscle function in dystrophic dogs. Nature 444(7119):574–579
Shimizu F et al (2008) Peripheral nerve pericytes originating from the blood-nerve barrier expresses tight junctional molecules and transporters as barrier-forming cells. J Cell Physiol 217(2):388–399
Stallcup WB (2018) The NG2 proteoglycan in Pericyte biology. Adv Exp Med Biol 1109:5–19
Takahashi T et al (1999) Ischemia- and cytokine-induced mobilization of bone marrow-derived endothelial progenitor cells for neovascularization. Nat Med 5(4):434–438
Tang W et al (2008) White fat progenitor cells reside in the adipose vasculature. Science 322(5901):583–586
Tomita Y et al (2019) Ninjurin 1 mediates peripheral nerve regeneration through Schwann cell maturation of NG2-positive cells. Biochem Biophys Res Commun 519(3):462–468
van Dijk CG et al (2015) The complex mural cell: pericyte function in health and disease. Int J Cardiol 190:75–89
Vanlandewijck M et al (2018) Author correction: a molecular atlas of cell types and zonation in the brain vasculature. Nature 560(7716):E3
Wang L et al (2012) Aorta-derived mesoangioblasts differentiate into the oligodendrocytes by inhibition of the Rho kinase signaling pathway. Stem Cells Dev 21(7):1069–1089
Yin H, Price F, Rudnicki MA (2013) Satellite cells and the muscle stem cell niche. Physiol Rev 93(1):23–67
Yoshida Y et al (2020) Capillary-resident EphA7(+) pericytes are multipotent cells with anti-ischemic effects through capillary formation. Stem Cells Transl Med 9(1):120–130

Chapter 9
Skeletal Muscle-Resident Pericyte Responses to Conditions of Use and Disuse

Michael Munroe, Svyatoslav Dvoretskiy, and Marni D. Boppart

Abstract Skeletal muscle structure and function are dependent on the presence or absence of mechanical cues in the tissue microenvironment. High intensity or sustained contractions can build muscle mass with exercise training, whereas bedrest or limb immobilization result in loss of muscle mass. Molecular signaling pathways, such as mTORC1 and induction of activating transcription factor 4 (ATF4), regulate intrinsic myofiber growth and atrophy respectively, yet the extent to which mononuclear cells outside the fiber contribute to muscle structure remodeling remains unclear. The skeletal muscle microenvironment is enriched with mononuclear cells, including progenitor and stem cells, fibroblasts, immune cells, and vascular stromal cells. All possess the potential to sense mechanical cues and contribute to structural remodeling, but the intricate positioning of vascular stromal cells around vessels make these cells particularly receptive to changes in mechanical stimuli due to alterations in both blood perfusion and contraction. Recent studies suggest that pericytes, a population of vascular stromal cells supporting capillaries and venules, are highly responsive to conditions of use and disuse. This chapter discusses the characteristics and behavior of skeletal muscle-derived pericytes as well as the specific responses to physiological stimuli with which muscle can be subjected. Determining how pericytes behave in situations of muscle activation and immobilization is important for the development of cell-based therapies for the treatment of skeletal muscle disorders.

Keywords Pericyte · Skeletal muscle · Activity · Disuse · Atrophy · Contraction · CD146 · NG2 · Injury · Recovery · Microvessels · Capillaries · Immobilization · ECM remodeling · Stromal cell

---

M. Munroe
Department of Medicine, Washington University in St. Louis, St. Louis, MO, USA

S. Dvoretskiy · M. D. Boppart (✉)
Department of Kinesiology and Community Health, University of Illinois at Urbana-Champaign, Urbana, IL, USA

Beckman Institute for Advanced Science and Technology, University of Illinois at Urbana-Champaign, Urbana, IL, USA
e-mail: mboppart@illinois.edu

A. Birbrair (ed.), *Biology of Pericytes – Recent Advances*, Stem Cell Biology and Regenerative Medicine 68, https://doi.org/10.1007/978-3-030-62129-2_9

9.1 Skeletal Muscle Mononuclear Cells

Human skeletal muscle is one of the largest tissue types within the body. Comprising approximately 40% of total body mass, muscle is responsible for multiple vital functions including force production and movement, metabolic substrate storage and use, organ support and stabilization, thermoregulation, and cytokine and growth factor production (Frontera and Ochala 2015; Karstoft and Pedersen 2016).

Skeletal muscle originates from the embryonic paraxial mesoderm with myogenic progenitor cells arising from the dermomyotome (Chal and Pourquié 2017; Endo 2015). These progenitor cells migrate, proliferate, differentiate into myoblasts, and finally fuse together to form myotubes which continue to mature into post-mitotic, multinucleated myofibers (Kim et al. 2015; Petrany and Millay 2019). Myofibers are bundled together along with the necessary vascularization, innervation, and extracellular matrix (ECM) support required for muscle growth and function (Gillies and Lieber 2011). Myonuclei localize to the periphery of individual myofibers and are responsible for its transcriptional output, primarily expression of myofibril (actin and myosin) and sarcomeric proteins (Kirby et al. 2016; Newlands et al. 1998; Pavlath et al. 1989).

Additional mononuclear cells reside adjacent to myofibers that can aid in overall repair and remodeling in adult skeletal muscle (Costamagna et al. 2015). The primary myogenic stem/progenitor cell is the satellite cell ($Pax7^+$; SC), which resides between the basal lamina and the myofiber sarcolemma (Forcina et al. 2019). While normally mitotically quiescent (i.e. undifferentiated) and transcriptionally inactive, in response to injury and myotrauma, SCs are activated and proliferate, co-expressing myogenic regulatory factors (MRFs), specifically MyoD and Myf5 (Brooks and Myburgh 2014; Yin et al. 2013). These activated cells have two possible fates: continue to express MRFs (MyoD and Myogenin) and thus commit to the myogenic lineage by differentiating into myoblasts, or, retain Pax7 expression and remain quiescent SCs (Dumont et al. 2015; Hawke and Garry 2001). Differentiated myoblasts either fuse with adjacent myoblasts to form a new myofiber, or in the case of injury, fuse with the damaged myofiber to allow for repair and recovery (Goh and Millay 2017; Relaix and Zammit 2012; Sambasivan et al. 2011). In addition to SCs, various other cell populations reside in close proximity to myofibers such as immune cells, $PDGFR\alpha^+$ fibroadipogenic progenitors (FAPs), $Tcf4^+$ fibroblasts, $Abcg2^+$ side population cells, and $NG2^+CD146^+$ pericytes (Boppart et al. 2013) (Fig. 9.1). Each of these populations present with a heterogenous mixture of nuclear and cell surface markers that do not overlap with SC marker expression (Pax7/CD56), making them distinguishable from SCs (Costamagna et al. 2015). Non-SC populations demonstrate some capacity for differentiation into mesodermal lineages (i.e., myogenic, osteogenic, chondrogenic, or adipogenic) in response to injury, yet their primary function in muscle appears to be stromal support (Caplan 2010; Jankowski et al. 2002; Otto et al. 2009). The direct and indirect contributions of non-SCs to muscle repair, regeneration and remodeling is not completely clear and additional studies are necessary to address the full extent of their therapeutic potential.

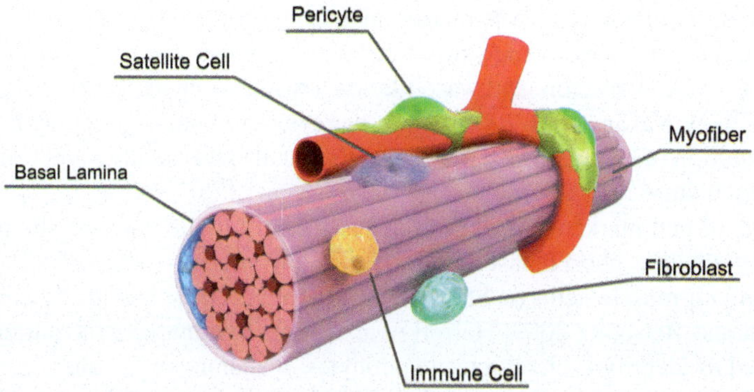

Fig. 9.1 Stem and stromal cells in the skeletal muscle microenvironment. The skeletal muscle microenvironment is enriched with stem and stromal cells, including satellite cells, pericytes, and fibroblasts. Satellite cells, localized underneath the basal lamina, are essential for repair and regeneration of muscle in response to injury. Fibroblasts (or fibroadipogenic progenitors), localized to the interstitium, support matrix turnover and indirectly influence satellite cell function. Pericytes share a basement membrane with endothelial cells in microvessels and regulate vascular function, including vasodilation/vasoconstriction and angiogenesis. The extent to which pericytes respond to muscle use (contraction) and disuse (inactivity) and ultimately impact myofiber structure and function is largely unexplored

9.2 Skeletal Muscle Pericytes

Skeletal muscle is an energetically demanding tissue, requiring significant vascularization to ensure proper nutrient deposition, gas exchange, and dissipation of metabolic heat. Pericytes and other vascular associated cells serve to regulate vessel structure and function, and are also appropriately dense within skeletal muscle. Pericytes possess cytoplasmic processes which extend across multiple endothelial cells of capillaries and venules, making direct, physical contact with endothelial cells via peg-socket contacts, gap junctions, and adhesion plaques. The unique structure of the pericyte makes them morphologically identifiable with electron microscopy (Armulik et al. 2005; Díaz-Flores et al. 2009). Pericytes are present in nearly all tissues of the body, yet pericyte density is highly variable between tissue types. Whereas the pericyte to endothelial cell ratio is 1:1 in the brain, the ratio is believed to range 1:10–1:100 in skeletal muscle (Armulik et al. 2011). Functionally, pericytes are required for local angiogenesis and blood flow regulation by providing critical structural and paracrine support to adjacent vessels. Angiogenesis is a highly complex, multi-step process that is mediated in part by pericyte activity. After detaching from their associated blood vessels, pericytes assist in the degradation and recycling of the local basement membrane, helping to initiate angiogenesis (Díaz-Flores et al. 1992). Pericytes must proliferate and migrate to the new capillary branches arising from vascular sprouting and maturation in order to re-stabilize the new microvessels (Díaz-Flores et al. 2009). Completion of these steps requires pericytes to synthesize and release various ECM remodeling proteins as well as growth factors that can interact in a paracrine manner with the local microenvironment. Post-angiogenesis, pericytes are key to controlling blood flow rates via their modulation of vasoconstriction and vasodilation (Armulik et al. 2005; Kutcher and Herman 2009; Nakano et al. 2000). Thus, pericytes are vital for not only the growth and maturation of newly formed microvessels, but also the stability and contractility of the vasculature.

A host of cell markers have been used to identify tissue-non-specific pericytes, including alkaline phosphatase (ALP/AP), PDGF receptor-beta (PDGFR-β), and alpha-smooth muscle actin (α-SMA) (Armulik et al. 2011; Dellavalle et al. 2011; Gerhardt and Betsholtz 2003; Harrell et al. 2018). Two markers have traditionally been used to identify skeletal muscle pericytes: melanoma cell adhesion molecule (MCAM; CD146) (Cappellari and Cossu 2013; Crisan et al. 2008; Sacchetti et al. 2016) and chondroitin sulfate proteoglycan, neural-glial antigen 2 (CSPG4; NG2) (Birbrair et al. 2013; Kostallari et al. 2015). Cells expressing either or both cell surface markers, and lacking expression of endothelial (CD31, CD34) and immune/hematopoietic (CD45) cell lineages, are closely associated with a traditional pericyte phenotype. Thus, these markers are highly useful in the identification and characterization of pericytes, especially within skeletal muscle.

Characterization of adult muscle pericytes has been completed by several groups. Pericytes derived from muscle are expandable *in vitro*, karyotypically normal, and enter proliferative senescence after approximately 15–20 population doublings (Crisan et al. 2008; Dellavalle et al. 2007). Gene expression profiles of muscle-derived pericytes are enriched for factors related to muscle development, muscle contraction, and energy metabolism (Sacchetti et al. 2016). Pericytes retain pericyte cell surface marker expression and remain distinct from SCs (i.e. $Pax7^-$, $MyoD^-$, $Myogenin^-$) during routine *in vitro* culture (Crisan et al. 2008; Dellavalle et al. 2007). However, human adult pericyte-derived clones spontaneously differentiate into myosin-positive myotubes (~20–40%) upon application of muscle differentiation medium (Dellavalle et al. 2007) Likewise, the prolonged culture of perivascular cells ($CD146^+CD56^-CD34^-CD45^-$) from human skeletal muscle for 2 weeks were shown to yield multinucleated myofibers expressing slow myosin heavy chain (MHC), desmin, and dystrophin (Crisan et al. 2008). Birbrair et al. (2013) further demonstrated specificity of Type 2 pericytes ($NG2^+/Nestin^+$) to differentiate into mature myofibers *in vitro*. Differentiation capacity appears to be unique to skeletal muscle-derived pericytes, as pericytes isolated from bone marrow, periosteum and cord blood do not possess the same potential (Sacchetti et al. 2016). In each of these sets of experiments, isolated pericytes remained characteristically unique from SCs indicating a SC-independent capacity for myogenic differentiation.

Overall, pericytes are enriched in nearly all tissues of the body and provide an essential role in regulating the vasculature. However, clonal analyses and tissue-specific differentiation potential suggest that pericytes may also represent a multipotent stem cell with unique capacity to regulate resident tissue morphology and function. The remainder of this chapter will provide insight into the capacity for pericytes to contribute to skeletal muscle recovery and remodeling.

9.3 Pericyte Contribution to Skeletal Muscle Development and Injury

A tempo-spatial relationship exists between muscle pericytes and development. Kostallari et al. (2015) demonstrated that $NG2^+$ pericytes (primarily $CD146^+$ and $Nestin^+$) not only associate with endothelial cells and growing microvessels but also progressively localize to SCs to provide support during post-natal development (Kostallari et al. 2015). They further show that pericytes ($CD146^+CD56^-$ with concurrent expression of NG2/PDGFR/αSMA) promote myotube formation *in vitro* by stimulating differentiation via paracrine factor secretion of IGF-1 while also inducing SC quiescence via ANGPT1 secretion. As SC expression of both growth factor receptors is high, it is likely that SCs are uniquely receptive to the pericyte secretome (Kostallari et al. 2015). Targeted ablation of $NG2^+$ pericytes using an inducible diphtheria toxin receptor mouse line ($Tg{:}NG2^{Cre/+}{::}R26R^{iDTR}$) resulted in a significant reduction to pericyte number, an increase in SC proliferation, and myofiber hypotrophy in post-natal (P13) mice (Kostallari et al. 2015). In adult mice, ablation of $NG2^+$ pericytes did not result in the same overall reduction to myofiber size despite ~50% reduction in pericyte quantity (Kostallari et al. 2015), demonstrating that mature muscle does not rely on pericyte-mediated myogenesis for the regulation of myofiber size as similarly observed by Sacchetti et al. (2016). However, given the essential role for satellite cells in repair and regeneration in response to injury, pericytes likely influence satellite cell contribution to recovery in adult skeletal muscle.

Injury and trauma to skeletal muscle has severe consequences for function. From relatively minor structural disruptions following exercise to the more drastic volumetric muscle loss (VML), any disruption to skeletal muscle tissue can impair movement and disrupt metabolic homeostasis. In general, damage to myofibers or the surrounding ECM initiates an immediate inflammatory response leading to macrophage accumulation and phagocytosis of damaged tissue (Souza and Gottfried 2013; Tidball 2005). Depending on the signaling cascade to follow, myofiber regeneration, fibrosis, or ectopic adipose accumulation will occur at the injury site in an attempt to morphologically recover the damaged tissue; however, non-myofiber infiltration will result in a significant loss to strength and function (Huard et al. 2002; Mann et al. 2011). Proper chemical and mechanical signaling is required for optimal skeletal muscle recovery post-trauma, which is mediated in part by the mononuclear stem/stromal cells residing adjacent to the myofibers. The role of SCs in the recovery process following injury is well acknowledged (Günther et al. 2013; Lepper et al. 2011; Murphy et al. 2011; Sambasivan et al. 2011). Lepper et al. (2011) observed that ablation of $Pax7^+$ SCs in mouse skeletal muscle completely prevents proper myogenesis and regeneration post-cardiotoxin injury, which has since been confirmed (Finnerty et al. 2017; Lepper et al. 2011; Murphy et al. 2011).

Given the necessity of endogenous SCs to muscle recovery post-injury, exogenous transplantation of SCs as a treatment for various skeletal muscle disorders has been of great interest. However, utilizing SCs therapeutically has been challenging and often unsuccessful (Montarras et al. 2005; Sampaolesi et al. 2003). Mode of delivery, quantity, and purity of transplantable cells as well as the functional characteristics of cultured SC all play a role in the efficacy of SC therapy (Motohashi et al. 2014). Recent studies have shown promising results for SC transplantation (Charville et al. 2015; Marg et al. 2019; Xu et al. 2015), but much more work is needed to develop methods capable of maintaining sufficient quantities of functional SCs that could be therapeutically useful. As such, other muscle resident progenitor cells, including pericytes, which are generally easier to isolate and maintain, have been identified as potential therapeutic alternatives to skeletal muscle stem cell transplantation.

As skeletal muscle-derived pericytes appear to be significantly myogenic *in vitro*, assessing their myogenic potential *in vivo* is crucial to determining their full therapeutic potential upon transplantation. In order to assess that potential as well as how exogenous pericyte transplantation compares to SC transplantation, multiple research groups have investigated pericyte myogenesis post-transplantation in conditions of myofiber damage. Dellavalle et al. (2007) investigated the pericyte (ALP^+) contribution to skeletal muscle repair under various conditions by isolating and culturing primary human skeletal muscle pericytes followed by a single intra-arterial (i.a.) transplantation of these pericytes into the femoral artery of dystrophic (*scid-mdx*) mice. Twenty-four hours post-transplantation, human pericytes were present in the downstream muscles, and following 7 days, human pericytes were primarily localized within the basal lamina of those myofibers. In a separate experiment, multiple i.a. injections of human-derived ALP^+ pericytes were performed in 2-month-old *scid-mdx* mice and examined one-month following the last injection. Various areas of the downstream tibialis anterior (TA) muscle contained human dystrophin$^+$ fibers, indicating that the exogenous pericytes are myogenic and can contribute to *de novo* myofiber synthesis in response to muscle damage (i.e. dystrophy) (Dellavalle et al. 2007). In contrast, i.a. administered SCs failed to cross the microvessel membrane, and thus, did not generate any new myofibers (Dellavalle et al. 2007). Similarly, it has been observed that intramuscular transplantation of human skeletal muscle-derived pericytes ($CD146^+$) into cardiotoxin-injured (CTX) muscle will result in muscle regeneration. Crisan et al. (2008) demonstrated that 3 weeks post-transplantation, the TA muscles of CTX injured mice presented with myonuclei expressing human spectrin as well as myofibers containing human dystrophin (Crisan et al. 2008). Likewise, transplantation of pericytes into CTX-injured TA muscles of either wild type or *scid-mdx* mice resulted in significant nuclei incorporation with dystrophin- and laminin-expressing myofibers in both healthy and dystrophic muscles (Sacchetti et al. 2016). These findings lend credence to the concept that exogenous pericytes adopt a myogenic fate in response to muscle damage.

Interestingly, subpopulations within pericytes might influence the specific fate of transplanted cells. Type 1 ($Nestin^-/NG2^+$) and Type 2 ($Nestin^+/NG2^+$) pericytes appear to be either adipogenic or myogenic, respectively (Birbrair et al. 2013; Birbrair et al. 2014; Gautam et al. 2017; Nirwane et al. 2017). This difference appears to influence their exogenous transplantation efficacy. Type 2 pericytes intramuscularly injected post-$BaCl_2$ injury integrate with the existing musculature and differentiate into myofibers, while Type 1 pericytes remain in the interstitium between damaged fibers and fail to directly contribute to muscle regeneration (Birbrair et al. 2013). However, in situations of specific basement membrane disruption via glycerol injection (where myofibers remain intact), Type 1 pericyte transplantation results in ectopic fat accumulation suggesting a greater adipogenic potential compared to Type 2 pericytes. However, in this study, ~18% of Type 1 pericytes expressed PDGFRα^+, a cell surface marker commonly used to identify fibroadipogenic progenitor (FAP) cells (Biferali et al. 2019; Joe et al. 2010; Uezumi et al. 2011; Wosczyna et al. 2019). While Type 1 pericytes did exhibit both *in vitro* and *in vivo* adipogenic potential, Type 1 pericytes lacking PDGFRα expression failed to differentiate into adipocytes *in vitro*. This suggests that a more specific subset of Type 1 pericytes (Nestin$^-$/NG2$^+$/PDGFRα^-), most likely FAPs, are responsible for adipose accumulation observed post-transplantation making interpretation of Type 1 pericyte behavior difficult to determine. However, the myogenic potential of Nestin$^+$/NG2$^+$ pericytes is convincing. Overall, the evidence suggests a potentially robust capacity for exogenous skeletal muscle-derived pericytes to improve skeletal muscle recovery post-injury either via paracrine factor secretion or direct myogenic incorporation. Continuing research and clinical trials into the utilization of pericyte transplantation therapies is paramount to developing alternative treatments for skeletal muscle disorders.

9.4 Impact of Skeletal Muscle Use (Contraction) on Pericyte Function

As a highly plastic tissue, skeletal muscle anatomy and physiology is influenced considerably by external stimuli. Exercise significantly alters muscle structure and functional capacity at both the macro- and microscopic level. Acute and chronic periods of resistance/strength training can increase total muscle mass and myofiber density, improve muscle strength via changes to muscle size and neuronal activation, as well as enhance anaerobic metabolism and glucose utilization. Endurance exercise results in more efficient oxygen consumption and utilization, expanded muscle vascularization, and improved aerobic metabolism. Likewise, muscle disuse, whether in response to whole body bed rest or isolated limb immobilization, will have a substantial detrimental impact on muscle structure and function. It has been demonstrated that even transient periods of inactivity can result in substantial and quantifiable losses to skeletal muscle mass and strength (English and Paddon-Jones 2010; Wall et al. 2013). The mechanistic basis by which mechanical activity regulates muscle structure and function is still an active area of investigation. However, recent studies suggest that pericytes are responsive to conditions of increased and decreased muscular effort (Fig. 9.2).

The skeletal muscle pericyte response to muscle contractions is a largely unstudied field. While most research is focused on the requirement of the SC for muscle

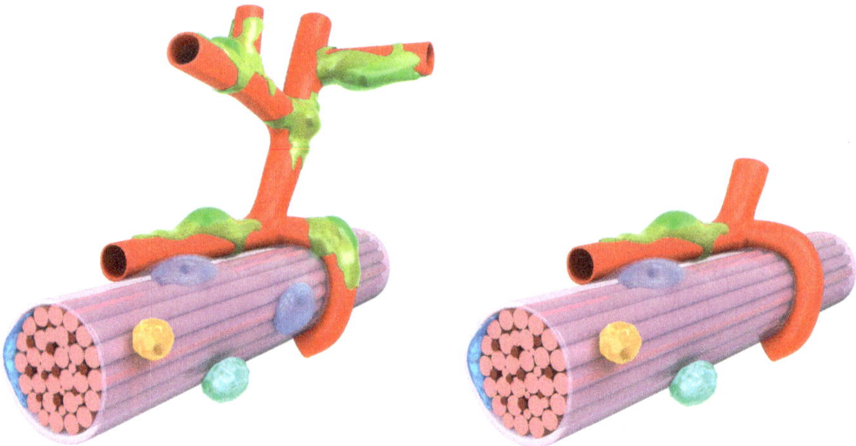

Fig. 9.2 Pericyte response to use (Contraction) and disuse (Inactivity). Skeletal muscle is highly sensitive to mechanical cues, increasing in structure and function in response to physical activity and atrophying in response to disuse. Capillary content also correlates with skeletal muscle morphology, such that higher capillary content is observed with use and decreased with inactivity. Pericytes are highly responsive to mechanical cues and likely regulate structural and functional changes observed in skeletal muscle with patterns of movement. Pericytes may, therefore, provide an important target in the design of novel strategies to prevent or treat loss of skeletal muscle mass and function following periods of disuse (extended bed rest, limb immobilization, microgravity)

hypertrophy, few researchers have investigated the contribution of non-SC cells to myofiber growth in response to muscle contraction/activation. Select research has demonstrated that myofiber hypertrophy in response to chronic overload requires direct SC involvement (Egner et al. 2016; Goh and Millay 2017), however, others have observed significant muscle growth even in the absence of SCs (Fry et al. 2014; McCarthy et al. 2011; Murach et al. 2017). This leaves room for other perivascular stem cells, like pericytes, to play a key role in modulating muscle hypertrophy.

Previous work by our group has investigated the impact of contraction on perivascular stem cell function in mice. We observed that muscle contraction and mechanical strain preferentially activates Sca-1$^+$CD45$^-$ mesenchymal stem cells possessing a pericyte signature (CD146$^+$/NG2$^+$/PDGFRβ$^+$) to upregulate various factors related to muscle growth and vascularization (Huntsman et al. 2013; Huntsman et al. 2018; Valero et al. 2012; Zou et al. 2015). In addition, intramuscular transplantation of perivascular stem cells (Sca-1$^+$CD45$^-$; CD146$^+$/NG2$^+$) in combination with eccentric exercise contributed to increased satellite cell proliferation, myofiber size and strength, and microvessel quantity and lumen diameter (Huntsman et al. 2013; Zou et al. 2015) suggesting that perivascular stem cell transplantation can augment the muscle strength and size gains observed with resistance training. Furthermore, we recently reported the extent to which skeletal muscle pericytes, including both $CD146^+CD31^-CD45^-$ [$CD31^+CD45^+$ = Lineage/Lin] and $NG2^+Lin^-$ populations, respond to acute and chronic muscle contractions via electrical stimulation (Dvoretskiy et al. 2019). Twenty-four hours after a single bout of muscle contraction, muscle resident pericyte quantity remained unchanged; however, population-specific changes to gene expression profiles were observed at 3 hours post-contraction (as measured by qPCR). $NG2^+Lin^-$ pericytes demonstrated minor alterations to gene expression, including *Fgf2* and *Ang*, while $CD146^+Lin^-$ pericytes demonstrated a robust response to electrical stimulation, including upregulation of *Fgf2* and various ECM remodeling factors (*Timp1, Timp2, Mmp2, Mmp14*, and *Col1α1*). Follow-up analysis with RNA-sequencing revealed that $CD146^+Lin^-$ pericytes synthesize RNAs important for ECM remodeling and immunomodulation (Dvoretskiy et al. 2019). Interestingly, exogenous $CD146^+Lin^-$ pericytes transplanted into hindlimb muscles of mice prior to 4 weeks of electrical stimulation training resulted in a significant increase to muscle capillarization and peak torque (independent of changes to myofiber CSA) compared to saline controls. These findings suggest pericytes, specifically $CD146^+Lin^-$, are highly responsive to skeletal muscle activation/contraction and help synthesize factors critical for ECM remodeling and muscle capillarization in order to aid in the muscle adaptation process.

Farup et al. (2015) investigated the impact of prolonged concentric and eccentric resistance exercise on pericyte quantity and proliferation in human skeletal muscle. Using the cell surface markers NG2 and ALP to identify pericytes, they observed variable effects on pericyte-SC interaction. Specifically, in response to concentric training, ALP^+ pericyte number decreased concurrent with an increase in overall $Pax7^+$ SC quantity. Following both concentric and eccentric training, $NG2^+$ pericyte number was reduced while the number of activated SC ($Pax7^+MyoD^+$) and total FAP ($PDGFR\alpha^+$) populations increased. This suggests a contraction-dependent pericyte transition where $NG2^+$ pericytes shift toward a more mesenchymal progenitor cell ($PDGFR\alpha^+$) as well as inducing SC activation. This concept is supported in part by the findings of Dellavalle et al. (2011) who observed similar pericyte-SC cellular changes in response to muscle injury. Resistance exercise training induces consistent, microscopic damage to myofibers, stimulating an injury response with which pericytes appear susceptible, thus possibly explaining the pericyte transition state observed by Farup et al. (2015). However, acute stimuli (e.g. single bouts of eccentric exercise) appear to have a more limited impact on pericyte quantity in humans, despite robustly inducing skeletal muscle damage. While neither Hyldahl et al. (2011) nor De Lisio et al. (2015) observed any change to pericyte quantity following a single, muscle damaging bout of eccentric exercise, it was observed that $NG2^+/ALP^+$ pericytes appear to be involved in the early, adaptive response to eccentric exercise via upregulation of NF-kβ transcription, which is critical for initiating the early immune response necessary for skeletal muscle turnover post-exercise (Hyldahl et al. 2011). Pericytes appear to be in a fluid-like state, transitioning between a vessel stabilizing cell structure to an immunomodulatory and local cell activating support population required for tissue recovery as previously proposed by Caplan and Correa (2011).

9.5 Impact of Skeletal Muscle Disuse on Pericyte Function

Skeletal muscle atrophy is a severe condition associated with general physical inactivity or muscle disuse. Frequently observed after bed rest, limb immobilization, spinal injury, or in situations of microgravity, the main consequences of disuse are significant losses to muscle mass and strength in addition to generalized muscle physiology dysfunction (Bodine 2013; Brooks and Myburgh 2014; Narici and de Boer 2011). Mechanistically, muscle atrophy is a multifaceted condition with disruptions to whole muscle protein balance being a major contributing factor to the rate and magnitude of muscle loss (Atherton et al. 2016). However, the severity and duration of disuse can also result in disruptions to the surrounding ECM and the cell populations which reside there, which can impact the time to recovery (Giannelli et al. 2005; Hammers et al. 2017; Li and Hu 2012).

Most research has focused on the SC response and contribution to disuse, which has yielded variable, and often conflicting results. Alterations to SC proliferation and overall quantity decrease (Arentson-Lantz et al. 2016), increase (Brooks et al. 2018; Suetta et al. 2013), or remain stable (Jackson et al. 2012; Snijders et al. 2014) during the atrophic process. Interestingly, the direct SC contribution to recovery post-disuse appears to be minimal (Itoh et al. 2014; Jackson et al. 2012). Jackson et al. (2012) utilized a conditional SC ablation model ($Pax7^{CreER/+};Rosa26^{DTA/+}$) to directly assess the SC role in disuse and recovery. They observed that the specific removal of skeletal muscle SCs before a period of hindlimb suspension yielded no significant difference in muscle weight, myofiber cross sectional area (CSA), or myofiber force production post-reload (Jackson et al. 2012). These findings suggest the SCs are not integral to the regrowth process associated with a return to activity following periods of disuse. As such, other skeletal muscle progenitor cells may be significant contributors to muscle regrowth in a SC-independent manner, particularly in mature muscle (Boppart et al. 2013; Murach et al. 2018).

The non-SC response to disuse, and subsequent contribution to recovery, has not previously been studied. Our group has initiated a series of studies to assess the extent to which pericytes respond to skeletal muscle immobilization and subsequently determine the impact of exogenous pericyte transplantation on skeletal muscle regrowth following disuse (Munroe et al. 2019). Concurrent with significant reductions to TA muscle weight following 2 weeks of hindlimb immobilization, we observed a reduction in $NG2^{+}Lin^{-}$ pericyte quantity and a trend toward a decrease in $CD146^{+}Lin^{-}$ pericyte content, both which were restored following 2 weeks of remobilization, despite a lack of muscle weight recovery (Munroe et al. 2019). Whether these cells are undergoing apoptosis (Mitchell and Pavlath 2004; Schwartz 2019) or are adopting an alternative fate remains to be elucidated, but it is clear that pericyte quantity is specifically impacted by 2 weeks of disuse. The pericyte gene expression profile appeared minimally altered in response to disuse (Munroe et al. 2019). However, our analysis was targeted and completion of a full-time course experiment is still necessary to thoroughly examine the pericyte response under the condition of disuse. The loss of pericyte quantity justified a transplantation study to determine the extent to which exogenous transplantation could restore myofiber size during the recovery period. Co-transplantation of both pericyte populations into the atrophied TA muscle followed by 2 weeks of hindlimb remobilization resulted in a significant restoration of myofiber CSA compared to non-treated controls (Munroe et al. 2019). In addition, we observed a significant increase in both capillary to fiber ratio and capillary density following transplantation suggesting a role for pericyte-mediated vascularization in myofiber recovery (Munroe et al. 2019). Both pericyte populations were fluorescently labeled to track cell fate within the TA muscle post-transplantation, and interestingly, neither population appeared to directly engraft with existing myofibers nor generate *de novo* fibers. The pericytes instead resided in the interstitial space between fibers, suggesting a supportive role in the regrowth process following disuse (Munroe et al. 2019). Identifying the underlying molecular basis for pericyte-mediated regrowth post-atrophy remains to be determined.

9.6 Conclusions and Future Directions

Pericytes classically serve as vascular stromal cells, providing an essential role in the regulation of vascularization and vessel tone. However, studies also demonstrate the capacity for pericytes to directly and indirectly contribute to endogenous tissue healing and remodeling. T-Box Transcription Factor 18 (Tbx18) was recently identified as a unique pericyte marker and Tbx18-CreERT2 lineage tracing experiments suggest limited potential for pericytes to differentiate in response to injury, high-fat feeding, or aging in a variety of tissues (Guimarães-Camboa et al. 2017). The Tbx18-CreERT2 mouse line now provides the opportunity to compare Tbx18$^+$ pericytes with other well characterized pericytes (NG2$^+$Nestin$^+$, CD146$^+$Lin$^-$) and subsequently determine the true extent to which these cells possess the capacity to renew and/or differentiate, both *in vitro* and *in vivo*. Consistent identification and replication of the resident pericyte will be essential for revealing the response to physiological stimuli and the ultimate potential for these cells to be used in the clinic to repair and restore muscle in the context of injury, disuse and neuromuscular disease.

References

Arentson-Lantz EJ, English KL, Paddon-Jones D, Fry CS (2016) Fourteen days of bed rest induces a decline in satellite cell content and robust atrophy of skeletal muscle fibers in middle-aged adults. J Appl Physiol 120:965–975

Armulik A, Abramsson A, Betsholtz C (2005) Endothelial/pericyte interactions. Circ Res 97:512–523

Armulik A, Genové G, Betsholtz C (2011) Pericytes: developmental, physiological, and pathological perspectives, problems, and promises. Dev Cell 21:193–215

Atherton PJ, Greenhaff PL, Phillips SM et al (2016) Control of skeletal muscle atrophy in response to disuse: clinical/preclinical contentions and fallacies of evidence. Am J Physiol Endocrinol Metab 311:594–604

Biferali B, Proietti D, Mozzetta C, Madaro L (2019) Fibro–adipogenic progenitors cross-talk in skeletal muscle: the social network. Front Physiol 10:1074

Birbrair A, Zhang T, Wang ZM et al (2013) Role of pericytes in skeletal muscle regeneration and fat accumulation. Stem Cells Dev 22:2298–2314

Birbrair A, Zhang T, Files DC et al (2014) Type-1 pericytes accumulate after tissue injury and produce collagen in an organ-dependent manner. Stem Cell Res Ther 5:122

Bodine SC (2013) Disuse-induced muscle wasting. Int J Biochem Cell Biol 45:2200–2208

Boppart MD, De Lisio M, Zou K, Huntsman HD (2013) Defining a role for non-satellite stem cells in the regulation of muscle repair following exercise. Front Physiol 4:310

Brooks NE, Myburgh KH (2014) Skeletal muscle wasting with disuse atrophy is multi-dimensional: the response and interaction of myonuclei, satellite cells and signaling pathways. Front Physiol 5:99

Brooks MJ, Hajira A, Mohamed JS, Alway SE (2018) Voluntary wheel running increases satellite cell abundance and improves recovery from disuse in gastrocnemius muscles from mice. J Appl Physiol 124:1616–1628

Caplan AI (2010) What's in a name? Tissue Eng Part A 16:2415–2417

Caplan AI, Correa D (2011) The MSC: an injury drugstore. Cell Stem Cell 9:11–15

Cappellari O, Cossu G (2013) Pericytes in development and pathology of skeletal muscle. Circ Res 113:341–347

Chal J, Pourquié O (2017) Making muscle: Skeletal myogenesis in vivo and in vitro. Development 144:2104–2122

Charville GW, Cheung TH, Yoo B et al (2015) Ex vivo expansion and in vivo self-renewal of human muscle stem cells. Stem Cell Rep 5:621–632

Costamagna D, Berardi E, Ceccarelli G, Sampaolesi M (2015) Adult stem cells and skeletal muscle regeneration. Curr Gene Ther 15:348–363

Crisan M, Yap S, Casteilla L et al (2008) A perivascular origin for mesenchymal stem cells in multiple human organs. Cell Stem Cell 3:301–313

De Lisio M, Farup J, Sukiennik RA et al (2015) The acute response of pericytes to muscle-damaging eccentric contraction and protein supplementation in human skeletal muscle. J Appl Physiol 119:900–907

De Souza J, Gottfried C (2013) Muscle injury: review of experimental models. J Electromyogr Kinesiol 23:1253–1260

Dellavalle A, Sampaolesi M, Tonlorenzi R et al (2007) Pericytes of human skeletal muscle are myogenic precursors distinct from satellite cells. Nat Cell Biol 9:255–267

Dellavalle A, Maroli G, Covarello D et al (2011) Pericytes resident in postnatal skeletal muscle differentiate into muscle fibres and generate satellite cells. Nat Commun 2:499

Díaz-Flores L, Gutiérrez R, Varela H (1992) Behavior of postcapillary venule pericytes during postnatal angiogenesis. J Morphol 213:33–45

Díaz-Flores L, Gutiérrez R, Madrid JF et al (2009) Pericytes. Morphofunction, interactions and pathology in a quiescent and activated mesenchymal cell niche. Hist Histopathol 24:909–969

Dumont NA, Bentzinger CF, Sincennes MC, Rudnicki MA (2015) Satellite cells and skeletal muscle regeneration. Compr Physiol 5:1027–1059

Dvoretskiy S, Garg K, Munroe M et al (2019) The impact of skeletal muscle contraction on $CD146^{+}Lin^{-}$ pericytes. Am J Physiol Cell Physiol 317:1011–1024

Egner IM, Bruusgaard JC, Gundersen K (2016) Satellite cell depletion prevents fiber hypertrophy in skeletal muscle. Development 143:2898–2906

Endo T (2015) Molecular mechanisms of skeletal muscle development, regeneration, and osteogenic conversion. Bone 80:2–13

English KL, Paddon-Jones D (2010) Protecting muscle mass and function in older adults during bed rest. Curr Opin Clin Nutr Metab Care 13:34–39

Farup J, De Lisio M, Klejs Rahbek S et al (2015) Pericyte response to contraction mode-specific resistance exercise training in human skeletal muscle. J Appl Physiol 119:1053–1063

Finnerty CC, McKenna CF, Cambias LA et al (2017) Inducible satellite cell depletion attenuates skeletal muscle regrowth following a scald-burn injury. J Physiol 595:6687–6701

Forcina L, Miano C, Pelosi L, Musarò A (2019) An overview about the biology of skeletal muscle satellite cells. Curr Genomics 20:24–37

Frontera WR, Ochala J (2015) Skeletal muscle: a brief review of structure and function. Behav Genet 45:183–195

Fry CS, Lee JD, Jackson JR et al (2014) Regulation of the muscle fiber microenvironment by activated satellite cells during hypertrophy. FASEB J 28:1654–1665

Gautam J, Nirwane A, Yao Y (2017) Laminin differentially regulates the stemness of type I and type II pericytes. Stem Cell Res Ther 8:1–16

Gerhardt H, Betsholtz C (2003) Endothelial-pericyte interactions in angiogenesis. Cell Tissue Res 314:15–23

Giannelli G, de Marzo A, Marinosci F, Antonaci S (2005) Matrix metalloproteinase imbalance in muscle disuse atrophy. Hist Histopathol 20:99–106

Gillies AR, Lieber RL (2011) Structure and function of the skeletal muscle extracellular matrix. Muscle Nerve 44:318–331

Goh Q, Millay DP (2017) Requirement of myomaker-mediated stem cell fusion for skeletal muscle hypertrophy. Elife 6

Guimarães-Camboa N, Cattaneo P, Sun Y et al (2017) Pericytes of multiple organs do not behave as mesenchymal stem cells in vivo. Cell Stem Cell 20:345–359

Günther S, Kim J, Kostin S et al (2013) Myf5-positive satellite cells contribute to Pax7-dependent long-term maintenance of adult muscle stem cells. Cell Stem Cell 13:590–601

Hammers DW, Merscham-Banda M, Hsiao JY et al (2017) Supraphysiological levels of GDF11 induce striated muscle atrophy. EMBO Mol Med 9:531–544

Harrell CR, Simovic Markovic B, Fellabaum C et al (2018) Molecular mechanisms underlying therapeutic potential of pericytes. J Biomed Sci 25:21

Hawke TJ, Garry DJ (2001) Myogenic satellite cells: physiology to molecular biology. J Appl Physiol 91:534–551

Huard J, Li Y, Fu FH (2002) Muscle injuries and repair: current trends in research. J Bone Joint Surg 84-A:822–832

Huntsman HD, Zachwieja N, Zou K et al (2013) Mesenchymal stem cells contribute to vascular growth in skeletal muscle in response to eccentric exercise. Am J Physiol Heart Circ Physiol 304:H72–H81

Huntsman HD, Rendeiro C, Merritt JR et al (2018) The impact of mechanically stimulated muscle-derived stromal cells on aged skeletal muscle. Exp Gerontol 103:35–46

Hyldahl RD, Xin L, Hubal MJ et al (2011) Activation of nuclear factor-κB following muscle eccentric contractions in humans is localized primarily to skeletal muscle-residing pericytes. FASEB J 25:2956–2966

Itoh Y, Hayakawa K, Mori T et al (2014) Stand-up exercise training facilitates muscle recovery from disuse atrophy by stimulating myogenic satellite cell proliferation in mice. Physiol Rep 2

Jackson JR, Mula J, Kirby TJ et al (2012) Satellite cell depletion does not inhibit adult skeletal muscle regrowth following unloading-induced atrophy. Am J Physiol Cell Physiol 303:854–861

Jankowski RJ, Deasy BM, Huard J (2002) Muscle-derived stem cells. Gene Ther 9:642–647

Joe AWB, Yi L, Natarajan A et al (2010) Muscle injury activates resident fibro/adipogenic progenitors that facilitate myogenesis. Nat Cell Biol 12:153–163

Karstoft K, Pedersen BK (2016) Skeletal muscle as a gene regulatory endocrine organ. Curr Opin Clin Nutr Metab Care 19:270–275

Kim JH, Jin P, Duan R, Chen EH (2015) Mechanisms of myoblast fusion during muscle development. Curr Opin Genet Dev 32:162–170

Kirby TJ, Patel RM, McClintock TS et al (2016) Myonuclear transcription is responsive to mechanical load and DNA content but uncoupled from cell size during hypertrophy. Mol Biol Cell 27:788–798
Kostallari E, Baba-Amer Y, Alonso-Martin S et al (2015) Pericytes in the myovascular niche promote post-natal myofiber growth and satellite cell quiescence. Development 142:1242–1253
Kutcher ME, Herman IM (2009) The pericyte: cellular regulator of microvascular blood flow. Microvasc Res 77:235–246
Lepper C, Partridge TA, Fan CM (2011) An absolute requirement for Pax7-positive satellite cells in acute injury-induced skeletal muscle regeneration. Development 138:3639–3646
Li S, Hu GF (2012) Emerging role of angiogenin in stress response and cell survival under adverse conditions. J Cell Physiol 227:2822–2826
Mann CJ, Perdiguero E, Kharraz Y et al (2011) Aberrant repair and fibrosis development in skeletal muscle. Skelet Muscle 1:21
Marg A, Escobar H, Karaiskos N et al (2019) Human muscle-derived CLEC14A-positive cells regenerate muscle independent of PAX7. Nat Commun 10:5776
Mccarthy JJ, Mula J, Miyazaki M et al (2011) Effective fiber hypertrophy in satellite cell-depleted skeletal muscle. Development 138:3657–3666
Mitchell PO, Pavlath GK (2004) Skeletal muscle atrophy leads to loss and dysfunction of muscle precursor cells. Am J Physiol Cell Physiol 287:1753–1762
Montarras D, Morgan J, Collins C et al (2005) Direct isolation of satellite cells for skeletal muscle regeneration. Science 309:2064–2067
Motohashi N, Asakura Y, Asakura A (2014) Isolation, culture, and transplantation of muscle satellite cells. J Vis Exp 86
Munroe M, Dvoretskiy S, Lopez A et al (2019) Pericyte transplantation improves skeletal muscle recovery following hindlimb immobilization. FASEB J 33:7694–7706
Murach KA, White SH, Wen Y et al (2017) Differential requirement for satellite cells during overload-induced muscle hypertrophy in growing versus mature mice. Skelet Muscle 7:14
Murach KA, Fry CS, Kirby TJ et al (2018) Starring or supporting role? Satellite cells and skeletal muscle fiber size regulation. Physiology 33:26–38
Murphy MM, Lawson JA, Mathew SJ et al (2011) Satellite cells, connective tissue fibroblasts and their interactions are crucial for muscle regeneration. Development 138:3625–3637
Nakano M, Atobe Y, Goris RC et al (2000) Ultrastructure of the capillary pericytes and the expression of smooth muscle α-actin and desmin in the snake infrared sensory organs. Anat Rec 260:299–307
Narici MV, de Boer MD (2011) Disuse of the musculo-skeletal system in space and on earth. Euro J Appl Physiol 111:403–420
Newlands S, Levitt LK, Robinson CS et al (1998) Transcription occurs in pulses in muscle fibers. Genes Dev 12:2748–2758
Nirwane A, Gautam J, Yao Y (2017) Isolation of type I and type II pericytes from mouse skeletal muscles. J Vis Exp 123
Otto A, Collins-Hooper H, Patel K (2009) The origin, molecular regulation and therapeutic potential of myogenic stem cell populations. J Anat 215:477–497
Pavlath GK, Rich K, Webster SG, Blau HM (1989) Localization of muscle gene products in nuclear domains. Nature 337:570–573
Petrany MJ, Millay DP (2019) Cell fusion: merging membranes and making muscle. Trends Cell Biol 29:964–973
Relaix F, Zammit PS (2012) Satellite cells are essential for skeletal muscle regeneration: the cell on the edge returns Centre stage. Development 139:2845–2856
Sacchetti B, Funari A, Remoli C et al (2016) No identical "mesenchymal stem cells" at different times and sites: human committed progenitors of distinct origin and differentiation potential are incorporated as adventitial cells in microvessels. Stem Cell Rep 6:897–913
Sambasivan R, Yao R, Kissenpfennig A et al (2011) Pax7-expressing satellite cells are indispensable for adult skeletal muscle regeneration. Development 138:3647–3656

Sampaolesi M, Torrente Y, Innocenzi A et al (2003) Cell therapy of alpha-sarcoglycan null dystrophic mice through intra-arterial delivery of mesoangioblasts. Science 301:487–492

Schwartz LM (2019) Skeletal muscles do not undergo apoptosis during either atrophy or programmed cell death-revisiting the myonuclear domain hypothesis. Front Physiol 9:1887

Snijders T, Wall BT, Dirks ML et al (2014) Muscle disuse atrophy is not accompanied by changes in skeletal muscle satellite cell content. Clin Sci 126:557–566

Suetta C, Frandsen U, Mackey AL et al (2013) Ageing is associated with diminished muscle re-growth and myogenic precursor cell expansion early after immobility-induced atrophy in human skeletal muscle. J Physiol 591:3789–3804

Tidball JG (2005) Inflammatory processes in muscle injury and repair. Am J Physiol Regul Integr Comp Physiol 288:345–353

Uezumi A, Ito T, Morikawa D et al (2011) Fibrosis and adipogenesis originate from a common mesenchymal progenitor in skeletal muscle. J Cell Sci 124:3654–3664

Valero MC, Huntsman HD, Liu J et al (2012) Eccentric exercise facilitates mesenchymal stem cell appearance in skeletal muscle. PLoS One 7:e29760

Wall BT, Dirks ML, van Loon LJC (2013) Skeletal muscle atrophy during short-term disuse: implications for age-related sarcopenia. Ageing Res Rev 12:898–906

Wosczyna MN, Konishi CT, Perez Carbajal EE et al (2019) Mesenchymal stromal cells are required for regeneration and homeostatic maintenance of skeletal muscle. Cell Rep 27:2029–2035

Xu X, Wilschut KJ, Kouklis G et al (2015) Human satellite cell transplantation and regeneration from diverse skeletal muscles. Stem Cell Reports 5:419–434

Yin H, Price F, Rudnicki MA (2013) Satellite cells and the muscle stem cell niche. Physiol Rev 93:23–67

Zou K, Huntsman HD, Carmen Valero M et al (2015) Mesenchymal stem cells augment the adaptive response to eccentric exercise. Med Sci Sports Exerc 47:315–325

Chapter 10
Pericytes in Myocardial Diseases

Linda Alex, Izabela Tuleta, and Nikolaos G. Frangogiannis

Abstract The adult mammalian heart contains a large number of pericytes that regulate blood flow, while supporting baseline microvascular structure and function. Considering their abundance, strategic peri-endothelial location, functional diversity and phenotypic plasticity, pericytes may be critically involved in a wide range of myocardial pathologic conditions. Following myocardial infarction, cardiac pericytes may regulate inflammatory, reparative, angiogenic and fibrogenic responses. Moreover, pericyte-mediated microvascular constriction may contribute to the pathogenesis of "no-reflow" in the ischemic and reperfused myocardium. Cell therapy with pericytes has been suggested to exert beneficial actions in experimental models of myocardial infarction. The underlying mechanisms of protection may involve modulation of inflammation, suppression of fibrosis and stimulation of angiogenesis. Pericytes may also be involved in the pathogenesis of heart failure by converting to myofibroblasts, by secreting fibrogenic and pro-inflammatory mediators, and my regulating microvascular blood flow. Unfortunately, limited information is currently available on the role of pericytes in myocardial diseases. There is an urgent need for systematic investigation of pericyte actions in the ischemic, infarcted and failing heart using experimental animal models and human studies.

Keywords Pericyte · Vascular smooth muscle cell · Endothelium · Myocardial ischemia · Fibroblast · Cytokine · Growth factors · TGF-β · PDGF · α-smooth muscle actin · No-reflow · Fibrosis · Heart failure · Diabetes · Cell therapy

L. Alex · I. Tuleta · N. G. Frangogiannis (✉)
The Wilf Family Cardiovascular Research Institute, Department of Medicine (Cardiology), Albert Einstein College of Medicine, Bronx, NY, USA
e-mail: nikolaos.frangogiannis@einsteinmed.org

A. Birbrair (ed.), *Biology of Pericytes – Recent Advances*, Stem Cell Biology and Regenerative Medicine 68, https://doi.org/10.1007/978-3-030-62129-2_10

10.1 Introduction

The adult mammalian heart is comprised of cardiomyocytes and a highly cellular interstitium. Fibroblasts (Tallquist 2020), immune cells (including macrophages (Epelman et al. 2014), mast cells (Gersch et al. 2002) and dendritic cells (Lee et al. 2018a)), and abundant vascular cells are enmeshed within the cardiac interstitium and may play a role in cardiac homeostasis. In young adult mice, vascular endothelial cells have been suggested to be the most abundant non-cardiomyocytes (Pinto et al. 2016). Considering that in all vascularized organs, a significant proportion of the endothelial abluminal surface is covered by mural cells (Sims 1986), it is not surprising that the myocardium contains a large number of pericytes (Nees et al. 2012; Pinto et al. 2016). In the adult heart, pericytes may regulate baseline microvascular function and permeability (Murray et al. 2017). Moreover, myocardial pericytes may be implicated in regulation of inflammatory, fibrotic and angiogenic responses following cardiac injury.

The adult mammalian heart has negligible regenerative capacity; thus repair following injury is dependent on activation of matrix-producing interstitial cells. In the infarcted heart, sudden death of a large number of cardiomyocytes stimulates expression of cytokines and chemokines leading to recruitment of inflammatory leukocytes, and subsequent expansion of interstitial cell populations that may play an important role in cardiac repair (Prabhu and Frangogiannis 2016). Moreover, most forms of chronic heart failure are associated with proliferation and stimulation of fibroblast-like interstitial cells that may initially transduce protective signals (Russo et al. 2019), but when chronically activated may contribute to dysfunction and adverse remodeling (Frangogiannis 2019). Considering their abundance, strategic location in the cardiac microvasculature, phenotypic plasticity and functional diversity, pericytes may be implicated in regulation of both reparative and injurious responses in myocardial diseases. This chapter discusses the role of pericytes in myocardial pathology, focusing on their involvement in regulating inflammatory, fibrogenic and angiogenic responses following cardiac injury, and on the potential significance of pericyte-mediated responses in the pathobiology of myocardial infarction and heart failure.

10.2 Characteristics of Cardiac Pericytes

Pericytes are defined as perivascular mesenchymal cells embedded within the vascular basement membrane (Armulik et al. 2011; Sims 1986). Ultrastructural studies show that pericytes have a rounded cell body and numerous projections that span the abluminal surface of several endothelial cells. It is widely accepted that pericytes share some common characteristics with vascular smooth muscle cells: both cell types are periendothelial and are part of the continuum of vascular mural cells. In the brain, the organizational hierarchy of mural cells has been well-described.

VSMCs are ring-shaped and cover arterioles with circumferential processes, whereas pericytes run longitudinally along capillaries, extending processes that form connections between pericytes and endothelial cells (Grant et al. 2019; Attwell et al. 2016). Between these mural cell types are transitional cells with features of both VSMCs and pericytes. Although pericyte populations in the myocardium have not been systematically characterized existing evidence suggests a hierarchy of mural cell cytoarchitecture that may resemble the cerebral microvasculature (O'Farrell et al. 2017).

Several investigations have suggested tissue-specific variations in endothelial coverage with pericytes. It has been suggested that the central nervous system vasculature exhibits the highest level of pericyte coverage (Mathiisen et al. 2010), whereas skeletal muscle has much less extensive pericyte coverage (Sims 1986; Diaz-Flores et al. 2009). There is conflicting evidence on the abundance of pericytes in the adult mammalian myocardium. Early electron microscopic studies identified a large population of pericytes in normal adult mammalian hearts (Forbes et al. 1977; Nag 1980). Some studies suggested that the extent of pericyte coverage in myocardial microvessels may be comparable to the cerebral microvasculature, and that pericytes may be the second most abundant cell population in mammalian hearts (Nees et al. 2012). In contrast, other studies characterizing the cellular composition of the mouse myocardium suggested a much lower pericyte content (Pinto et al. 2016). Recent investigations in both mice and rats suggested that the majority of coronary capillaries were in contact with pericytes (O'Farrell et al. 2017). Conflicting data regarding the extent of pericyte coverage in mammalian hearts may reflect the absence of specific and reliable indicators for pericytes. No single molecular marker can identify all pericytes; thus, several different markers are typically used (reviewed in Table 10.1). The proteoglycan neural/glial2 (NG2)/chondroitin sulfate proteoglycan 4(CSPG4), and platelet derived growth factor β (PDGFRβ) are the most reliable markers for cardiac pericytes. Although NG2 is also expressed by cardiomyocytes in embryonic hearts (Ozerdem et al. 2001), in the adult mammalian myocardium NG2 expression specifically labels mural cells, both pericytes and vascular smooth muscle cells (Fig. 10.1). In normal hearts, PDGFRβ expression also seems to be restricted to vascular mural cells (Chen et al. 2016). α-smooth muscle actin (SMA), desmin, alkaline phosphatase, CD146, CD13, CD73, the intermediate filament protein nestin, and the zinc finger protein Glioma-associated oncogene protein 1 (Gli1) (Chen et al. 2015; Kramann et al. 2015) have also been used as markers for pericytes. Both α-SMA and desmin lack specificity and sensitivity as markers of myocardial pericytes. Pericapillary pericytes do not express α-SMA (Fig. 10.1b) (Birbrair et al. 2015), whereas vascular smooth muscle cells and injury-site myofibroblasts (Frangogiannis et al. 2000a) exhibit high levels of α-SMA expression. Desmin on the other hand, is not expressed by many cardiac pericytes (Chen et al. 2015) and is abundantly expressed in cardiomyocytes.

A growing body of evidence suggests that cardiac pericytes exhibit significant heterogeneity. In rat hearts, the majority of myocardial pericytes were found to co-express PDGFRβ and NG2, while a much smaller subset (~6%) of PDGFRβ+ cells were NG2-negative (O'Farrell et al. 2017). Other studies classified mouse pericytes

Table 10.1 Markers of cardiac pericytes

Marker	Properties	Sensitivity for myocardial pericytes	Specificity for myocardial pericytes
NG2 (neural/glial antigen 2)/chondroitin sulfate proteoglycan 4 (CSPG4)	Transmembrane molecule with a role in cell adhesion, cell communication and cell polarity (Biname 2014).	Expressed by pericytes in fetal, neonatal and adult hearts in several different species, including humans (Chen et al. 2015; Chintalgattu et al. 2013; Nees et al. 2012; Lee et al. 2019; Ozerdem et al. 2001; Crisan et al. 2008; Avolio et al. 2015b; O'Farrell et al. 2017). 95% of neonatal cardiac pericytes were NG2+ (Avolio et al. 2015b).	In embryonic hearts, NG2 is also expressed by cardiomyocytes. In adult hearts, NG2 is expressed by all mural cells (both vascular smooth muscle cells and pericytes).
PDGFR-β (platelet-derived growth factor receptor-β)	Tyrosine kinase receptor involved in PDGF signaling, promoting mural cell differentiation, migration and survival.	Expressed by pericytes in fetal, neonatal and adult hearts in several different species, including humans (Chen et al. 2015; Chintalgattu et al. 2010, 2013; Nees et al. 2012; Lee et al. 2019; Ozerdem et al. 2001; Crisan et al. 2008; Avolio et al. 2015b; O'Farrell et al. 2017). 58% of neonatal cardiac pericytes expressed PDGFR-β (Avolio et al. 2015b). Reported to be expressed by virtually all human ventricular pericytes (both fetal and adult) (Chen et al. 2015).	Expressed by both vascular smooth muscle cells and pericytes. Under conditions of stress (such as left ventricular pressure overload), enhanced PDGFR-β expression was also noted in other myocardial cell types (including cardiomyocytes) (Chintalgattu et al. 2010).

(continued)

Table 10.1 (continued)

Marker	Properties	Sensitivity for myocardial pericytes	Specificity for myocardial pericytes
α-smooth muscle actin (α-SMA)/ Acta2	Actin isoform, found predominantly in vascular smooth muscle cells, but also expressed by pericyte subsets and by injury-site myofibroblasts.	α-SMA expression has been reported in a subset of cardiac pericytes (O'Farrell et al. 2017; Chen et al. 2015; Chintalgattu et al. 2010, 2013; Nees et al. 2012; Lee et al. 2019; Ozerdem et al. 2001; Crisan et al. 2008; Avolio et al. 2015b; O'Farrell et al. 2017).	In embryonic hearts, α-SMA is also expressed by cardiomyocytes (Ya et al. 1997; Sawtell and Lessard 1989). In adult hearts, α-SMA is highly expressed by vascular smooth muscle cells at levels much higher than in pericytes (Nees et al. 2012). Myofibroblasts infiltrating infarcted or pressure-overloaded hearts also express large amounts of α-SMA (Shinde et al. 1863b).
Desmin	Intermediate filament protein expressed in cardiac, skeletal and smooth muscles.	Whereas pericytes in other organs are typically desmin-positive, few cardiac pericytes were found to express desmin (Birbrair et al. 2015; Chen et al. 2016).	Desmin is highly expressed by cardiomyocytes, representing the major component of intermediate filaments (Milner et al. 1996).
CD13 Membrane alanyl aminopeptidase	CD13 functions as a peptidase, a viral receptor and a signal transduction molecule.	Expressed by pericytes in fetal hearts (Crisan et al. 2008).	Lacks specificity. In the infarcted heart, CD13 is also expressed on F4/80-positive macrophages (high expression), CD11c+dendritic cells and some endothelial cells (Pereira et al. 2013).
CD44	Ubiquitously distributed glycoprotein that mediates a wide variety of cell-cell and cell-matrix interactions, CD44 influences inflammatory and fibrotic processes (Huebener et al. 2008).	Expressed by most neonatal cardiac pericytes and by human ventricular pericytes in fetal and adult hearts (Chen et al. 2015; Avolio et al. 2015b).	Lacks specificity. CD44 is ubiquitously expressed. In the infarcted myocardium, CD44 is upregulated and is localized on infiltrating leukocytes, wound myofibroblasts, and vascular cells (Huebener et al. 2008).

(continued)

Table 10.1 (continued)

Marker	Properties	Sensitivity for myocardial pericytes	Specificity for myocardial pericytes
CD73 Ecto-5′-nucleotidase, NT5E; CALJA; eN; eNT	Ectonucleotidase that converts AMP to adenosine. This enzyme is used as a marker of lymphocyte differentiation and may have effects in immunity, inflammation, and cancer.	Expressed by fetal and adult pericytes (Chen et al. 2015; Crisan et al. 2008).	Lacks specificity. CD73 is expressed by cardiomyocytes, vascular smooth muscle cells, lymphocytes and endothelial cells in normal mouse hearts (Minor et al. 2019).
CD105, Endoglin	CD105 is a type III auxiliary receptor for the transforming growth factor (TGF)-β superfamily.	Expressed by neonatal fetal and adult cardiac pericytes (Chen et al. 2015; Crisan et al. 2008; Avolio et al. 2015b).	Lacks specificity. Highly expressed in cardiac fibroblasts and endothelial cells (Kapur et al. 2012; Tual-Chalot et al. 2020).
CD90, Thy-1 cell surface antigen	CD90 plays role in inflammation, wound healing, cell–matrix and cell–cell adhesion.	Expressed by subsets of neonatal, fetal and adult cardiac pericytes (Chen et al. 2015; Crisan et al. 2008; Avolio et al. 2015b).	Lacks specificity. Also expressed by subsets of fibroblasts, immune cells and endothelial cells in normal and injured hearts (Ali et al. 2014; Li et al. 2020).
Vimentin	Vimentin is a type III intermediate filament (IF) protein expressed in all mesenchymal cells.	Expressed by neonatal, fetal and adult cardiac pericytes (Chen et al. 2015; Crisan et al. 2008; Avolio et al. 2015b).	Lacks specificity. Expressed in large amounts by all cells of mesenchymal origin, including fibroblasts and vascular smooth muscle cells (Villalobos et al. 2019; Ivey and Tallquist 2016).

(continued)

Table 10.1 (continued)

Marker	Properties	Sensitivity for myocardial pericytes	Specificity for myocardial pericytes
Nestin	Intermediate filament protein, mostly expressed in nerve cells.	Nestin marks subpopulations of pericytes in several different tissues, including the brain, bone marrow, liver and skeletal muscle. Normal hearts contain both nestin-positive and nestin-negative pericytes; however, only nestin-expressing cells expand following injury (Birbrair et al. 2014).	Lacks specificity. Cardiac resident neural progenitor/stem cells constitutively express nestin. A subset of normal adult ventricular fibroblasts expresses nestin. Nestin is upregulated in reparative and reactive fibrosis. Nestin upregulation has also been reported in cardiomyocytes of the infarct border zone (Meus et al. 2017; Hertig et al. 2017).
CD146	Adhesion molecule expressed by vascular cells, including endothelial and mural cells (Leroyer et al. 2019).	Expressed by human ventricular pericytes from fetal and adult hearts (Chen et al. 2015; Crisan et al. 2008).	Lacks specificity. Expressed in the endothelium of infarct microvessels (Ren et al. 2002).
ALP (alkaline phosphatase)	Enzyme	Expressed by human ventricular pericytes from fetal and adult hearts (Chen et al. 2015; Crisan et al. 2008; Nees et al. 2012).	Expressed in pericytes, but not in vascular smooth muscle cells (Nees et al. 2012). Reported to be expressed by many other cell types, including endothelial cells and fibroblasts (Hui and Tenenbaum 1998).
Gli-1 (glioma-associated oncogene 1)	Effector of hedgehog signaling with a role in embryogenesis. Gli-1-expressing perivascular cells are often referred to as "pericytes" (Kramann et al. 2016).	Myocardial perivascular Gli1+ cells have been reported to contribute to fibrosis in the heart and in other organs by converting to myofibroblasts (Kramann et al. 2015).	Systematic characterization of cardiac Gli1+ cells and study of their potential overlap with cardiac fibroblast and pericyte populations identified with other markers is lacking.

(continued)

Table 10.1 (continued)

Marker	Properties	Sensitivity for myocardial pericytes	Specificity for myocardial pericytes
Calponin	Calcium binding protein which inhibits the ATPase activity of myosin in smooth muscles.	Coronary microvascular pericytes from 8 different species (including humans) expressed calponin at a concentration six times higher than venular endothelial cells (Nees et al. 2012).	Vascular smooth muscle cells of aortic or arteriolar have ten-fold higher expression of calponin than pericytes (Nees et al. 2012).
3G5	The antigen recognized by 3G5 is a cell surface ganglioside, characterized originally on vascular pericytes from bovine retina (Nayak et al. 1988).	Coronary microvascular pericytes from 8 spezies, including humans were positive for 3G5 (Nees et al. 2012). In the pressure-overloaded heart 3G5 expression was noted in pericyte-like cells (Souders et al. 2012).	Some studies have challenged the specificity of 3G5, suggesting that is may be expressed predominantly by mast cells (Gushi et al. 2008). Expression by bone marrow-derived mesenchymal cells has also been reported (Khan et al. 2010).
Tbx18 (T-box transcription factor gene 18)	Transcription factor with a crucial role in embryonic development.	Tbx18 has been reported to selectively mark pericytes and vascular smooth muscle cells in multiple organs, including the heart of adult mice (Guimaraes-Camboa et al. 2017).	Pluripotent Tbx18-expressing cardiac progenitor cells from epicardium give rise to myocytes in the ventricular septum and the atrial and ventricular walls, cardiac fibroblasts and coronary smooth muscle cells (Cai et al. 2008). Tbx18-derived fibroblasts are noted in pressure-overloaded hearts (Ali et al. 2014).

on the basis of nestin expression. Normal hearts contained both nestin-positive and nestin-negative pericytes; however, only nestin-expressing cells expanded following injury (Birbrair et al. 2014). Identification of new markers and use of single cell transcriptomics will likely expand our understanding of pericyte heterogeneity. Whether specific subsets of cardiac pericytes exhibit distinct functional properties remains unknown.

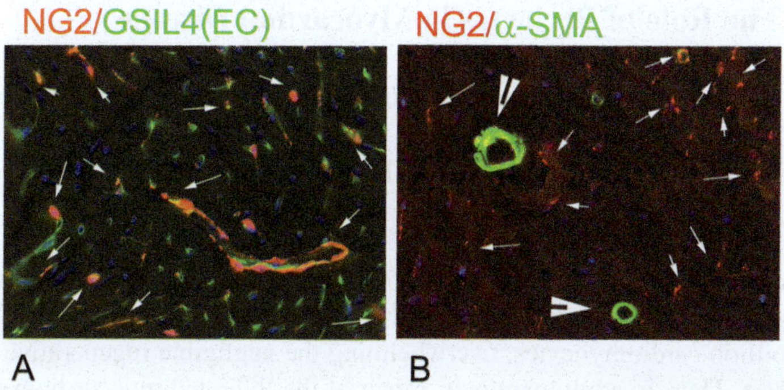

Fig. 10.1 Pericytes in adult mouse hearts, identified using NG2^{dsred} reporter mice. (**a**) Dual staining for NG2 and endothelial cell (EC) labeling with Griffonia simplicifolia isolectin 4 (GSIL4) identifies myocardial NG2+ pericytes as peri-endothelial cells (arrows). (**b**) Dual fluorescence for NG2 and α-SMA shows that the majority of pericapillary pericytes do not express α-SMA (arrows). Arteriolar NG2+/α-SMA+ vascular smooth muscle cells are also identified (arrowheads)

10.3 The Role of Pericytes in Cardiac Homeostasis

The functional role of pericytes in cardiac homeostasis remains poorly understood. In several different tissues, pericytes have been suggested to play an important role in vascular patterning and maturation (Simonavicius et al. 2012). Moreover, pericytes support endothelial cell survival, restrain endothelial cell proliferation, and regulate microvascular permeability (Edelman et al. 2006). The significance of pericyte-mediated regulation of endothelial function in cardiac homeostasis is supported through indirect evidence from several different studies. First, mice with pericyte loss due to absence of PDGF-B or PDGFRβ had perturbations in vascular development and formed microaneurysms in many organs, including the heart (Hellstrom et al. 1999). Second, administration of tyrosine kinase inhibitors (such as the anticancer agent sunitinib) that target PDGF receptors was found to induce cardiotoxicity, associated with pericyte loss and microvascular dysfunction (Chintalgattu et al. 2013). Third, mice with endothelial cell-specific overexpression of angiopoietin-2 had marked loss of pericytes associated with development of cardiomyopathy (Ziegler et al. 2013). In all these cases, the associative data support an important role for pericytes in cardiac homeostasis, but do not document direct causative effects.

Subsets of pericytes express contractile proteins, such as α-SMA and may regulate capillary constriction and physiologic changes in blood flow. A recently published study suggested that cardiomyocytes may transduce signals that regulate pericyte contraction, thus regulating blood flow in response to their metabolic needs (Zhao et al. 2020).

10.4 The Role of Pericytes in Myocardial Disease

10.4.1 Pericytes in Myocardial ischemia and Infarction

10.4.1.1 The Cell Biology of Cardiac Repair: From Inflammation to Fibrosis

Myocardial infarction typically results from sudden occlusion of an epicardial coronary vessel, leading to cessation of blood flow. Prolonged ischemia causes death of up to a billion cardiomyocytes, overwhelming the negligible regenerative capacity of the heart. Thus, in adult mammals, repair of the infarcted myocardium is dependent on a well-orchestrated inflammatory/reparative response that ultimately leads to formation of a collagen-based scar (Frangogiannis 2014). The cell biological response following myocardial infarction can be divided into 3 distinct, but overlapping phases: the inflammatory phase, the proliferative phase and the maturation phase. Release of damage-associated molecular patterns (DAMPs) from dying cardiomyocytes and degraded extracellular matrix, activates Toll-like receptor (TLR) pathways in cardiomyocytes, interstitial cells and immune cells, inducing cytokine and chemokine expression (Arslan et al. 2010). Chemokines recruit leukocytes in the infarcted heart, contributing to the accumulation of professional phagocytes that clears the wound from dead cells and matrix fragments, while setting the stage for expansion of reparative cells through macrophage and lymphocyte-mediated secretion of growth factors (Dutta and Nahrendorf 2015; Chen and Frangogiannis 2017). Secretion of anti-inflammatory mediators, such as IL-10 and TGF-βs, suppresses inflammation, but also activates fibroblasts and vascular cells (Frangogiannis et al. 2000b; Frangogiannis 2017a; Hanna and Frangogiannis 2019) leading to expansion of mesenchymal cells, fibroblast to myofibroblast conversion and enhanced angiogenesis (Frangogiannis et al. 2000a; Shinde et al. 1863a). These events mark the transition to the proliferative phase of infarct healing. Activated myofibroblasts secrete both structural and matricellular extracellular matrix proteins, preserving the structural integrity of the infarcted heart and protecting from cardiac rupture (Kong et al. 2018; Frangogiannis 2017b), but may also contribute to chronic dysfunction and heart failure. Deposition of matrix proteins stimulates yet unidentified suppressive signals that reduce fibroblast activity leading to the maturation phase of infarct repair. Activated myofibroblasts become quiescent (Fu et al. 2018), and secretion of cross-linking enzymes results in formation of a dense collagenous network. As the scar matures, the cellular responses associated with chronic remodeling predominantly involve the non-infarcted segments. It has been suggested that hypertrophic and fibrotic remodeling of viable myocardium may contribute to adverse remodeling, worsening dysfunction and development of post-infarction heart failure (Frangogiannis 2015). Considering their strategic periendothelial location and their functional diversity, pericytes may play an active role in regulation of inflammatory, fibrogenic and angiogenic responses following myocardial infarction (Alex and Frangogiannis 2019).

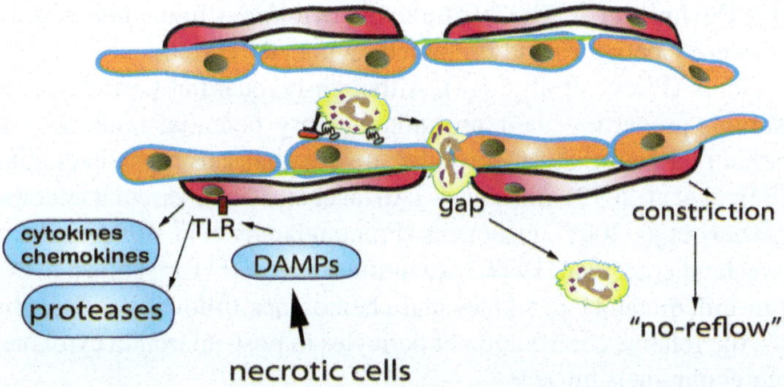

Fig. 10.2 Pericytes during the inflammatory phase of infarct healing. Necrotic cardiomyocytes release damage-associated molecular patterns (DAMPs) that activate Toll-like receptor (TLR) pathways in all cells involved in cardiac repair. DAMPs may activate a pro-inflammatory phenotype in pericytes inducing chemokine and cytokine secretion. Considering that several other cell types are capable of synthesizing and secreting pro-inflammatory mediators in the infarcted heart (including cardiomyocytes, endothelial cells, macrophages and fibroblasts), the relative significance of pericyte-mediated inflammatory actions remains unknown. Inflammatory leukocytes are recruited in the infarct through adhesive interactions with endothelial cells. Formation of gaps between pericytes may be important in leukocyte transmigration. It has been suggested that pericyte constriction in the reperfused ischemic myocardium may be implicated in the pathogenesis of "no reflow", the failure to reperfuse despite restoration of blood flow in the epicardial coronary

10.4.1.2 Pericytes in the Inflammatory Phase of Cardiac Repair (Fig. 10.2)

The role of pericytes in regulation of microvascular permeability is well-established (Bhowmick et al. 2019; Bischoff et al. 2017). Following myocardial infarction, induction of pro-inflammatory cytokines increases microvascular permeability (Gao et al. 2017). Experimental evidence suggests that pericyte perturbations may be involved in accentuation of microvascular permeability in the infarcted heart. Although the molecular signals involved in modulation of pericyte function have not been systematically characterized, published evidence suggests a role for nerve growth factor (NGF) and angiopoietin-2 in pericyte-mediated vascular hyperpermeability. The injury-induced cytokine pro-NGF is rapidly upregulated in ischemic cardiomyocytes and activates the p75 receptor in pericytes promoting microvascular dysfunction that extends the area of infarction (Siao et al. 2012). On the other hand, angiopoietin-2 is released by ischemic endothelial cells and perturbs endothelial-pericyte connections, promoting pericyte detachment, and increasing microvascular permeability (Lee et al. 2018b).

In addition to their effects in regulation of microvascular function, pericytes in the infarcted heart could also serve as inflammatory cells, producing and secreting cytokines and chemokines. DAMPs released in the infarcted area by dying cardiomyocytes may activate TLR signaling in pericytes promoting a pro-inflammatory phenotype. Brain and lung pericytes are known to express TLRs and can activate inflammatory cascades in response to stimulation with TLR ligands (Nyul-Toth et al. 2017; Edelman et al. 2007). Moreover, cytokine stimulation induced chemokine synthesis by pericytes and promoted leukocyte recruitment in many different in vitro models (Pieper et al. 2013). Although myocardial pericytes are likely to exhibit similar properties, their pro-inflammatory potential remains unknown. It should be emphasized that several other myocardial cell types, including macrophages (Bajpai et al. 2019), fibroblasts (Anzai et al. 2017), vascular endothelial cells (Frangogiannis et al. 2001), mast cells (Frangogiannis et al. 1998) and cardiomyocytes (Gwechenberger et al. 1999; Mezzaroma et al. 2011) are known to express and secrete pro-inflammatory cytokines and chemokines following myocardial infarction. Thus, the relative contribution of pericytes in post-infarction cytokine and chemokine upregulation is unclear.

Due to their strategic peri-endothelial location, pericytes may serve as critical regulators of leukocyte transendothelial trafficking cascades. Recruitment of neutrophils, monocytes and lymphocytes in sites of injury is driven by adhesive interactions between endothelial cells and circulating leukocytes (Vestweber 2015). Intravital microscopy experiments in cremaster muscle demonstrated that cytokine stimulation induces transition of pericytes to an inflammatory phenotype, characterized by expression of adhesion molecules and chemoattractant mediators that provide molecular cues to transmigrating leukocytes. Moreover, inflammatory activation causes shape changes in pericytes, promoting enlargement of gaps between adjacent pericytes. It has been proposed that leukocytes may use the enlarged gaps as exit points during transmigration into the injured tissue (Proebstl et al. 2012). Whether morphologic alterations and adhesive interactions between pericytes and leukocytes are involved in inflammatory cell trafficking in infarcted hearts remains unknown.

Degradation of the native extracellular matrix by activated proteases is a prominent early event following myocardial infarction (Frangogiannis 2017b). In addition to their potential role in activation of leukocyte trafficking and in post-infarction inflammation, pericytes may also play an important role in degradation of the pericellular extracellular matrix in the healing infarct. Pericyte-mediated breakdown of the basement membrane and detachment from endothelial cells is an important early step in angiogenesis (van Hinsbergh and Koolwijk 2008).

10.4.1.3 The Role of Pericytes in the Pathogenesis of No-Reflow

The "no-reflow phenomenon" is a term used to describe the failure to perfuse the ischemic myocardium despite restoration of coronary flow (Kloner et al. 1974). In experimental models of myocardial ischemia/reperfusion, endothelial injury, and leukocyte plugging have been suggested to cause microvessel occlusion, contributing to coronary "no-reflow", and accentuating ischemic injury. In human patients, thrombi and atherosclerotic debris may embolize microvessels further reducing microvascular flow (Rezkalla et al. 2017). A recently published study in a rat model of myocardial ischemia and reperfusion suggested that pericyte constriction contributes to no-reflow by causing microvascular obstruction (O'Farrell et al. 2017;

O'Farrell and Attwell 2014). Pericyte relaxation through adenosine treatment markedly increased myocardial perfusion. Although this finding may have potential therapeutic implications for patients with myocardial infarction, the molecular signals that promote pericyte constriction following reperfusion remain poorly understood (Costa et al. 2018).

10.4.1.4 The Contribution of Pericytes in Fibroblast Activation Following Myocardial Infarction (Fig. 10.3)

Transition to the proliferative phase of cardiac repair is preceded by suppression of pro-inflammatory signaling cascades (Chen et al. 2012). Induction and activation of TGF-β in the healing infarct suppresses pro-inflammatory signaling in infarct macrophages, while activating fibroblasts (Hanna and Frangogiannis 2019; Chen et al. 2019; Kong et al. 2018). Fibroblasts become activated and acquire a myofibroblast phenotype, incorporating α-SMA into cytoskeletal stress fibers, and producing large amounts of extracellular matrix proteins (Willems et al. 1994; Cleutjens et al. 1995; Frangogiannis et al. 2000a). Myofibroblast activation is accompanied by an intense

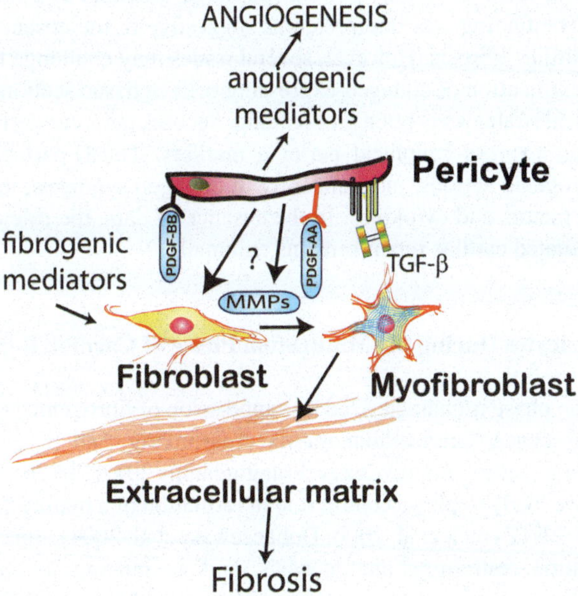

Fig. 10.3 Pericytes during the proliferative phase of cardiac repair. During the proliferative phase of cardiac repair, pericytes may act as regulators of fibrosis and angiogenesis. Pericytes may contribute to formation of the scar by converting to myofibroblasts, or by producing fibroblast-activating mediators. The relative contribution of pericytes in expansion of infarct myofibroblasts remains unknown due to the absence of robust data. Moreover, pericytes may regulate angiogenic responses in the infarcted heart. Pericyte activation in the healing infarct may reflect actions of growth factors, such as PDGFs and TGF-βs

angiogenic response, that may serve to enhance perfusion of the hypercellular, metabolically active infarct, ensuring supply of oxygen and nutrients (Ren et al. 2002). Limited experimental evidence is available on the potential role of pericytes in the proliferative phase of cardiac repair. Considering their abundance, phenotypic plasticity and functional diversity, pericytes may contribute to several of the cellular responses associated with cardiac repair.

Experimental studies exploring the cell biology of fibrotic remodeling in the kidney (Leaf et al. 2017), lung (Sava et al. 2017) and skin (Sundberg et al. 1996), suggest that pericytes may directly contribute to myofibroblast activation and to deposition of collagenous matrix. However, evidence documenting effects of pericytes in activation of fibrogenic pathways in the infarcted heart is lacking. Recent studies using robust lineage tracing approaches suggest that resident Tcf21+ fibroblast populations account for the expansion of myofibroblasts in the infarcted heart (Kanisicak et al. 2016). To what extent, cardiac pericytes may undergo conversion to myofibroblast-like cells remains unknown. Kramann et al. found that a large number of myofibroblasts in remodeling and infarcted hearts are derived from Gli1+ perivascular cells (Kramann et al. 2016). The potential overlap of these cells with bona fide pericytes is unclear. Studies using pericyte-reporter mice suggested that, although myocardial infarction triggers expansion of a nestin-negative/NG2+ pericyte population, these cells may not express large amounts of collagen (Birbrair et al. 2014). The finding was interpreted as suggestive of the absence of a role for pericytes in cardiac fibrosis. However, several issues may challenge this interpretation. First, identification of collagen-expressing cells in tissue sections using immunohistochemical strategies is often problematic. Second, pericyte-derived fibroblasts may no longer express canonical pericyte markers. Third, pericytes may exert important fibrogenic actions, independently of collagen synthesis, through synthesis of growth factors and cytokines. In the absence of data, the role of pericytes in fibroblast-mediated cardiac repair remains enigmatic.

10.4.1.5 Pericytes During the Maturation Phase of Cardiac Repair

The maturation phase is characterized by suppression of fibrogenic responses, myofibroblast quiescence, extracellular matrix crosslinking, and by acquisition of a mural cell coat by infarct microvessels (Frangogiannis 2012, 2017b). Both pericytes and VSMCs are likely to play a central role in formation of a mature scar, by coating infarct neovessels (Zymek et al. 2006; Dobaczewski et al. 2004), thus stabilizing the microvasculature, restraining angiogenesis, and attenuating inflammation. In a mouse model of reperfused myocardial infarction, inhibition of the PDGF-BB/PDGFRβ axis perturbed coating of infarct neovessels with mural cells and prolonged hemorrhagic and inflammatory changes in the healing infarct (Zymek et al. 2006). Thus, mural cells may play a critical role in orchestration of cardiac repair by preventing excessive or prolonged pro-inflammatory and angiogenic activation of endothelial cells.

10.4.2 Pericytes in Chronic Heart Failure

In contrast to the sudden and dramatic loss of cardiomyocytes following myocardial infarction, chronic heart failure is associated with insidious onset of cardiac dysfunction. In human patients, chronic heart failure may be caused by a wide range of pathophysiologic perturbations, including pressure overload (such as in patients with hypertension or critical aortic stenosis), volume overload (typically observed in patients with valvular regurgitant lesions), diabetes, obesity and metabolic dysfunction, genetic cardiomyopathies, myocarditis, etc. Two distinct clinical phenotypes of heart failure have been identified. About 50% of heart failure patients have reduced ejection fraction (Heart Failure with reduced Ejection Fraction – HFrEF), reflecting systolic dysfunction, whereas the other half has preserved ejection fraction (Heart Failure with preserved Ejection Fraction – HFpEF) (Borlaug and Redfield 2011). Cell biological alterations in both cardiomyocytes and non-cardiomyocytes may be implicated in the pathogenesis of heart failure. The role of pericytes in chronic heart failure is unknown. Considering their known properties as fibrogenic cells and as regulators of microvascular function and angiogenesis, pericytes may contribute to the pathogenesis of heart failure through several different mechanisms.

First, cardiac pericytes may contribute to the development of fibrosis (Fig. 10.4). Expansion of the cardiac interstitium due to net accumulation of extracellular matrix proteins is a common pathophysiologic companion in most cardiac pathologic conditions (Frangogiannis 2019; Berk et al. 2007; Kong et al. 2014). Although in many

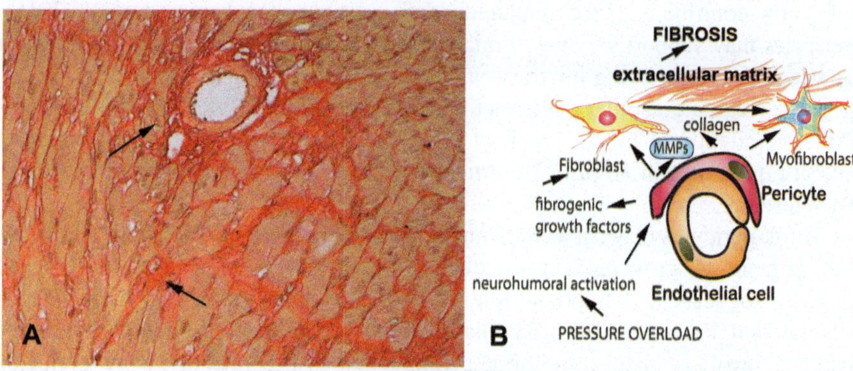

Fig. 10.4 The potential role of pericytes in myocardial fibrosis associated with chronic heart failure. (**a**) Picrosirius red staining shows collagen deposition in the pressure-overloaded heart. Seven days after transverse aortic constriction, fibrosis is noted in the interstitium and in perivascular areas. The presence of significiant perivascular fibrosis (arrows) may suggest an involvement of mural cells in collagen deposition. (**b**) Pericytes may contribute to fibrosis in the pressure-overloaded heart through several different mechanisms. First, pericytes may convert to myofibroblasts. Second, pericytes may produce fibrogenic growth factors or extracellular matrix proteins (structural and matricellular). Third, pericytes may secrete matrix metalloproteinases (MMPs), contributing to matrix remodeling

cases, cardiac fibrosis is the result of a reparative process that is activated in response to cardiomyocyte necrosis (for example, the typical reparative response noted in the infarcted myocardium), in many pathophysiologic conditions, insidious interstitial or perivascular fibrosis develops in the absence of significant cardiomyocyte replacement. In these cases, fibrotic remodeling may contribute to ventricular dysfunction. Activation of resident cardiac fibroblasts and conversion to activated myofibroblasts has been suggested as the predominant cellular mechanism involved in fibrosis of the pressure-overloaded heart (Ali et al. 2014; Moore-Morris et al. 2014). Whether pericytes may also contribute to the expansion of activates myofibroblasts remains unknown. In non-cardiac tissues (including the kidney and the lung), pericytes have been suggested as an important source of activated fibroblasts in both injury models and in human fibrotic conditions (Sava et al. 2017; Lin et al. 2008). However, a study using lineage tracing of Tbx18+ "pericytes" in the pressure-overloaded myocardium demonstrated that these cells maintain their identity and do not significantly contribute to other lineages (Guimaraes-Camboa et al. 2017). Considering the presence of significant perivascular fibrosis in experimental models of left ventricular pressure overload may reflect direct conversion of mural cells to collagen-secreting fibroblasts, or may result from induction and secretion of fibrogenic mediators by activated pericytes.

Second, perturbed pericyte function may promote dysfunction in the failing heart through actions on the microvasculature. Pathologic cardiac hypertrophy is typically associated with microvascular dysfunction and capillary rarefaction (McConkey et al. 2019; Camici et al. 2012, 2020) that may exacerbate systolic function or promote ischemia, even in the absence of epicardial coronary stenosis. HFpEF patients exhibit significant coronary microvascular rarefaction that may significantly contribute to the impaired cardiac reserve (Mohammed et al. 2015). Pericytes may support survival of microvascular endothelial cells under conditions of stress, thus preventing microvascular loss and preserving function. In addition, pericytes regulate capillary constriction (Methner et al. 2019). Impaired pericyte-driven regulation of blood flow may be involved in the pathogenesis of impaired exercise tolerance in HFpEF. Unfortunately, these intriguing hypotheses have not been tested.

Third, pericytes may act as inflammatory cells contributing to dysfunction in the failing heart. A growing body of evidence in experimental models and human patients suggests an important role for microvascular inflammation in HFpEF (Paulus and Tschope 2013). To what extent, inflammatory activation in HFpEF patients involves acquisition of a cytokine-expressing phenotype by pericytes remains unknown.

10.4.2.1 Pericytes in Diabetic Cardiomyopathy

Epidemiologic studies have demonstrated that diabetes, obesity and metabolic dysfunction are associated with an increased incidence of heart failure (Gilbert and Krum 2015). Histopathologic studies in myocardial samples from diabetics (Rubler et al. 1972) suggested a distinct entity, termed "diabetic cardiomyopathy" that may be independent of coronary disease or other concomitant conditions, and may contribute to the pathogenesis of heart failure. Obesity on the other hand, has been suggested to cause a distinct clinical phenotype of HFpEF, with worse exercise capacity, and the presence of cardiac inflammation, interstitial fibrosis and microvascular dysfunction (Obokata et al. 2017). Both diabetes and obesity are associated with fibrotic remodeling of the ventricle and with significant diastolic dysfunction (Russo and Frangogiannis 2016; Alex et al. 2018; Cavalera et al. 2014). Pericyte dysfunction is prominently noted in diabetic patients and in experimental models of diabetes, and has been implicated in the pathogenesis of diabetic complications. In diabetic retinopathy loss of pericytes through apoptosis or necrosis may abrogate key homeostatic signals that stabilize the endothelium, promoting neovessel formation (Feenstra et al. 2013). The role of myocardial pericytes in the pathogenesis of diabetic cardiomyopathy remains poorly understood. Whether pericyte to fibroblast conversion, or pericyte-derived growth factor secretion accounts for fibrotic remodeling of diabetic hearts has not been investigated. Moreover, diabetes-associated pericyte dysfunction may play a critical role in the pathogenesis of microvascular dysfunction and impaired angiogenesis in diabetic hearts. A recent study showed that myocardial samples from patients with diabetes exhibit capillary rarefaction, associated with pericyte loss (Hinkel et al. 2017). Moreover, in a diabetic pig model of myocardial ischemia, impaired responses to angiogenic factor stimulation were associated with perturbed pericyte recruitment (Hinkel et al. 2017).

10.5 Pericytes as Therapeutic Targets in Myocardial Disease

Due to the limited evidence on the functional role of pericytes in myocardial disease, development of therapeutic approaches targeting specific pericyte-mediated actions remains challenging. Inhibition of pericyte constriction has been suggested as a promising strategy for protection of the ischemic and reperfused myocardium from "no-reflow" (O'Farrell et al. 2017; O'Farrell and Attwell 2014). Whether approaches targeting endogenous pericyte-mediated actions may protect from adverse post-infarction remodeling through effects on inflammatory or fibrotic responses remains unknown. Targeting pericyte functions may perturb vascular maturation in the healing infarct (Zymek et al. 2006), thus causing formation of defective, hyperpermeable vessels that may result in impaired cardiac repair.

Considering their established role in angiogenesis, and suggestions for potential regenerative properties, cell therapy with pericytes has been considered an attractive strategy to improve repair of the infarcted heart. Several experimental studies have reported that intramyocardial delivery of pericytes may protect the infarcted myocardium from adverse remodeling through angiogenic and anti-apoptotic actions (Avolio and Madeddu 2016; Avolio et al. 2015a; Cathery et al. 2018). Beneficial effects were attributed to paracrine effects of pericytes on cardiomyocytes, fibroblasts and endothelial cells. In a mouse model of myocardial infarction, transplantation of saphenous vein-derived pericytes into the peri-infarct zone ameliorated systolic dysfunction and attenuated adverse remodeling. The protective effects were attributed to secretion of angiogenic mediators and to release of microRNAs (Katare et al. 2011). In another study, transplantation of human skeletal muscle pericytes into immunocompromised mice subjected to coronary artery ligation improved dysfunction, attenuating cardiac fibrosis and reducing macrophage infiltration (Chen et al. 2013). Studies in large animal models have also suggested beneficial actions of pericyte cell therapy following infarction. In a swine model of reperfused myocardial infarction, transplantation of human adventitial pericytes improved capillary dysfunction and reduced fibrosis (Alvino et al. 2018) without affecting ventricular function. Some investigations have suggested that subsets of pericytes may also have limited cardiomyogenic potential (Chen et al. 2015). Although this is an intriguing possibility, documentation of pericyte-driven remuscularization in vivo is lacking.

10.6 Conclusions and Future Directions

Despite their abundance, their strategic location around vessels, and their diverse functions in regulation of microvascular blood flow, vascular permeability, inflammation, fibrosis and angiogenesis, cardiac pericytes have been ignored by myocardial biologists. There is an urgent need for studies systematically characterizing cardiac pericyte subpopulations using single cell transcriptomics, identifying subsets with distinct functional properties, and investigating the functional role of pericytes in experimental models of heart failure. Moreover, there is a need to study the molecular and functional profile of pericytes in human hearts, systematically characterizing their properties in normal and diseased human myocardium.

Acknowledgments Dr. Frangogiannis' laboratory is supported by NIH grants R01 HL76246 and R01 HL85440, and by Department of Defense grants PR151029 and PR181464. Dr. Tuleta is supported by a grant from the Deutsche Forschungsgemeinschaft (TU 632/1-1).

References

Alex L, Frangogiannis NG (2019) Pericytes in the infarcted heart. Vasc Biol 1:H23–H31

Alex L, Russo I, Holoborodko V, Frangogiannis NG (2018) Characterization of a mouse model of obesity-related fibrotic cardiomyopathy that recapitulates features of human heart failure with preserved ejection fraction. Am J Physiol Heart Circ Physiol 315:H934–H949

Ali SR, Ranjbarvaziri S, Talkhabi M, Zhao P, Subat A, Hojjat A, Kamran P, Muller AM, Volz KS, Tang Z, Red-Horse K, Ardehali R (2014) Developmental heterogeneity of cardiac fibroblasts does not predict pathological proliferation and activation. Circ Res 115:625–635

Alvino VV, Fernandez-Jimenez R, Rodriguez-Arabaolaza I, Slater S, Mangialardi G, Avolio E, Spencer H, Culliford L, Hassan S, Sueiro Ballesteros L, Herman A, Ayaon-Albarran A, Galan-Arriola C, Sanchez-Gonzalez J, Hennessey H, Delmege C, Ascione R, Emanueli C, Angelini

GD, Ibanez B, Madeddu P (2018) Transplantation of allogeneic pericytes improves myocardial vascularization and reduces interstitial fibrosis in a swine model of reperfused acute myocardial infarction. J Am Heart Assoc 7:e006727

Anzai A, Choi JL, He S, Fenn AM, Nairz M, Rattik S, McAlpine CS, Mindur JE, Chan CT, Iwamoto Y, Tricot B, Wojtkiewicz GR, Weissleder R, Libby P, Nahrendorf M, Stone JR, Becher B, Swirski FK (2017) The infarcted myocardium solicits GM-CSF for the detrimental oversupply of inflammatory leukocytes. J Exp Med 214:3293–3310

Armulik A, Genove G, Betsholtz C (2011) Pericytes: developmental, physiological, and pathological perspectives, problems, and promises. Dev Cell 21:193–215

Arslan F, Smeets MB, O'Neill LA, Keogh B, McGuirk P, Timmers L, Tersteeg C, Hoefer IE, Doevendans PA, Pasterkamp G, de Kleijn DP (2010) Myocardial ischemia/reperfusion injury is mediated by leukocytic toll-like receptor-2 and reduced by systemic administration of a novel anti-toll-like receptor-2 antibody. Circulation 121:80–90

Attwell D, Mishra A, Hall CN, O'Farrell FM, Dalkara T (2016) What is a pericyte? J Cereb Blood Flow Metab 36:451–455

Avolio E, Madeddu P (2016) Discovering cardiac pericyte biology: from physiopathological mechanisms to potential therapeutic applications in ischemic heart disease. Vasc Pharmacol 86:53–63

Avolio E, Meloni M, Spencer HL, Riu F, Katare R, Mangialardi G, Oikawa A, Rodriguez-Arabaolaza I, Dang Z, Mitchell K, Reni C, Alvino VV, Rowlinson J, Livi U, Cesselli D, Angelini G, Emanueli C, Beltrami AP, Madeddu P (2015a) Combined intramyocardial delivery of human pericytes and cardiac stem cells additively improves the healing of mouse infarcted hearts through stimulation of vascular and muscular repair. Circ Res 116:e81–e94

Avolio E, Rodriguez-Arabaolaza I, Spencer HL, Riu F, Mangialardi G, Slater SC, Rowlinson J, Alvino VV, Idowu OO, Soyombo S, Oikawa A, Swim MM, Kong CH, Cheng H, Jia H, Ghorbel MT, Hancox JC, Orchard CH, Angelini G, Emanueli C, Caputo M, Madeddu P (2015b) Expansion and characterization of neonatal cardiac pericytes provides a novel cellular option for tissue engineering in congenital heart disease. J Am Heart Assoc 4:e002043

Bajpai G, Bredemeyer A, Li W, Zaitsev K, Koenig AL, Lokshina I, Mohan J, Ivey B, Hsiao HM, Weinheimer C, Kovacs A, Epelman S, Artyomov M, Kreisel D, Lavine KJ (2019) Tissue resident CCR2- and CCR2+ cardiac macrophages differentially orchestrate monocyte recruitment and fate specification following myocardial injury. Circ Res 124:263–278

Berk BC, Fujiwara K, Lehoux S (2007) ECM remodeling in hypertensive heart disease. J Clin Invest 117:568–575

Bhowmick S, D'Mello V, Caruso D, Wallerstein A, Abdul-Muneer PM (2019) Impairment of pericyte-endothelium crosstalk leads to blood-brain barrier dysfunction following traumatic brain injury. Exp Neurol 317:260–270

Biname F (2014) Transduction of extracellular cues into cell polarity: the role of the transmembrane proteoglycan NG2. Mol Neurobiol 50:482–493

Birbrair A, Zhang T, Files DC, Mannava S, Smith T, Wang ZM, Messi ML, Mintz A, Delbono O (2014) Type-1 pericytes accumulate after tissue injury and produce collagen in an organ-dependent manner. Stem Cell Res Ther 5:122

Birbrair A, Zhang T, Wang ZM, Messi ML, Mintz A, Delbono O (2015) Pericytes at the intersection between tissue regeneration and pathology. Clin Sci (Lond) 128:81–93

Bischoff FC, Werner A, John D, Boeckel JN, Melissari MT, Grote P, Glaser SF, Demolli S, Uchida S, Michalik KM, Meder B, Katus HA, Haas J, Chen W, Pullamsetti SS, Seeger W, Zeiher AM, Dimmeler S, Zehendner CM (2017) Identification and functional characterization of hypoxia-induced endoplasmic reticulum stress regulating lncRNA (HypERlnc) in pericytes. Circ Res 121:368–375

Borlaug BA, Redfield MM (2011) Diastolic and systolic heart failure are distinct phenotypes within the heart failure spectrum. Circulation 123:2006–2013. discussion 2014

Cai CL, Martin JC, Sun Y, Cui L, Wang L, Ouyang K, Yang L, Bu L, Liang X, Zhang X, Stallcup WB, Denton CP, McCulloch A, Chen J, Evans SM (2008) A myocardial lineage derives from Tbx18 epicardial cells. Nature 454:104–108

Camici PG, Olivotto I, Rimoldi OE (2012) The coronary circulation and blood flow in left ventricular hypertrophy. J Mol Cell Cardiol 52:857–864

Camici PG, Tschope C, Di Carli MF, Rimoldi O, Van Linthout S (2020) Coronary microvascular dysfunction in hypertrophy and heart failure. Cardiovasc Res 116:806–816

Cathery W, Faulkner A, Maselli D, Madeddu P (2018) Concise review: the regenerative journey of pericytes toward clinical translation. Stem Cells 36:1295–1310

Cavalera M, Wang J, Frangogiannis NG (2014) Obesity, metabolic dysfunction, and cardiac fibrosis: pathophysiological pathways, molecular mechanisms, and therapeutic opportunities. Transl Res 164:323–335

Chen B, Frangogiannis NG (2017) Immune cells in repair of the infarcted myocardium. Microcirculation 24:e12305

Chen W, Saxena A, Li N, Sun J, Gupta A, Lee DW, Tian Q, Dobaczewski M, Frangogiannis NG (2012) Endogenous IRAK-M attenuates postinfarction remodeling through effects on macrophages and fibroblasts. Arterioscler Thromb Vasc Biol 32:2598–2608

Chen CW, Okada M, Proto JD, Gao X, Sekiya N, Beckman SA, Corselli M, Crisan M, Saparov A, Tobita K, Peault B, Huard J (2013) Human pericytes for ischemic heart repair. Stem Cells 31:305–316

Chen WC, Baily JE, Corselli M, Diaz ME, Sun B, Xiang G, Gray GA, Huard J, Peault B (2015) Human myocardial pericytes: multipotent mesodermal precursors exhibiting cardiac specificity. Stem Cells 33:557–573

Chen Q, Zhang H, Liu Y, Adams S, Eilken H, Stehling M, Corada M, Dejana E, Zhou B, Adams RH (2016) Endothelial cells are progenitors of cardiac pericytes and vascular smooth muscle cells. Nat Commun 7:12422

Chen B, Huang S, Su Y, Wu YJ, Hanna A, Brickshawana A, Graff J, Frangogiannis NG (2019) Macrophage Smad3 protects the infarcted heart, stimulating phagocytosis and regulating inflammation. Circ Res 125:55–70

Chintalgattu V, Ai D, Langley RR, Zhang J, Bankson JA, Shih TL, Reddy AK, Coombes KR, Daher IN, Pati S, Patel SS, Pocius JS, Taffet GE, Buja LM, Entman ML, Khakoo AY (2010) Cardiomyocyte PDGFR-beta signaling is an essential component of the mouse cardiac response to load-induced stress. J Clin Invest 120:472–484

Chintalgattu V, Rees ML, Culver JC, Goel A, Jiffar T, Zhang J, Dunner K Jr, Pati S, Bankson JA, Pasqualini R, Arap W, Bryan NS, Taegtmeyer H, Langley RR, Yao H, Kupferman ME, Entman ML, Dickinson ME, Khakoo AY (2013) Coronary microvascular pericytes are the cellular target of sunitinib malate-induced cardiotoxicity. Sci Transl Med 5:187–169

Cleutjens JP, Verluyten MJ, Smiths JF, Daemen MJ (1995) Collagen remodeling after myocardial infarction in the rat heart. Am J Pathol 147:325–338

Costa MA, Paiva AE, Andreotti JP, Cardoso MV, Cardoso CD, Mintz A, Birbrair A (2018) Pericytes constrict blood vessels after myocardial ischemia. J Mol Cell Cardiol 116:1–4

Crisan M, Yap S, Casteilla L, Chen CW, Corselli M, Park TS, Andriolo G, Sun B, Zheng B, Zhang L, Norotte C, Teng PN, Traas J, Schugar R, Deasy BM, Badylak S, Buhring HJ, Giacobino JP, Lazzari L, Huard J, Peault B (2008) A perivascular origin for mesenchymal stem cells in multiple human organs. Cell Stem Cell 3:301–313

Diaz-Flores L, Gutierrez R, Madrid JF, Varela H, Valladares F, Acosta E, Martin-Vasallo P, Diaz-Flores L Jr (2009) Pericytes. Morphofunction, interactions and pathology in a quiescent and activated mesenchymal cell niche. Histol Histopathol 24:909–969

Dobaczewski M, Akrivakis S, Nasser K, Michael LH, Entman ML, Frangogiannis NG (2004) Vascular mural cells in healing canine myocardial infarcts. J Histochem Cytochem 52:1019–1029

Dutta P, Nahrendorf M (2015) Monocytes in myocardial infarction. Arterioscler Thromb Vasc Biol 35:1066–1070

Edelman DA, Jiang Y, Tyburski J, Wilson RF, Steffes C (2006) Pericytes and their role in microvasculature homeostasis. J Surg Res 135:305–311

Edelman DA, Jiang Y, Tyburski JG, Wilson RF, Steffes CP (2007) Lipopolysaccharide activation of pericyte's toll-like receptor-4 regulates co-culture permeability. Am J Surg 193:730–735

Epelman S, Lavine KJ, Beaudin AE, Sojka DK, Carrero JA, Calderon B, Brija T, Gautier EL, Ivanov S, Satpathy AT, Schilling JD, Schwendener R, Sergin I, Razani B, Forsberg EC, Yokoyama WM, Unanue ER, Colonna M, Randolph GJ, Mann DL (2014) Embryonic and adult-derived resident cardiac macrophages are maintained through distinct mechanisms at steady state and during inflammation. Immunity 40:91–104

Feenstra DJ, Yego EC, Mohr S (2013) Modes of retinal cell death in diabetic retinopathy. J Clin Exp Ophthalmol 4:298

Forbes MS, Rennels ML, Nelson E (1977) Ultrastructure of pericytes in mouse heart. Am J Anat 149:47–70

Frangogiannis NG (2012) Regulation of the inflammatory response in cardiac repair. Circ Res 110:159–173

Frangogiannis NG (2014) The inflammatory response in myocardial injury, repair, and remodelling. Nat Rev Cardiol 11:255–265

Frangogiannis NG (2015) Pathophysiology of myocardial infarction. Compr Physiol 5:1841–1875

Frangogiannis NG (2017a) The role of transforming growth factor (TGF)-beta in the infarcted myocardium. J Thorac Dis 9:S52–S63

Frangogiannis NG (2017b) The extracellular matrix in myocardial injury, repair, and remodeling. J Clin Invest 127:1600–1612

Frangogiannis NG (2019) Cardiac fibrosis: cell biological mechanisms, molecular pathways and therapeutic opportunities. Mol Asp Med 65:70–99

Frangogiannis NG, Lindsey ML, Michael LH, Youker KA, Bressler RB, Mendoza LH, Spengler RN, Smith CW, Entman ML (1998) Resident cardiac mast cells degranulate and release preformed TNF-alpha, initiating the cytokine cascade in experimental canine myocardial ischemia/reperfusion. Circulation 98:699–710

Frangogiannis NG, Michael LH, Entman ML (2000a) Myofibroblasts in reperfused myocardial infarcts express the embryonic form of smooth muscle myosin heavy chain (SMemb). Cardiovasc Res 48:89–100

Frangogiannis NG, Mendoza LH, Lindsey ML, Ballantyne CM, Michael LH, Smith CW, Entman ML (2000b) IL-10 is induced in the reperfused myocardium and may modulate the reaction to injury. J Immunol 165:2798–2808

Frangogiannis NG, Mendoza LH, Lewallen M, Michael LH, Smith CW, Entman ML (2001) Induction and suppression of interferon-inducible protein 10 in reperfused myocardial infarcts may regulate angiogenesis. FASEB J 15:1428–1430

Fu X, Khalil H, Kanisicak O, Boyer JG, Vagnozzi RJ, Maliken BD, Sargent MA, Prasad V, Valiente-Alandi I, Blaxall BC, Molkentin JD (2018) Specialized fibroblast differentiated states underlie scar formation in the infarcted mouse heart. J Clin Invest 128:2127–2143

Gao XM, Wu QZ, Kiriazis H, Su Y, Han LP, Pearson JT, Taylor AJ, Du XJ (2017) Microvascular leakage in acute myocardial infarction: characterization by histology, biochemistry, and magnetic resonance imaging. Am J Physiol Heart Circ Physiol 312:H1068–H1075

Gersch C, Dewald O, Zoerlein M, Michael LH, Entman ML, Frangogiannis NG (2002) Mast cells and macrophages in normal C57/BL/6 mice. Histochem Cell Biol 118:41–49

Gilbert RE, Krum H (2015) Heart failure in diabetes: effects of anti-hyperglycaemic drug therapy. Lancet 385:2107–2117

Grant RI, Hartmann DA, Underly RG, Berthiaume AA, Bhat NR, Shih AY (2019) Organizational hierarchy and structural diversity of microvascular pericytes in adult mouse cortex. J Cereb Blood Flow Metab 39:411–425

Guimaraes-Camboa N, Cattaneo P, Sun Y, Moore-Morris T, Gu Y, Dalton ND, Rockenstein E, Masliah E, Peterson KL, Stallcup WB, Chen J, Evans SM (2017) Pericytes of multiple organs do not behave as mesenchymal stem cells in vivo. Cell Stem Cell 20:345–359. e345

Gushi A, Tanaka M, Tsuyama S, Nagai T, Kanzaki T, Kanekura T, Matsuyama T (2008) The 3G5 antigen is expressed in dermal mast cells but not pericytes. J Cutan Pathol 35:278–284
Gwechenberger M, Mendoza LH, Youker KA, Frangogiannis NG, Smith CW, Michael LH, Entman ML (1999) Cardiac myocytes produce interleukin-6 in culture and in viable border zone of reperfused infarctions. Circulation 99:546–551
Hanna A, Frangogiannis NG (2019) The role of the TGF-beta superfamily in myocardial infarction. Front Cardiovasc Med 6:140
Hellstrom M, Kalen M, Lindahl P, Abramsson A, Betsholtz C (1999) Role of PDGF-B and PDGFR-beta in recruitment of vascular smooth muscle cells and pericytes during embryonic blood vessel formation in the mouse. Development 126:3047–3055
Hertig V, Tardif K, Meus MA, Duquette N, Villeneuve L, Toussaint F, Ledoux J, Calderone A (2017) Nestin expression is upregulated in the fibrotic rat heart and is localized in collagen-expressing mesenchymal cells and interstitial CD31(+)- cells. PLoS One 12:e0176147
Hinkel R, Howe A, Renner S, Ng J, Lee S, Klett K, Kaczmarek V, Moretti A, Laugwitz KL, Skroblin P, Mayr M, Milting H, Dendorfer A, Reichart B, Wolf E, Kupatt C (2017) Diabetes mellitus-induced microvascular destabilization in the myocardium. J Am Coll Cardiol 69:131–143
Huebener P, Abou-Khamis T, Zymek P, Bujak M, Ying X, Chatila K, Haudek S, Thakker G, Frangogiannis NG (2008) CD44 is critically involved in infarct healing by regulating the inflammatory and fibrotic response. J Immunol 180:2625–2633
Hui M, Tenenbaum HC (1998) New face of an old enzyme: alkaline phosphatase may contribute to human tissue aging by inducing tissue hardening and calcification. Anat Rec 253:91–94
Ivey MJ, Tallquist MD (2016) Defining the cardiac fibroblast. Circ J 80:2269–2276
Kanisicak O, Khalil H, Ivey MJ, Karch J, Maliken BD, Correll RN, Brody MJ, SC JL, Aronow BJ, Tallquist MD, Molkentin JD (2016) Genetic lineage tracing defines myofibroblast origin and function in the injured heart. Nat Commun 7:12260
Kapur NK, Wilson S, Yunis AA, Qiao X, Mackey E, Paruchuri V, Baker C, Aronovitz MJ, Karumanchi SA, Letarte M, Kass DA, Mendelsohn ME, Karas RH (2012) Reduced endoglin activity limits cardiac fibrosis and improves survival in heart failure. Circulation 125:2728–2738
Katare R, Riu F, Mitchell K, Gubernator M, Campagnolo P, Cui Y, Fortunato O, Avolio E, Cesselli D, Beltrami AP, Angelini G, Emanueli C, Madeddu P (2011) Transplantation of human pericyte progenitor cells improves the repair of infarcted heart through activation of an angiogenic program involving micro-RNA-132. Circ Res 109:894–906
Khan WS, Adesida AB, Tew SR, Lowe ET, Hardingham TE (2010) Bone marrow-derived mesenchymal stem cells express the pericyte marker 3G5 in culture and show enhanced chondrogenesis in hypoxic conditions. J Orthop Res 28:834–840
Kloner RA, Ganote CE, Jennings RB (1974) The "no-reflow" phenomenon after temporary coronary occlusion in the dog. J Clin Invest 54:1496–1508
Kong P, Christia P, Frangogiannis NG (2014) The pathogenesis of cardiac fibrosis. Cell Mol Life Sci 71:549–574
Kong P, Shinde AV, Su Y, Russo I, Chen B, Saxena A, Conway SJ, Graff JM, Frangogiannis NG (2018) Opposing actions of fibroblast and cardiomyocyte Smad3 signaling in the infarcted myocardium. Circulation 137:707–724
Kramann R, Schneider RK, DiRocco DP, Machado F, Fleig S, Bondzie PA, Henderson JM, Ebert BL, Humphreys BD (2015) Perivascular Gli1+ progenitors are key contributors to injury-induced organ fibrosis. Cell Stem Cell 16:51–66
Kramann R, Wongboonsin J, Chang-Panesso M, Machado FG, Humphreys BD (2016) Gli1+ Pericyte loss induces capillary rarefaction and proximal tubular injury. J Am Soc Nephrol 25:1924–1931
Leaf IA, Nakagawa S, Johnson BG, Cha JJ, Mittelsteadt K, Guckian KM, Gomez IG, Altemeier WA, Duffield JS (2017) Pericyte MyD88 and IRAK4 control inflammatory and fibrotic responses to tissue injury. J Clin Invest 127:321–334
Lee JS, Jeong SJ, Kim S, Chalifour L, Yun TJ, Miah MA, Li B, Majdoubi A, Sabourin A, Keler T, Guimond JV, Haddad E, Choi EY, Epelman S, Choi JH, Thibodeau J, Oh GT, Cheong C

(2018a) Conventional dendritic cells impair recovery after myocardial infarction. J Immunol 201:1784–1798
Lee SJ, Lee CK, Kang S, Park I, Kim YH, Kim SK, Hong SP, Bae H, He Y, Kubota Y, Koh GY (2018b) Angiopoietin-2 exacerbates cardiac hypoxia and inflammation after myocardial infarction. J Clin Invest 128:5018–5033
Lee LL, Khakoo AY, Chintalgattu V (2019) Isolation and purification of murine cardiac pericytes. J Vis Exp 2019:e53208
Leroyer AS, Blin MG, Bachelier R, Bardin N, Blot-Chabaud M, Dignat-George F (2019) CD146 (cluster of differentiation 146). Arterioscler Thromb Vasc Biol 39:1026–1033
Li Y, Song D, Mao L, Abraham DM, Bursac N (2020) Lack of Thy1 defines a pathogenic fraction of cardiac fibroblasts in heart failure. Biomaterials 236:119824
Lin SL, Kisseleva T, Brenner DA, Duffield JS (2008) Pericytes and perivascular fibroblasts are the primary source of collagen-producing cells in obstructive fibrosis of the kidney. Am J Pathol 173:1617–1627
Mathiisen TM, Lehre KP, Danbolt NC, Ottersen OP (2010) The perivascular astroglial sheath provides a complete covering of the brain microvessels: an electron microscopic 3D reconstruction. Glia 58:1094–1103
McConkey HZR, Marber M, Chiribiri A, Pibarot P, Redwood SR, Prendergast BD (2019) Coronary microcirculation in aortic stenosis. Circ Cardiovasc Interv 12:e007547
Methner C, Mishra A, Golgotiu K, Li Y, Wei W, Yanez ND, Zlokovic B, Wang RK, Alkayed NJ, Kaul S, Iliff JJ (2019) Pericyte constriction underlies capillary derecruitment during hyperemia in the setting of arterial stenosis. Am J Physiol Heart Circ Physiol 317:H255–H263
Meus MA, Hertig V, Villeneuve L, Jasmin JF, Calderone A (2017) Nestin expressed by pre-existing cardiomyocytes recapitulated in part an embryonic phenotype; suppressive role of p38 MAPK. J Cell Physiol 232:1717–1727
Mezzaroma E, Toldo S, Farkas D, Seropian IM, Van Tassell BW, Salloum FN, Kannan HR, Menna AC, Voelkel NF, Abbate A (2011) The inflammasome promotes adverse cardiac remodeling following acute myocardial infarction in the mouse. Proc Natl Acad Sci U S A 108:19725–19730
Milner DJ, Weitzer G, Tran D, Bradley A, Capetanaki Y (1996) Disruption of muscle architecture and myocardial degeneration in mice lacking desmin. J Cell Biol 134:1255–1270
Minor M, Alcedo KP, Battaglia RA, Snider NT (2019) Cell type- and tissue-specific functions of ecto-5′-nucleotidase (CD73). Am J Physiol Cell Physiol 317:C1079–C1092
Mohammed SF, Hussain S, Mirzoyev SA, Edwards WD, Maleszewski JJ, Redfield MM (2015) Coronary microvascular rarefaction and myocardial fibrosis in heart failure with preserved ejection fraction. Circulation 131:550–559
Moore-Morris T, Guimaraes-Camboa N, Banerjee I, Zambon AC, Kisseleva T, Velayoudon A, Stallcup WB, Gu Y, Dalton ND, Cedenilla M, Gomez-Amaro R, Zhou B, Brenner DA, Peterson KL, Chen J, Evans SM (2014) Resident fibroblast lineages mediate pressure overload-induced cardiac fibrosis. J Clin Invest 124:2921–2934
Murray IR, Baily JE, Chen WCW, Dar A, Gonzalez ZN, Jensen AR, Petrigliano FA, Deb A, Henderson NC (2017) Skeletal and cardiac muscle pericytes: functions and therapeutic potential. Pharmacol Ther 171:65–74
Nag AC (1980) Study of non-muscle cells of the adult mammalian heart: a fine structural analysis and distribution. Cytobios 28:41–61
Nayak RC, Berman AB, George KL, Eisenbarth GS, King GL (1988) A monoclonal antibody (3G5)-defined ganglioside antigen is expressed on the cell surface of microvascular pericytes. J Exp Med 167:1003–1015
Nees S, Weiss DR, Senftl A, Knott M, Forch S, Schnurr M, Weyrich P, Juchem G (2012) Isolation, bulk cultivation, and characterization of coronary microvascular pericytes: the second most frequent myocardial cell type in vitro. Am J Physiol Heart Circ Physiol 302:H69–H84
Nyul-Toth A, Kozma M, Nagyoszi P, Nagy K, Fazakas C, Hasko J, Molnar K, Farkas AE, Vegh AG, Varo G, Galajda P, Wilhelm I, Krizbai IA (2017) Expression of pattern recognition receptors

and activation of the non-canonical inflammasome pathway in brain pericytes. Brain Behav Immun 64:220–231

O'Farrell FM, Attwell D (2014) A role for pericytes in coronary no-reflow. Nat Rev Cardiol 11:427–432

O'Farrell FM, Mastitskaya S, Hammond-Haley M, Freitas F, Wah WR, Attwell D (2017) Capillary pericytes mediate coronary no-reflow after myocardial ischaemia. elife 6:e29280

Obokata M, Reddy YNV, Pislaru SV, Melenovsky V, Borlaug BA (2017) Evidence supporting the existence of a distinct obese phenotype of heart failure with preserved ejection fraction. Circulation 136:6–19

Ozerdem U, Grako KA, Dahlin-Huppe K, Monosov E, Stallcup WB (2001) NG2 proteoglycan is expressed exclusively by mural cells during vascular morphogenesis. Dev Dyn 222:218–227

Paulus WJ, Tschope C (2013) A novel paradigm for heart failure with preserved ejection fraction: comorbidities drive myocardial dysfunction and remodeling through coronary microvascular endothelial inflammation. J Am Coll Cardiol 62:263–271

Pereira FE, Cronin C, Ghosh M, Zhou SY, Agosto M, Subramani J, Wang R, Shen JB, Schacke W, Liang B, Yang TH, McAulliffe B, Liang BT, Shapiro LH (2013) CD13 is essential for inflammatory trafficking and infarct healing following permanent coronary artery occlusion in mice. Cardiovasc Res 100:74–83

Pieper C, Pieloch P, Galla HJ (2013) Pericytes support neutrophil transmigration via interleukin-8 across a porcine co-culture model of the blood-brain barrier. Brain Res 1524:1–11

Pinto AR, Ilinykh A, Ivey MJ, Kuwabara JT, D'Antoni ML, Debuque R, Chandran A, Wang L, Arora K, Rosenthal NA, Tallquist MD (2016) Revisiting cardiac cellular composition. Circ Res 118:400–409

Prabhu SD, Frangogiannis NG (2016) The biological basis for cardiac repair after myocardial infarction: from inflammation to fibrosis. Circ Res 119:91–112

Proebstl D, Voisin MB, Woodfin A, Whiteford J, D'Acquisto F, Jones GE, Rowe D, Nourshargh S (2012) Pericytes support neutrophil subendothelial cell crawling and breaching of venular walls in vivo. J Exp Med 209:1219–1234

Ren G, Michael LH, Entman ML, Frangogiannis NG (2002) Morphological characteristics of the microvasculature in healing myocardial infarcts. J Histochem Cytochem 50:71–79

Rezkalla SH, Stankowski RV, Hanna J, Kloner RA (2017) Management of no-reflow phenomenon in the catheterization laboratory. JACC Cardiovasc Interv 10:215–223

Rubler S, Dlugash J, Yuceoglu YZ, Kumral T, Branwood AW, Grishman A (1972) New type of cardiomyopathy associated with diabetic glomerulosclerosis. Am J Cardiol 30:595–602

Russo I, Frangogiannis NG (2016) Diabetes-associated cardiac fibrosis: cellular effectors, molecular mechanisms and therapeutic opportunities. J Mol Cell Cardiol 90:84–93

Russo I, Cavalera M, Huang S, Su Y, Hanna A, Chen B, Shinde AV, Conway SJ, Graff J, Frangogiannis NG (2019) Protective effects of activated myofibroblasts in the pressure-overloaded myocardium are mediated through smad-dependent activation of a matrix-preserving program. Circ Res 124:1214–1227

Sava P, Ramanathan A, Dobronyi A, Peng X, Sun H, Ledesma-Mendoza A, Herzog EL, Gonzalez AL (2017) Human pericytes adopt myofibroblast properties in the microenvironment of the IPF lung. JCI Insight 2:e96352

Sawtell NM, Lessard JL (1989) Cellular distribution of smooth muscle actins during mammalian embryogenesis: expression of the alpha-vascular but not the gamma-enteric isoform in differentiating striated myocytes. J Cell Biol 109:2929–2937

Shinde AV, Humeres C, Frangogiannis NG (1863a) The role of alpha-smooth muscle actin in fibroblast-mediated matrix contraction and remodeling. Biochim Biophys Acta 2017:298–309

Shinde AV, Humeres C, Frangogiannis NG (1863b) The role of alpha-smooth muscle actin in fibroblast-mediated matrix contraction and remodeling. Biochim Biophys Acta 2016:298–309

Siao CJ, Lorentz CU, Kermani P, Marinic T, Carter J, McGrath K, Padow VA, Mark W, Falcone DJ, Cohen-Gould L, Parrish DC, Habecker BA, Nykjaer A, Ellenson LH, Tessarollo L, Hempstead

BL (2012) ProNGF, a cytokine induced after myocardial infarction in humans, targets pericytes to promote microvascular damage and activation. J Exp Med 209:2291–2305
Simonavicius N, Ashenden M, van Weverwijk A, Lax S, Huso DL, Buckley CD, Huijbers IJ, Yarwood H, Isacke CM (2012) Pericytes promote selective vessel regression to regulate vascular patterning. Blood 120:1516–1527
Sims DE (1986) The pericyte--a review. Tissue Cell 18:153–174
Souders CA, Borg TK, Banerjee I, Baudino TA (2012) Pressure overload induces early morphological changes in the heart. Am J Pathol 181:1226–1235
Sundberg C, Ivarsson M, Gerdin B, Rubin K (1996) Pericytes as collagen-producing cells in excessive dermal scarring. Lab Investig 74:452–466
Tallquist MD (2020) Cardiac fibroblast diversity. Annu Rev Physiol 82:63–78
Tual-Chalot S, Garcia-Collado M, Redgrave RE, Singh E, Davison B, Park C, Lin H, Luli S, Jin Y, Wang Y, Lawrie A, Jakobsson L, Arthur HM (2020) Loss of endothelial endoglin promotes high-output heart failure through peripheral arteriovenous shunting driven by VEGF signaling. Circ Res 126:243–257
van Hinsbergh VW, Koolwijk P (2008) Endothelial sprouting and angiogenesis: matrix metalloproteinases in the lead. Cardiovasc Res 78:203–212
Vestweber D (2015) How leukocytes cross the vascular endothelium. Nat Rev Immunol 15:692–704
Villalobos E, Criollo A, Schiattarella GG, Altamirano F, French KM, May HI, Jiang N, Nguyen NUN, Romero D, Roa JC, Garcia L, Diaz-Araya G, Morselli E, Ferdous A, Conway SJ, Sadek HA, Gillette TG, Lavandero S, Hill JA (2019) Fibroblast primary cilia are required for cardiac fibrosis. Circulation 139:2342–2357
Willems IE, Havenith MG, De Mey JG, Daemen MJ (1994) The alpha-smooth muscle actin-positive cells in healing human myocardial scars. Am J Pathol 145:868–875
Ya J, Markman MW, Wagenaar GT, Blommaart PJ, Moorman AF, Lamers WH (1997) Expression of the smooth-muscle proteins alpha-smooth-muscle actin and calponin, and of the intermediate filament protein desmin are parameters of cardiomyocyte maturation in the prenatal rat heart. Anat Rec 249:495–505
Zhao G, Joca HC, Nelson MT, Lederer WJ (2020) ATP- and voltage-dependent electro-metabolic signaling regulates blood flow in heart. Proc Natl Acad Sci U S A 117(13):7461–7470
Ziegler T, Horstkotte J, Schwab C, Pfetsch V, Weinmann K, Dietzel S, Rohwedder I, Hinkel R, Gross L, Lee S, Hu J, Soehnlein O, Franz WM, Sperandio M, Pohl U, Thomas M, Weber C, Augustin HG, Fassler R, Deutsch U, Kupatt C (2013) Angiopoietin 2 mediates microvascular and hemodynamic alterations in sepsis. J Clin Invest 123:3436–3445
Zymek P, Bujak M, Chatila K, Cieslak A, Thakker G, Entman ML, Frangogiannis NG (2006) The role of platelet-derived growth factor signaling in healing myocardial infarcts. J Am Coll Cardiol 48:2315–2323

Chapter 11
Adventitial and Skeletal Muscle Pericytes in Health and Ischemic Tissue Regeneration

A. Rampin, P. Campagnolo, V. Mastrullo, P. Madeddu, and G. Spinetti

Abstract
Introduction
Adventitial and skeletal muscle pericytes possess intrinsic reparative capacity, both *in situ* where they participate to tissue repair and remodelling and when used as a cell therapy option. Understanding their reparative potential and how pathological conditions associated with ischemia affect these cells is crucial in order to fully exploit their therapeutic potential.

The following chapter will therefore address the biological characteristics of adventitial and skeletal muscle pericytes in health and tissue-specific diseases. We will focus on the latest molecular discoveries, discussing their potential in particular relating to the still unmet clinical need of restoring blood flow in peripheral arterial disease and critical limb ischemia.

Methods
Extensive literature search has been performed using PubMed and Google Scholar online databases using key words such as 'adventitial progenitors', 'skeletal muscle pericytes', 'pericytes and diabetes', 'pericytes and atherosclerosis', 'pericytes and ischemia'. Information has been gathered and summarised regarding their biology, development, regenerative potential and their roles in disease and repair.

A. Rampin and P. Campagnolo contributed equally.

A. Rampin · G. Spinetti (✉)
Laboratory of Cardiovascular Research, IRCCS, MultiMedica, Milan, Italy
e-mail: gaia.spinetti@multimedica.it

P. Campagnolo · V. Mastrullo
Department of Biochemical Sciences, University of Surrey, Guildford, UK

P. Madeddu (✉)
Laboratory of Experimental Cardiovascular Medicine, University of Bristol, Bristol, UK
e-mail: paolo.madeddu@bristol.ac.uk

A. Birbrair (ed.), *Biology of Pericytes – Recent Advances*, Stem Cell Biology and Regenerative Medicine 68, https://doi.org/10.1007/978-3-030-62129-2_11

Results
Recent evidence demonstrates pericytes residing in the tunica adventitia of large vessels are endowed with pro-angiogenic activity. Adventitial pericytes share markers of mesenchymal progenitors and contributes to the remodelling of the vessel wall in pathological conditions. Similarly, skeletal muscle pericytes can support endothelial cell organization during angiogenesis and vasculogenesis, but can also differentiate in skeletal myocytes. Skeletal muscle pericytes also play central roles in the development, homeostasis and ageing of their native tissue and play a central role in diabetes-related endothelial dysfunction. Both cell populations have been extensively and successfully used in pre-clinical models to promote the regeneration of post-ischemic tissues and combined in tissue engineering approaches to enhance their performance.

Conclusions
Their specific biological features render these two subclasses of pericytes especially attractive tools in the development of tissue regeneration strategies to treat ischemic tissues damage.

Keywords Adipogenesis · Adventitia · Angiogenesis · Blood flow · Critical limb ischemia · Diabetes mellitus · Molecular mechanisms · Multilineage differentiation · Non-coding RNAs · Peripheral arterial disease · Regeneration · Skeletal muscle · Tissue engineering · Vasculogenesis · Vessel wall

Abbreviations

Ang-1	Angiopoietin-1
Ang-2	Angiopoietin-2
AP	Alkaline phosphatase
bFGF-2	Basic fibroblasts growth factor – 2
BM	Basement membrane
CLI	Critical limb ischemia
ECs	Endothelial Cells
EV	Extracellular vesicle
hiPSCs	Human induced pluripotent stem cells
IGF-1	Insulin-like growth factor – 1
ITGBL1	Integrin subunit βlike – 1
MBs	Mesoangioblasts
MSCs	Mesenchymal stem cells
Myf-5	Myogenic factor-5
NCCs	Neural crest cells
NF-kB	Nuclear factor kappa-light-chain-enhancer of activated B cells
NG-2	Neural/glial antigen 2
$p75^{NTR}$	Neurotrophine receptor p75

PAD	Peripheral arterial disease
PAX7	Paired box protein 7
PDGFR-β	Platelet-derived growth factor receptor β
PEG	Polyethylene glycol
α-SG	A-sarcoglycan
PACs	Pro-angiogenic cells
PKC	Protein kinase C
SCs	Satellite cells
SMPs	Skeletal muscle pericytes
SHP1	Src homology region 2 domain-containing phosphatase-1
TGF-β	Transforming growth factor-β
T1D	Type 1 diabetes
T2DM	Type 2 diabetes mellitus
VCAF	Vascular calcification associated factor
VEGF	Vascular endothelial growth factor
VEZF1	Vascular endothelial zinc finger-1
VSMCs	Vascular smooth muscle cells

11.1 Introduction

Upon ischemia, the condition of low oxygen and starvation ignites the process of tissue repair aimed at re-establishing blood flow and replacing damaged/dead cells. First in line to respond to the insult are endothelial cells (ECs) via proliferation, migration, and secretion of pro-angiogenic molecules. However, to accomplish the difficult task of growing new vessels, ECs require the support of circulating pro-angiogenic cells (PACs) and tissue resident cells, such as pericytes.

The first description of pericytes dates back to the late 1800s when Eberth and Rouget described them as adventitial/mural cells juxtaposed to capillary ECs (Rouget 1873). In 1923 Zimmerman coined the term 'pericyte' referring to their location adjacent to capillaries and embedded within the same basement membrane (BM) (Zimmermann 1923).

The origin of pericytes has been largely debated during the last one hundred years due to both the mosaic-like embryonic vasculogenesis and the similarity in cell phenotypes (pericytes, vascular smooth muscle cells (VSMCs) and various stromal cells) observed throughout the vessel walls, which makes their ontological classification challenging. One of the first observations relating to the origin of pericytes dates back in 1925, when Eliot R. and Eleanor L. Clark observed that mural cells in amphibian larvae originate from a mesodermic population clustering around the yolk sack, during the capillary plexus formation (Clark and Clark 1925). However, it gradually became evident that pericytes and other mural cells have, and most likely share, at least four different origins, depending on their location. These have been identified in the splancnic mesoderm, neural crest cells (NCCs), second heart field and somites (Armulik et al. 2011). Evidence from markers expression, topology and morphology suggests, but does not necessarily implies, a common origin with VSMCs. An interesting study conducted on chicken embryos by Bergwerff and colleagues in 1998 can help us understand the intertwined paths of these two cell types. By using lacZ reporter vectors coupled with α-SMA immunofluorescence, NCCs were tracked down to the development sites of some important vessels, where the authors observed they originally differentiated into VSMCs, and later into cells of the intima as well as in fibroblasts of the adventitia (Bergwerff et al. 1998). However, it is important to note that the early identification as VSMCs was by the use of a monoclonal antibody recognizing a muscle specific actin reported to also be expressed in pericytes Moreover, differentiating VMSCs were apparently in close contact with the endothelium, therefore suggesting these differentiating mural cells might be a common progenitor for adult VMSCs and pericytes (Bergwerff et al. 1998; Tsukada et al. 1987). In support of this hypothesis, the accumulation of de-differentiated, actively proliferating VSMCs that is observed in the intima as a consequence of injury or atherosclerosis has been attributed at least in part to vessel wall derived progenitor cells that show characteristics not dissimilar to pericytes (Wang et al. 2015). Moreover, a common origin has been suggested also for VSMCs and pericytes comprising the walls of vessels originating from mesoderm. In an *in vitro* study conducted on human induced pluripotent stem cells (hiPSCs) mesenchyme angioblast, basic fibroblasts growth factor-2(bFGF-2) dependent colonies were identified as common clonal precursors for VSMCs and pericytes (Kumar et al. 2017). Pericytes have recently attracted increasing attention since they are both able to modulate ECs functions (migration, proliferation, permeability and contractility) via direct and paracrine interactions, and possess multilineage differentiation ability (Campagnolo et al. 2010; Dulmovits and Herman 2012). Therefore pericytes may represent a plastic tool for vascular regeneration strategies. However, although pericytes can participate in vascular and muscular regeneration in ischemic conditions, the cardio-metabolic diseases often associated with ischemia negatively impact on their functional capabilities. This is particularly relevant in the case of diabetes mellitus, a disease frequently associated with cardiovascular ischemic complications. Of the 425 million diabetic patients in the world (1/11 people), 20–30% have macrovascular complications, including ischemic heart disease, cerebrovascular disease and peripheral arterial disease (PAD). Critical limb ischemia (CLI), the most advanced form of PAD, requires foot amputation in 25% of cases within one year from diagnosis. Furthermore, 25% of patients with CLI will die during the same period (Faglia et al. 2006; Hirsch et al. 2006). Poor revascularization results in more severe outcomes in terms of major amputation and mortality rates. Thus, researchers are focusing on understanding the biology of diabetic pericytes in order to exploit them in new vascular regenerative strategies for diabetic CLI (Fig. 11.1).

In this chapter, we will describe the features of pericytes residing in the adventitia of large vessels and surrounding capillaries in the skeletal muscles, their origin and distinctive characteristics, and their potential use in regenerative medicine.

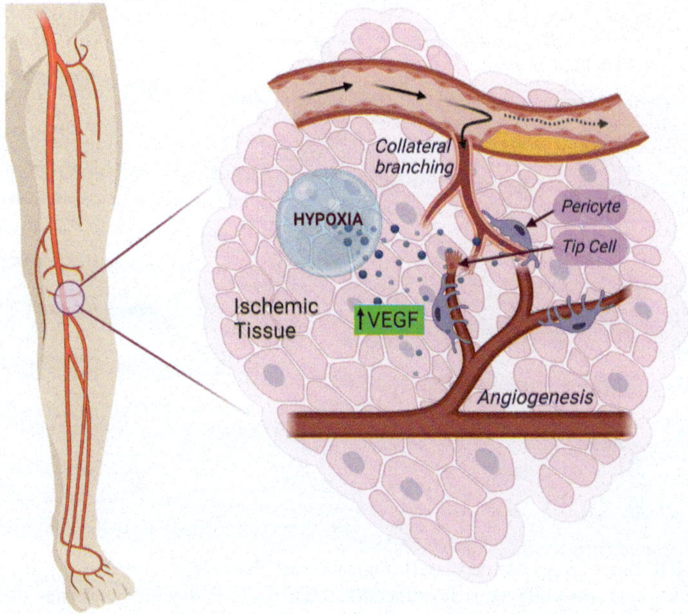

Fig. 11.1 Critical Limb Ischemia (CLI) impairs blood flow in peripheral tissues. The drop in oxygen diffusion, beyond physiological level (hypoxia) triggers the formation of new blood vessels to compensate the nutrients supply. Pericytes play a fundamental role in capillary formation and therefore may constitute a potential target for regenerative medicine. (Created with Biorender.com)

11.2 Adventitial Pericytes

11.2.1 Introduction

Large arteries and veins are complex structures formed by precisely organised concentric layers: tunica intima, media and adventitia. The tunica intima is composed of a single impermeable layer of ECs, strictly mediating the exchanges of fluids, gasses, nutrients and cells between the blood and the organism. Separated by the lamina intima, the tunica media is mainly composed of an aligned array of VSMCs maintaining the vascular tone in response to stimuli and changes in blood pressure and blood flow requirements. The outermost layer forming the wall of large blood vessels is the tunica adventitia, a loose network of extracellular matrix proteins, primarily collagens, containing a variety of cell types, such as macrophages and myofibroblasts (Mazurek et al. 2017). The tunica adventitia also contains the *vasa vasorum*, a thick network of microvessels providing oxygen and nutrients to cells within the vessel wall. This layer has been defined the 'vasculogenic zone' due to the presence of several progenitor cell populations that have been described by

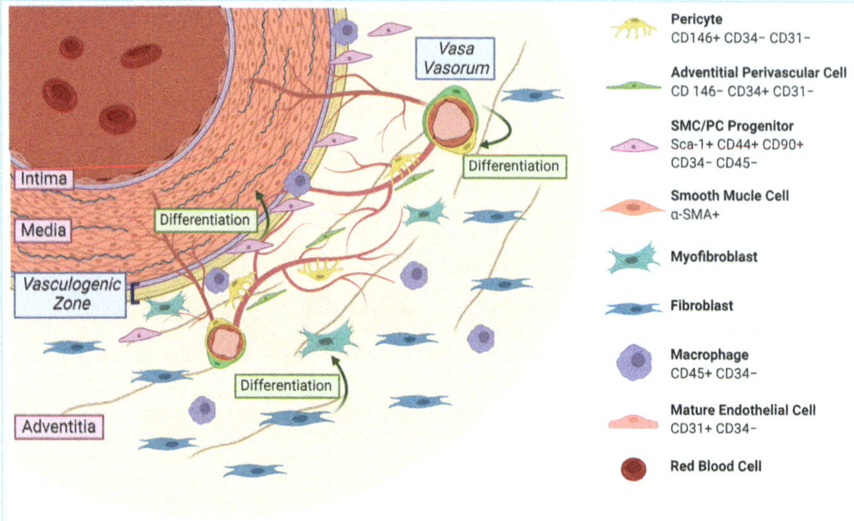

Fig. 11.2 The outer layer of large vessels (tunica adventitia) contains a heterogeneous population of cells embedded in a collagen rich extracellular matrix. The adventitial contains the "vasculogenic zone", a dynamic area containing progenitor cells involved in several processes such as migration, angiogenesis and remodelling of the vascular wall, and the vasa vasorum, a network of small vessels with the function of supplying blood nutrients to the surrounding and the tunica media. (Created with Biorender.com)

different groups to promote angiogenesis and to participate to the remodelling of the vascular wall (Fig. 11.2) (Zengin et al. 2006).

Among those, different populations of pericyte-like cells have been characterised either for their expression of typical pericyte markers or for their perivascular location within the *vasa vasorum*. Others have been described as 'adventitial progenitor' or 'mesenchymal progenitors' with no direct reference to their location but are nonetheless characterised by largely similar phenotypes and will therefore be discussed here.

11.2.2 A Brief Definition of Adventitial 'Pericyte-Like' Populations

As in all other tissues, pericytes in large vessels elude precise definition based on surface markers alone and appear as a heterogeneous population broadly presenting mesenchymal-like characteristics *in vitro*. CD146+ CD34- cells have been identified as bona fide pericytes residing in a perivascular position around microvessels across many human organs (Crisan et al. 2008), including the *vasa vasorum* of large vessels (Billaud et al. 2017). CD34+ CD146- cells were also identified in the adventitia of large vessels such as the aorta and the saphenous vein (Campagnolo et al.

2010; Billaud et al. 2017; Howson et al. 2005; Pasquinelli et al. 2007; Corselli et al. 2012) and visualized wrapping around the microvasculature of the *vasa vasorum*. The current view of the field is that these two distinct populations are of the same origin, with the CD34+ population representing the more undifferentiated of the two (Corselli et al. 2012).

Several other surface markers have been used to identify mesenchymal cells from the adventitia, although their perivascular localisation has not always been confirmed. The cell populations of the adventitia and their characteristic markers most relevant to this chapter are summarised in Table 11.1.

11.2.3 Adventitial Pericytes in Pathology

The vascular wall is predominantly composed of fully differentiated cells which stay quiescent, unless stimulated or damaged. Vascular remodelling is a process of repair and regeneration of the blood vessel wall after acute injury, but it also defines the slower continuous changes occurring in reaction to chronic insults, leading to atherosclerosis and calcification (Lusis 2000). Examples of acute injuries are endovascular interventions, that can either damage the endothelium alone by removing part of the monolayer, or affect the tunica media too, for example, revascularization through balloon inflation (Kim and Dean 2011). The causes of chronic injuries are instead less distinct and often rather defined as 'risk factors'. Common examples are hypercholesterolemia, diabetes and disturbed flow downstream of a stent (Lusis 2000).

In response to damage, the endothelium becomes activated, expresses pro-inflammatory proteins promoting leukocyte transmigration and represses the release of nitric oxide (NO). Activation of VSMCs follows and is characterised by increased proliferation and migration as well as a reduction of their contractile capacity. The uncontrolled proliferation of VSMC leads to their invasion into the luminal space and formation of the 'neointima' layer which reduces the lumen available to the blood flow (Kim and Dean 2011). In chronic disruption of blood vessel homeostasis, the thickening of the intima can be accompanied by the formation of a lipid and macrophage rich atherosclerotic plaque (Lusis 2000). While the role of the intimal and medial layer has been characterised since the 1970s, the participation of the adventitial cells is still poorly understood and hotly debated. In 2013, Tiggers et al. definitively reported the post-injury activation of neural/glial antigen-2 (NG-2) + and platelet derived growth factor receptor-β (PDGFR-β) + adventitial pericytes; these cells were then detected in the neointimal layer at later time points, demonstrating their direct contribution to remodelling (Tigges et al. 2013). The contribution of adventitial pericytes to vascular pathology can also be linked to their broad differentiation capacity, which includes osteogenesis (Campagnolo et al. 2010; Crisan et al. 2008; Corselli et al. 2012). Indeed, in the pro-atherosclerotic environment of ApoE−/− mice 3G5+ pericytes from the vascular wall formed robust and mature bone, indicating that they might play a significant role in the calcification of large

Table 11.1 Summary of the principal markers used to identify mesenchymal cells

Marker expression	Perivascular location	Vasa vasorum	Mesenchymal differentiation	Pro-angiogenic in vitro	Pro-angiogenic in vivo
CD146+	Corselli et al. (2012), Tsang et al. (2013), James et al. (2012a), Tigges et al. (2013), Robin et al. (2009)	Billaud et al. (2017)	Crisan et al. (2008), Tsang et al. (2013), Schwab and Gargett (2007), Deasy et al. (2009), Montemurro et al. (2011), Li et al. (2019)	Wang et al. (2019)	Maier et al. (2010)
CD34+	Pasquinelli et al. (2007), Torsney et al. (2007)	Campagnolo et al. (2010), Zengin et al. (2006), Chang et al. (2012)	Zimmerlin et al. (2010), Mekala et al. (2018)	Campagnolo et al. (2010), Howson et al. (2005), Invernici et al. (2007)	Campagnolo et al. (2010), Chang et al. (2012), Caporali et al. (2015), Alvino et al. (2018)
Sca1+	Hu et al. (2004), Passman et al. (2008), Ieronimakis et al. (2013), Psaltis et al. (2014)	Chen et al. (2013), Toledo-Flores et al. (2019)	Klatte-Schulz et al. (2012), Tsai et al. (2012), Wong et al. (2013), Majesky et al. (2017)	Toledo-Flores et al. (2019), Sainz et al. (2006)	Toledo-Flores et al. (2019)
3G5	Castrechini et al. (2010), Khan et al. (2008), Leszczynska et al. (2016)		Doherty et al. (1998), Farrington-Rock et al. (2004)		
CD44+	Klein et al. (2011)	Garoffolo et al. (2020)	Covas et al. (2005), Sarugaser et al. (2005)		
c-kit	Aicher et al. (2007), Blazquez-Martinez et al. (2014)	Davie et al. (2004), Montani et al. (2011)	Blazquez-Martinez et al. (2014), Ni et al. (2019)		Fang et al. (2012), Russell and Brown (2014)
α-SMA	Kim et al. (2020)			Benjamin et al. (1998)	
CD13+			De Palma et al. (2005)		
Gli1	Kramann et al. (2015)		Kramann et al. (2016)		
CD90+	Michelis et al. (2018)		Michelis et al. (2018)		

vessels (Leszczynska et al. 2016). Furthermore, vascular calcification associated factor (VCAF)-positive sporadic adventitial cells co-express CD34 and CD146 and contribute to atherosclerotic plaque formation and calcification (Wilkinson et al. 2007). Additionally, the expression of VCAF had been previously implicated in the pericyte calcification process *in vitro* (Alexander et al. 2005).

Given the discussed overlap in marker expression of adventitial populations and their cross-differentiation capacity (Corselli et al. 2012), the distinct contribution of adventitial pericytes to vascular remodelling is somewhat difficult to discern from that of other adventitial progenitor populations. For example Sca-1+ adventitial progenitor cells (Hu et al. 2004; Chen et al. 2013) and adventitial myofibroblasts (Shi et al. 1996), both of them characterized as heterogeneous populations, were shown to contribute to neointimal formation. Additionally, CD44+ cells from the adventitia of grafted saphenous vein are known to proliferate and participate to the recipient vessel remodeling (Garoffolo et al. 2020).

Together these findings demonstrate adventitial pericyte-like cells play an active role in the repair and pathogenesis of the vessel wall and can therefore represent a therapeutic target to reduce the occurrence of remodelling.

11.2.4 Adventitial Pericytes in Repair

Adventitial cells have been shown to participate in the vascular repopulation process, however the role of the pericytic subpopulation is less studied (Campagnolo et al. 2010, 2015a; Chen et al. 2013). Additionally, adventitial pericytes and mesenchymal cells display a remarkable pro-angiogenic capacity, promoting the formation and maturation of endothelial-formed capillaries through direct cell-cell contact and the release of paracrine factors (Campagnolo et al. 2010; Klein et al. 2011). *In vivo*, CD34+ progenitors from the foetal (Invernici et al. 2007) and adult (Campagnolo et al. 2010; Katare et al. 2011) vasculature were able to stimulate angiogenesis and regeneration in ischemic tissues. In particular, adult saphenous vein-derived adventitial pericytes increased angiogenesis and arteriogenesis promoting the recovery of function after hindlimb ischemia (Campagnolo et al. 2010) and myocardial infarction (Alvino et al. 2018; Katare et al. 2011), by secreting pro-angiogenic factors and non-coding RNAs of the microRNA subclass (e.g. miR-132) and physically interacting with the forming vasculature. Their regenerative potential after myocardial infarction was improved by the combined delivery of cardiac stem cells due to the synergistic stimulation of angiogenesis and muscular repair, and correlated with their epigenetic status and level of DNA methylation (Avolio et al. 2015).

11.2.5 Adventitial Pericytes in Tissue Engineering

Tissue engineering combines progenitor cells and three-dimensional biomaterial-based scaffolds to form tissue-like constructs with the aim of providing novel solutions for organ repair. Pro-angiogenic capacities (Campagnolo et al. 2010; Katare et al. 2011) and osteo-chondrogenic differentiation potential (Corselli et al. 2012; Leszczynska et al. 2016) of adventitial pericytes are particularly appealing to the tissue engineering field.

Tissue engineering approaches have been combined with adventitial pericytes to improve their retention at the site of ischemia, to prolong and maximize their well characterized pro-angiogenic activity. For example,. adventitial progenitors seeded onto a synthetic electrospun scaffold that was then shaped as a cuff and implanted around the occluded femoral artery in a model of limb ischemia yielded outstanding blood flow recovery (Carrabba et al. 2016).

Adventitial pericytes have also been utilised in vascular tissue engineering applications due to their ability to support the function of the endothelium. The combination of saphenous vein-derived pericytes with an advanced biomodified electrospun scaffold was employed to produce an innovative vessel graft solution. The scaffold functionalities synergistically enhanced the pro-angiogenic properties of the cells and ultimately improved endothelial coverage of the scaffold by stimulating migration of ECs (Campagnolo et al. 2016). In a previous paper, adventitial progenitor cells were seeded on a decellularized vascular scaffold. The presence of the cells and their *in situ* differentiation significantly improved the graft patency and performance after *in vivo* implantation (Campagnolo et al. 2015b).

Other applications of adventitial pericytes related to tissue engineering have focused on the osteo-chondrogenic interface. Lipoaspirate-derived CD146+ and CD34+ adventitial pericyte populations can be isolated consistently from patients, independently of age and gender and have been demonstrated to provide an easily accessible source of cells for bone regeneration (James et al. 2012b). While CD146+ cells possess superior bone healing capacity (James et al. 2012b), CD34+ cells release potent paracrine factor combinations. Together the combination of both two populations resulted in a greater effective response (Wang et al. 2019). Interestingly, both these cell populations were able to support chondrogenesis on electrospun scaffold material either through direct differentiation or through paracrine stimulation of chondrocytes (Zhang et al. 2015). However, this reparative potential was not translated into the histological repair of Achille's tendon defect, despite improvements in the overall mechanical properties being observed (Devana et al. 2018) (Fig. 11.3).

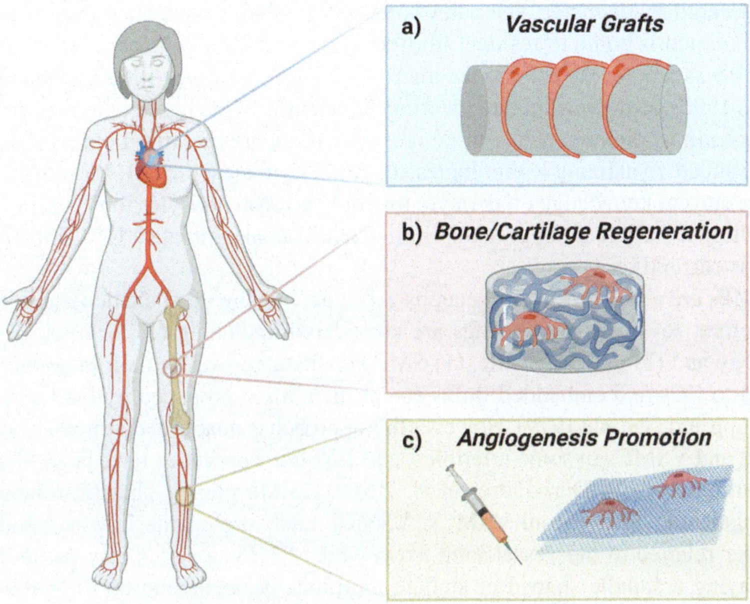

Fig. 11.3 Applications of pericytes in tissue engineering: (**a**) Adventitial pericytes seeded on a vessel shaped-scaffold to derive a vascular graft; (**b**) pericytes embedded in a scaffold for osteochondro regeneration; (**c**) pericytes injected onto a synthetic scaffold at the site of ischemia to promote angiogenesis and blood flow recovery. (Created with Biorender.com)

11.3 Skeletal Muscle Pericytes

11.3.1 Introduction

Skeletal muscle accounts for approximately 40% of the adult human body weight (Hoehn and Marieb 2012). Such a vast structure is fed by epimysial arteries, which branch into smaller arterioles running in close proximity to every muscle fascicle, eventually feeding smaller arcade arterioles that penetrate the fascicles and branch into a finely organised capillary network (Gitiaux et al. 2013). Within this vast network, skeletal muscle pericytes (SMPs) have been identified by a number of studies over the past 40 years (Bruns and Palade 1968; Tilton et al. 1979; Shepro and Morel 1993). In an observational study conducted by Burns and Palade in 1968, SMPs in several skeletal muscle tissues from Sprague Dawley rats were carefully observed and described with the aid of electron microscopy. The authors referred to them as a cell type very similar to the VSMCs that reside around wider blood vessels (Bruns and Palade 1968). SMPs, as is the case for pericytes, lie embedded in the BM they share with underlying ECs. This membrane, composed of collagen IV, nidogen, laminin and proteoglycans, forms a thin layer which envelopes SMPs (Paulson 1992; Allsopp and Gamble 1979). From within this location SMPs can, depending on their morphological subtype, radiate numerous filopodia that can embrace the whole capillary diameter. The outer border of the BM is embedded into connective tissue, which is home to resident fibroblasts, mast cells and macrophages. Initially 1 SMP was reported to exist in the microvasculature for every 100 ECs (Shepro and Morel 1993). However, such pioneering calculations were based on observations of phase contrast pictures, while more recent quantifications carried out with the help of advanced visualizing technologies suggest a ratio closer to 1:1. Moreover, based on the current knowledge on pericyte function, a 1:100 pericytes to ECs ratio would not allow a sufficient support for ECs. Therefore an estimated 1:10 SMP to ECs ratio is currently accepted.

SMPs are a rare and heterogeneous cell type, making them challenging to define Therefore, several characteristics are considered, including (1) location, (2) morphology and (3) marker profile. (1) SMP are characterised by their location, juxtaposed to ECs and embedded in basement membrane produced by both cell types (Allsopp and Gamble 1979). However this approach is unable to distinguish between SMPs and VSMCs in some arterioles, and BM free capillaries have been identified (Tilton et al. 1979; Díaz-Flores et al. 2009). (2) Morphology has been helpful in distinguishing SMPs from VSMCs. VSMCs tend to elongate in a perpendicular manner relative to the vessel long axes, whilst SMPs generally lay parallel to it, displaying a spindle shaped or stellate morphology depending on the vessel type (Shepro and Morel 1993; Van Dijk et al. 2015). Recent observations have highlighted SMPs processes can cross the entire EC and open up into the lumen, allowing for a direct contact between SMPs and the blood (Baum et al. 2020). When observed through electron microscopy, SMPs reveal a round nucleus with little cytoplasm. They display a branching morphology, allowing connections with ECs through connexin 43 expressing peg-and-socket junctions, or through integrin bound fibronectin plaques. While the former connection type is responsible for chemical communication, the latter provides a tool for mechanical transmission (Van Dijk et al. 2015; Larson et al. 1987). (3) SMPs can be also defined by their distinctive marker expression profile. Several markers have been identified for this cell type; however, none of them are specific alone. Many markers are expressed in a dynamic pattern possibly describing a differential behaviour of SMPs in different conditions that may depend on chemical or mechanical signalling (Table 11.2). This chapter section addresses the biology of SMPs, focusing on their suggested roles in various diseases affecting skeletal muscle tissues, and on their potential role as a target for novel treatments.

11.3.2 Skeletal Muscle Pericytes in Muscle Development and Homeostasis

After the formation of the primordial vessels through vasculogenesis, angiogenesis gradually becomes the only mechanism for vessel network expansion. It is fascinating how this process is intertwined with skeletal muscle development and

Table 11.2 Common SMPs markers and their expression in other cell types residing around the muscle vessel walls

Markers	SMPs	T1 SMPs	T2 SMPs	VSMCs	ECs	MSCs	MyoT	MyoF
NG-2	Armulik et al. (2011), Mazurek et al. (2017), Baum et al. (2020), Larson et al. (1987), Ozerdem et al. (2001)	Larson et al. (1987), Morgan et al. (1993)	Larson et al. (1987), Morgan et al. (1993)	Armulik et al. (2011)			Armulik et al. (2011)	
PDGFR-β	Armulik et al. (2011), Crisan et al. (2008), Dellavalle et al. (2011)					Armulik et al. (2011)		Armulik et al. (2011)
CD146	Crisan et al. (2008), Kostallari et al. (2015)			Allsopp and Gamble (1979)				
αSMA	Armulik et al. (2011), Kostallari et al. (2015), Nehls and Drenckhahn (1991, 1993)			Armulik et al. (2011), Kostallari et al. (2015)				Armulik et al. (2011)
CD56							Kostallari et al. (2015), Dellavalle et al. (2007)	
Desmin	Armulik et al. (2011), Ozerdem et al. (2001), Kostallari et al. (2015), Nehls et al. (1992)			Allsopp and Gamble (1979)				
Vimentin	Armulik et al. (2011), Ozerdem et al. (2001)			Allsopp and Gamble (1979)				
CD144					Mazurek et al. (2017), Ozerdem et al. (2001), Nehls and Drenckhahn (1993)			

(continued)

Table 11.2 (continued)

Markers	SMPs	T1 SMPs	T2 SMPs	VSMCs	ECs	MSCs	MyoT	MyoF
CD31					Kostallari et al. (2015), Dellavalle et al. (2007)			
CD34					Mazurek et al. (2017), Ozerdem et al. (2001), Nehls and Drenckhahn (1993)			
Annexin V	Dellavalle et al. (2011), Brachvogel et al. (2005), Carmeille et al. (2016).						Dellavalle et al. (2007)	
Nestin			Larson et al. (1987), Morgan et al. (1993)					
PDGFR-α		Larson et al. (1987), Morgan et al. (1993)						
AP	Mazurek et al. (2017), Birbrair et al. (2013a), Nehls and Drenckhahn (1993)				Dellavalle et al. (2007, 2011)			

MyoT myotubes, *MyoF* myofibroblasts, *AP* alkaline phosphatase

differentiation. During the early post-natal period, skeletal muscle undergoes a profound maturation that includes a rapid growth of its myofibers, which is supported by the concurrent development of a wide vessel network (Gitiaux et al. 2013; Kostallari et al. 2015; Sallum et al. 2013; Bloch and Iberall 1982). At the beginning of this process, in post-natal day 1 (P1), the SMPs to ECs microvascular ratio has been estimated as 78:100. Very few SMPs are closely associated with the myofibers, with only 24% located at less than 5 μm from the nearest fibre (Kostallari et al. 2015). At this stage myofibers average width is estimated as 5 μm. Concurrently, satellite cells (SCs), (a population of cells resident inside the endomysial membrane and responsible for myofibre growth and regeneration), are actively replicating and are therefore strongly positive for the Ki67 marker. During the following developmental stages myofibres grow wider as they include an increasing number of SCs derived nuclei (White et al. 2010). Meanwhile the capillary network grows quickly in order to comply with the increasing blood volume required by the tissue. This drastically reduces the average distance between capillaries and myofibres, resulting in approximately 73% of SMPs being found at less than 5 μm from the closest myofiber. At the end of the angiogenic sprout SMPs to ECs ratio increases to approximately 9:10, most likely as a consequence of the arrest in ECs proliferation and abundant secretion of Angiopoietin-2 (Ang-2), driving SMPs capillary coverage (Armulik et al. 2011). During the final steps of this developmental process, SCs proliferation gradually slows down, eventually reaching quiescence. This drastic reduction in SC proliferation has been attributed to the increased expression of Angiopoietin-1 (Ang-1) by the approaching SMPs (Fig. 11.4). In SCs as well as ECs, Tie-2 receptor activation by Ang-1 triggers tight junction formation, survival, self-renewal and quiescence (Kostallari et al. 2015; Abou-Khalil et al. 2009).

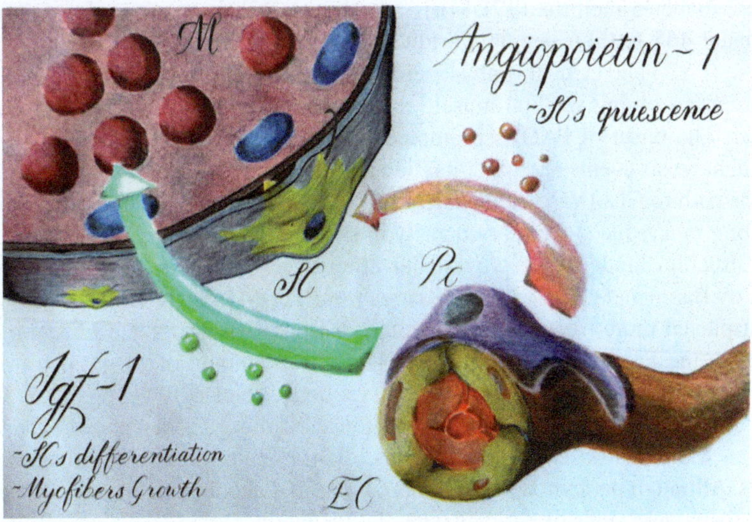

Fig. 11.4 Pericytes induce satellite cells quiescence and differentiation. *M* myofiber, *SC* satellite cell, *Pc* pericyte, *EC* endothelial cell

Moreover, SMPs have been demonstrated to play a pivotal role both *in vitro* (isolated by adherence selection and CD146+ CD56- sorting) and *in vivo*, on the differentiation and quiescence induction of SCs derived myogenic precursors. This is through the secretion of insulin-like growth factor – 1 (IGF-1) and Ang-1 respectively (Kostallari et al. 2015). In an interesting study conducted on mice at early post-natal development stage, SMPs were identified as AP+ CD31- mural cells and subjected to lineage tracking after pharmacologically induced SMPs depletion (Dellavalle et al. 2011). The authors demonstrated SMPs participated in both muscle development and regeneration. SMPs were shown to migrate inside the endomysium, differentiate into SCs, then migrate and differentiate into muscle fibers. However, this SMP differentiation capability was drastically reduced after the first week of life, suggesting SMPs can participate in the quick post-natal muscle growth previously described. Nevertheless, small amounts of SMPs derived cells were still found in myofibers, suggesting SMPs retain a myogenic potential throughout adulthood. Another study highlighted that the selective ablation of paired box protein 7 (PAX7) + SCs permanently precluded muscle regeneration in adult mice (Lepper et al. 2011). Taken together these results could lead to more than one possible conclusion about the role of SMPs in adult muscle regeneration: (1) SMPs can only participate in early post-natal muscle development, or (2) SCs are required for SMPs recruitment and/or activation (Sambasivan et al. 2011).

11.3.3 Skeletal Muscle Pericytes in Diabetes and Critical Limb Ischemia

Type 2 diabetes mellitus (T2DM) is a widespread disease currently affecting an estimated 463 million people worldwide (Saeedi et al. 2019). Peripheral arterial disease (PAD) and critical limb ischemia (CLI) are common diabetes-related complications involving skeletal muscle tissue, the latter being the 'end stage' of the former. The onset of PAD is prompted by insufficient blood supply, resulting in topical necrosis events and inflammation that can become chronic, eventually opening the path to CLI (Varu et al. 2010). In T2DM patients angiogenesis is impaired, resulting in a reduced capillary density in skeletal muscle, coupled with an altered wall structure hindering microvascular efficiency (Vono et al. 2016). In a recent work by Baum and colleagues, capillaries from T2DM human *vastus lateralis* muscle displayed more frequently gaps in the BM between SMPs and ECs, a reduced length in the ECs luminal processes, a significant amount of empty peg – socket junctions and, confirming previous observations, a thicker BM (Fig. 11.5) (Baum et al. 2020; Tilton et al. 1985). In 1985 Tilton and colleagues first showed that, similarly to what was observed in the retina, diabetes may have a negative effect on SMPs. Although pericyte coverage was similar in diabetic patients and control subjects, there were signs of pericyte degradation and increased acellular capillaries in feet (Tilton et al. 1985). Thus, uncovering the molecular basis of how diabetes and

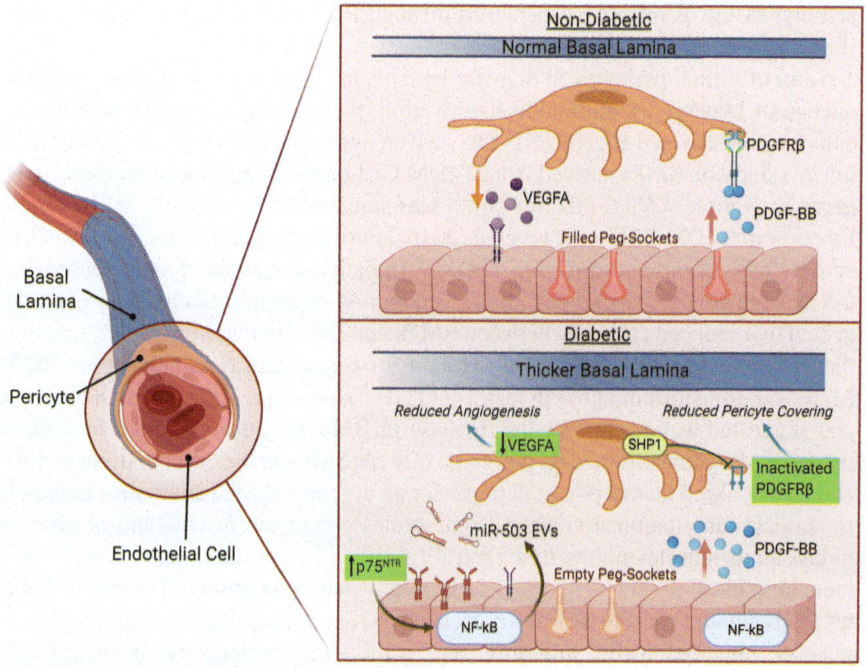

Fig. 11.5 Schematic of the impairment in cross-talk between ECs and SMPs in diabetes. (Created with Biorender.com)

ischemia affect SMPs function and the cross talk between SMPs, ECs, and circulating PACs could help develop new strategies to maintain tissue integrity in patients. This is of crucial importance since the skeletal muscle is more than an organ of motion. Indeed, being the largest site for glucose uptake following insulin stimulation, it is also involved in the regulation of glucose metabolism. However, under hyperglycemic conditions, continuous insulin stimulation induces its own receptor desensitization. Thus, a preferential uptake of free fatty acids together with a reduced rate of fat oxidation promotes the accumulation of intramuscular adipose tissue, and hence the muscle ability to control blood glucose uptake is compromised. As a consequence, muscle goes toward a condition referred to as diabetic myopathy characterized by metabolic impairment, decreased vascularization and fiber atrophy (D'Souza et al. 2013). Since Tilton and colleagues' observations on diabetes-related SMPs loss, evidence has accumulated for an impaired crosstalk between ECs and SMPs in diabetic conditions. This includes impaired expression of vascular endothelial growth factor (VEGF), PDGF-BB and Ang-2 (Hoong et al. 2004), in association with an hyperglycemia- triggered downregulation of PDGFR-β. In T2DM retinal pericytes hyperglycaemia-induced activation of protein kinase C (PKC) results in the Src homology region 2 domain-containing phosphatase-1 (SHP1) induced inactivation of PDGFR-β, reducing pro-survival signals (Geraldes et al. 2009). Conversely, possibly as an attempt to compensate for PDGFR-β loss of activity, PDGF-B mRNA is increased through PKC activity in the diabetic retina (Fig. 11.5) (Yokota et al. 2003).

Loss of retinal pericytes in diabetic patients has also been observed as a consequence of hyperglycaemia-induced activation of nuclear factor kappa-light-chain-enhancer of activated B cells (NF-kB) -driven apoptosis (Romeo et al. 2002). In an *in vitro* study on SMPs derived from T2DM CLI biopsies, an oxidative stress – triggered activation of PKC effector $p66^{shc}$ was suggested to play a role in the loss of function of T2DM SMPs (Vono et al. 2016). Furthermore, impaired mitochondrial activity was recently described in T2DM-CLI patients in terms of an increased proton leakage coupled with reduced ATP production efficiency. Such deficit was connected to a reduced crosstalk between SMPs and ECs, and was partially rescued by the *in vitro* administration of anti-oxidant 2-oxoglutarate (Faulkner et al. 2020). Moreover, transforming growth factor-β (TGF-β) overexpression has been observed and suggested as a reason for the increase in BM deposition observed in diabetic patients. Altogether, these data point to a role for SMPs in the onset of diabetic PAD and CLI, making them a potential target for an urgently needed alternative treatment to surgical amputation. Evidence from studies conducted *in vivo* and *in vitro* on mouse limb muscles points to a central role for the tyrosine kinase neurotrophin receptor p75 ($p75^{NTR}$) Overexpression of $p75^{NTR}$ has been observed in type 1 diabetes (T1D) mouse muscle ECs and correlated to reduced capillary density and angiogenesis, impaired SMPs migration and capillary coverage, and increased ECs apoptosis (Caporali et al. 2008). The $p75^{NTR}$ mechanism of action has been investigated through genome expression profiling, revealing the downregulation of several genes pivotal for angiogenesis and survival of ECs, including VEGF-A, integrin subunit βlike – 1 (ITGBL1) and vascular endothelial zinc finger-1 (VEZF1). Moreover, hyperglycaemia was shown to activate a series of events eventually leading to the onset of metabolic memory: a status of permanent epigenetic modifications causing deregulation of expression of several non-coding RNAs (Prattichizzo et al. 2016; Ceriello 2012). The transcription and extracellular vesicle (EV) -driven paracrine delivery of anti-angiogenic miR-503 generated by NF-kB activation in ECs of T2DM post ischemic mice limb muscle serves as an example of miRNA deregulation in T2DM CLI (Caporali et al. 2011). EVs mediated delivery of miR-503 caused SMPs to downregulate VEGF expression, impairing crosstalk with ECs and angiogenesis (Fig. 11.5) (Caporali et al. 2015). Many other microRNAs have been involved in the onset and perpetuation of T2DM and CLI (Spinetti et al. 2013, 2020), however, gaps are still present in our current knowledge of the complex networks involved in the deregulation of angiogenesis in diabetes.

11.3.4 Skeletal Muscle Pericytes in Regeneration and Regenerative Medicine

Although SCs are recognized as the main cell type responsible for adult skeletal muscle regeneration, there are some important limitations hindering their use as preferred cell type for skeletal muscle regenerative medicine. These include their incapability to cross the myofiber's basal lamina, as well as a complicated isolation procedure and poor survival after transplantation (Dellavalle et al. 2007; Morgan et al. 1993). SMPs attracted growing interest since they were shown to regenerate into muscle *in vivo*. On such basis, SMPs have recently gained credit as a cell source for regenerative medicine applications in several tissues. SMPs express typical markers of mesenchymal stem cells (MSCs) and display a MSC-like morphology when cultured *in vitro*. Moreover, SMPs can differentiate towards adipogenic, skeletal muscle and osteogenic lineages under appropriate *in vitro* conditions (Birbrair et al. 2014). It is important to mention that the exclusive isolation of such a rare cell type from skeletal muscle is hindered by the heterogeneity of SMPs and by the lack of specific markers (Armulik et al. 2011). Moreover, separation from their *in vivo* milieu, including their close contact and crosstalk with ECs, is likely to affect their phenotype, including marker expression. It is therefore reasonable to consider *in vitro* SMPs as a heterogeneous and dynamic population of actively proliferating mesenchymal cells, enriched in pericytes. Isolation method is a key aspect for the development of an *in vitro* pericyte culture, along with culture conditions. In their first application relating to muscle regeneration, SMPs were isolated by dissecting and plating the vessel containing interstitial tissue from human bicep muscles (Dellavalle et al. 2007). After an initial sprouting of fibroblasts, a second population of small, round and refractive cells was observed which adhered poorly to the dish surface. These cells were harvested by pipetting and re-plated, yet rapidly underwent senescence in standard conditions for MSCs and SCs culture. However, using low serum supplemented with bFGF-2 they were expanded for 20 subculture passages. The expanded cells were positive for alkaline phosphatase (AP). In the same study, SMPs were also isolated by digestion to a single cell suspension and sorting for AP+ and CD56- (a myoblast marker) as a comparison. 2–4% of the isolated cells were AP+ CD56-. After seeding at cloning density, clones were almost uniformly AP+ and myogenic factor-5(Myf-5)- after 5 days. However, their phenotype changed after 20 days, when only 20–60% of the clones were AP+ and 10–30% were Myf-5+, with few showing double positivity. These observations suggest skeletal muscle mural cells rapidly lose their phenotype in culture, possibly towards the myogenic fate. SMP-derived cells isolated by pipetting of interstitial tissue outgrowths were shown to regenerate healthy myofibers in a dystrophic mouse model (Dellavalle et al. 2007). Two different SMPs subtypes were identified on the basis of the expression of neural marker nestin, adipogenic progenitor marker PDGFR-α, and pericyte marker NG2 (Birbrair et al. 2011). In this study, cells were isolated through digestion of dissected murine hindlimb muscles and sorting for NG2 and nestin. Type 1 SMPs were described as nestin-, PDGFR-α + and NG2+, and are thought to participate in age-related muscle loss of function due to adipose tissue accumulation. Interestingly, type 1 and not type 2 SMPs were shown to participate in adipose tissue formation in mice. Type 2 pericytes are nestin+, PDGFR-α- and NG2+ and display the ability to undergo myogenic differentiation *in vitro* and *in vivo* (Birbrair et al. 2013a) (Fig. 11.6).

Systemic transplantation of both wild type mesoangioblasts (a type of foetal mural vessel stem cells considered a subset of pericytes, MB) and MBs transduced with α-sarcoglycan (α-SG) into α-SG null dystrophic mice resulted in their participation in muscle reconstitution and amelioration of the dystrophic condition of downstream muscles (Sampaolesi et al. 2003; Tedesco et al. 2012). Similar results were obtained through the systemic transplantation of human SMPs sorted as AP+ CD56- into a dystrophic mouse model (Dellavalle et al. 2007). SMPs sorted from human adult skeletal muscles as CD146+, CD56-, CD34- and CD144-, showed an

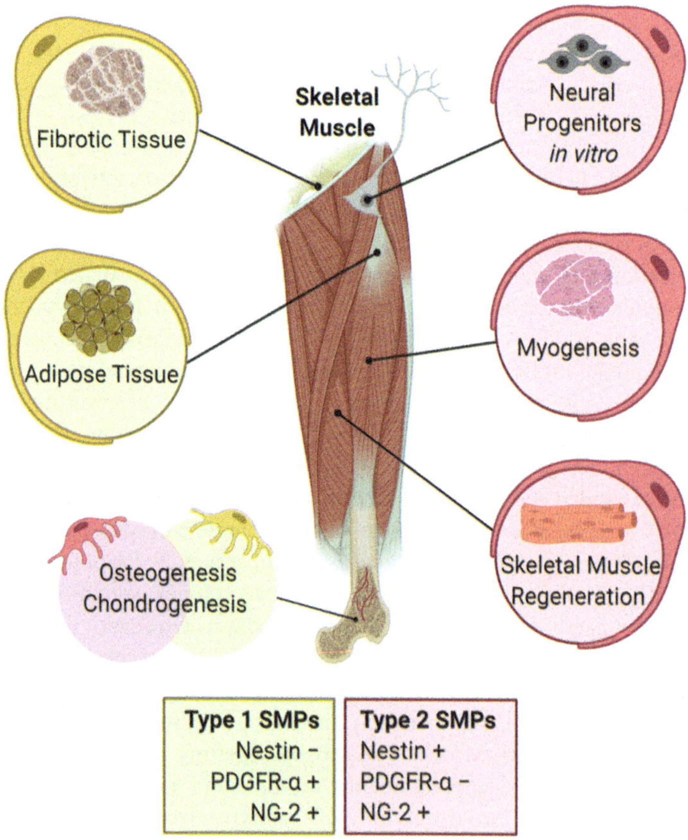

Fig. 11.6 SMPs classification in type 1 and type 2. While type 1 SMPs display fibrotic phenotype and adipogenic differentiation potential and could be involved in the onset of sarcopeny and muscle degeneration, type 2 SMPs display myogenic potential and could be involved in normal skeletal muscle regeneration. (Created with Biorender.com)

even higher myogenic potential *in vivo* compared to CD146-, CD56+, CD34-, CD144- myoblasts (Crisan et al. 2008). In a swine model it was observed that 3D culture within polyethylene glycol (PEG) hydrogels helped adult SMPs to recover from their reduced capacity to form myotubes *in vitro* compared to younger counterparts (Fuoco et al. 2014). Type 2 SMPs are also able to generate glial precursors *in vitro* (Birbrair et al. 2013b). Considering that non identified skeletal muscle cells are able to facilitate re-innervation in a denervated skeletal muscle mouse model (Lavasani et al. 2014), it would be interesting to understand whether SMPs are involved in this process *in vivo*, and to evaluate their regenerative potential as a therapy for the several pathologies involving muscle denervation (Birbrair et al. 2014). SMPs also possess the ability to differentiate toward the chondrogenic and osteogenic lineage, and can produce calcification under appropriate conditions *in vitro* (Crisan et al. 2008; Farrington-Rock et al. 2004). However, ectopic bone formation in skeletal muscle can occur due to pathological conditions such as fibrodysplasia ossificans progressive and progressive osseous heteroplasia. The exact cell source for ectopic bone formation has not been identified as yet, but evidence collected so far suggests the exclusion of macrophages and B cells, and points towards a phenotype that closely resembles that of SMPs (Birbrair et al. 2014). All these observations suggest SMPs as a very promising cell source for regenerative medicine purposes, due to their multipotency and relative ease of delivery. At present, the mechanisms involved in their adipogenic, chondrogenic and osteogenic transdifferentiation *in vivo* still needs to be completely elucidated and could provide an alternative target for the treatment of the pathologies involved.

11.4 Future Perspectives

Adventitial pericyte-like cells and SMPs represent a viable source of adult multipotent stem cells, with potential for use as an autologous treatment for several angiogenesis-related diseases. A growing number of studies have helped shed light on the vast network of signalling pathways involved in the onset and continuation of chronic angiogenesis-related pathologies such as atherosclerosis, T2DM and CLI. These studies highlight the important role played by pericytes and endothelial progenitor cells in these conditions. Future studies are needed in the context of T2DM and CLI in order to completely unravel the complex crosstalk between ECs, pericytes and the microenvironment they share. Areas for further investigation include secreted signals such as cytokines, oxidative stressors and non-coding RNAs. Identification of key pathways could provide valuable hints for *ex vivo* genetic correction of autologous pericytes, turning them into precious tools in the quest for effective treatments for diseases such as T2DM and its related complications. Moreover, the involvement of pericytes in the aberrant angiogenesis and inflammation related events that sustain the growth of a large proportion of human cancers also demands further investigation. Studies on ageing of connective tissues will also benefit from the study of adventitial derived pericytes – like cells and

SMPs biology. This is due to their acknowledged involvement in ectopic calcification and adipogenesis, two mechanisms considered strongly involved in age-related loss of function at both tissue and organ level.

Acknowledgments We thank Dr. Sadie Slater, University of Bristol, for proofreading the manuscript.

Compliance with Ethical Standards

Disclosure of Interests: All authors declare they have no conflict of interest.

References

Abou-Khalil R, Le Grand F, Pallafacchina G et al (2009) Autocrine and paracrine angiopoietin 1/Tie-2 signaling promotes muscle satellite cell self-renewal. Cell Stem Cell. https://doi.org/10.1016/j.stem.2009.06.001

Aicher A, Rentsch M, Sasaki KI et al (2007) Nonbone marrow-derived circulating progenitor cells contribute to postnatal neovascularization following tissue ischemia. Circ Res 100:581–589. https://doi.org/10.1161/01.RES.0000259562.63718.35

Alexander MY, Wilkinson FL, Kirton JP et al (2005) Identification and characterization of vascular calcification-associated factor, a novel gene upregulated during vascular calcification in vitro and in vivo. Arterioscler Thromb Vasc Biol. https://doi.org/10.1161/01.ATV.0000175750.94742.46

Allsopp G, Gamble HJ (1979) An electron microscopic study of the pericytes of the developing capillaries in human fetal brain and muscle. J Anat 128(Pt 1):155–168

Alvino VV, Fernández-Jiménez R, Rodriguez-Arabaolaza I et al (2018) Transplantation of allogeneic pericytes improves myocardial vascularization and reduces interstitial fibrosis in a swine model of reperfused acute myocardial infarction. J Am Heart Assoc 7. https://doi.org/10.1161/JAHA.117.006727

Armulik A, Genové G, Betsholtz C (2011) Pericytes: developmental, physiological, and pathological perspectives, problems, and promises. Dev Cell 21(2):193–215

Avolio E, Meloni M, Spencer HL et al (2015) Combined intramyocardial delivery of human pericytes and cardiac stem cells additively improves the healing of mouse infarcted hearts through stimulation of vascular and muscular repair. Circ Res. https://doi.org/10.1161/CIRCRESAHA.115.306146

Baum O, Bernd J, Becker S et al (2020) Structural microangiopathies in skeletal muscle related to systemic vascular pathologies in humans. Front Physiol. https://doi.org/10.3389/fphys.2020.00028

Benjamin LE, Hemo I, Keshet E (1998) A plasticity window for blood vessel remodelling is defined by pericyte coverage of the preformed endothelial network and is regulated by PDGF-B and VEGF. Development 125:1591–1598

Bergwerff M, Verberne ME, DeRuiter MC et al (1998) Neural crest cell contribution to the developing circulatory system implications for vascular morphology? Circ Res. https://doi.org/10.1161/01.RES.82.2.221

Billaud M, Donnenberg VS, Ellis BW et al (2017) Classification and functional characterization of vasa vasorum-associated perivascular progenitor cells in human aorta. Stem Cell Reports. https://doi.org/10.1016/j.stemcr.2017.04.028

Birbrair A, Wang ZM, Messi ML et al (2011) Nestin-GFP transgene reveals neural precursor cells in adult skeletal muscle. PLoS One 6. https://doi.org/10.1371/journal.pone.0016816

Birbrair A, Zhang T, Wang ZM et al (2013a) Role of pericytes in skeletal muscle regeneration and fat accumulation. Stem Cells Dev. https://doi.org/10.1089/scd.2012.0647

Birbrair A, Zhang T, Wang ZM et al (2013b) Skeletal muscle neural progenitor cells exhibit properties of NG2-glia. Exp Cell Res. https://doi.org/10.1016/j.yexcr.2012.09.008

Birbrair A, Zhang T, Wang ZM et al (2014) Pericytes: multitasking cells in the regeneration of injured, diseased, and aged skeletal muscle. Front Aging Neurosci 6(119):245

Blazquez-Martinez A, Chiesa M, Arnalich F et al (2014) C-kit identifies a subpopulation of mesenchymal stem cells in adipose tissue with higher telomerase expression and differentiation potential. Differentiation 87:147–160. https://doi.org/10.1016/j.diff.2014.02.007

Bloch EH, Iberall AS (1982) Toward a concept of the functional unit of mammalian skeletal muscle. Am J Phys. https://doi.org/10.1152/ajpregu.1982.242.5.r411

Brachvogel B, Moch H, Pausch F et al (2005) Pervascular cells expressing annexin A5 define a novel mesenchymal stem cell-like population with the capacity to differentiate into multiple mesenchymal lineages. Development. https://doi.org/10.1242/dev.01846

Bruns RR, Palade GE (1968) Studies on blood capillaries. I. General organization of blood capillaries in muscle. J Cell Biol. https://doi.org/10.1083/jcb.37.2.244

Campagnolo P, Cesselli D, Al Haj Zen A et al (2010) Human adult vena saphena contains perivascular progenitor cells endowed with clonogenic and proangiogenic potential. Circulation. https://doi.org/10.1161/CIRCULATIONAHA.109.899252

Campagnolo P, Hong X, Di Bernardini E et al (2015a) Resveratrol-induced vascular progenitor differentiation towards endothelial lineage via miR-21/Akt/β-catenin is protective in vessel graft models. PLoS One. https://doi.org/10.1371/journal.pone.0125122

Campagnolo P, Tsai TN, Hong X et al (2015b) C-kit+ progenitors generate vascular cells for tissue-engineered grafts through modulation of the Wnt/Klf4 pathway. Biomaterials. https://doi.org/10.1016/j.biomaterials.2015.04.055

Campagnolo P, Gormley AJ, Chow LW et al (2016) Pericyte seeded dual peptide scaffold with improved endothelialization for vascular graft tissue engineering. Adv Healthc Mater. https://doi.org/10.1002/adhm.201600699

Caporali A, Pani E, Horrevoets AJG et al (2008) Neurotrophin p75 receptor (p75NTR) promotes endothelial cell apoptosis and inhibits angiogenesis: implications for diabetes-induced impaired neovascularization in ischemic limb muscles. Circ Res. https://doi.org/10.1161/CIRCRESAHA.108.177386

Caporali A, Meloni M, Völlenkle C et al (2011) Deregulation of microRNA-503 contributes to diabetes mellitus-induced impairment of endothelial function and reparative angiogenesis after limb ischemia. Circulation. https://doi.org/10.1161/CIRCULATIONAHA.110.952325

Caporali A, Meloni M, Nailor A et al (2015) P75NTR-dependent activation of NF-κB regulates microRNA-503 transcription and pericyte-endothelial crosstalk in diabetes after limb ischaemia. Nat Commun. https://doi.org/10.1038/ncomms9024

Carmeille R, Bouvet F, Tan S et al (2016) Membrane repair of human skeletal muscle cells requires Annexin-A5. Biochim Biophys Acta, Mol Cell Res. https://doi.org/10.1016/j.bbamcr.2016.06.003

Carrabba M, De Maria C, Oikawa A et al (2016) Design, fabrication and perivascular implantation of bioactive scaffolds engineered with human adventitial progenitor cells for stimulation of arteriogenesis in peripheral ischemia. Biofabrication. https://doi.org/10.1088/1758-5090/8/1/015020

Castrechini NM, Murthi P, Gude NM et al (2010) Mesenchymal stem cells in human placental chorionic villi reside in a vascular niche. Placenta 31:203–212. https://doi.org/10.1016/j.placenta.2009.12.006

Ceriello A (2012) The emerging challenge in diabetes: the "metabolic memory". Vasc Pharmacol 57(5–6):133–138

Chang HN, Pang JHS, Chen CPC et al (2012) The effect of aging on migration, proliferation, and collagen expression of tenocytes in response to ciprofloxacin. J Orthop Res 30(5):764–768

Chen Y, Wong MM, Campagnolo P et al (2013) Adventitial stem cells in vein grafts display multilineage potential that contributes to neointimal formation. Arterioscler Thromb Vasc Biol 33:1844–1851. https://doi.org/10.1161/ATVBAHA.113.300902

Clark ER, Clark EL (1925) A. The development of adventitial (Rouget) cells on the blood capillaries of amphibian larvae. Am J Anat. https://doi.org/10.1002/aja.1000350205

Corselli M, Chen CW, Sun B et al (2012) The tunica adventitia of human arteries and veins as a source of mesenchymal stem cells. Stem Cells Dev. https://doi.org/10.1089/scd.2011.0200

Covas DT, Piccinato CE, Orellana MD et al (2005) Mesenchymal stem cells can be obtained from the human saphena vein. Exp Cell Res 309:340–344. https://doi.org/10.1016/j.yexcr.2005.06.005

Crisan M, Yap S, Casteilla L et al (2008) A perivascular origin for mesenchymal stem cells in multiple human organs. Cell Stem Cell. https://doi.org/10.1016/j.stem.2008.07.003

D'Souza DM, Al-Sajee D, Hawke TJ (2013) Diabetic myopathy: impact of diabetes mellitus on skeletal muscle progenitor cells. Front Physiol 88:158–164

Davie NJ, Crossno JT, Frid MG et al (2004) Hypoxia-induced pulmonary artery adventitial remodeling and neovascularization: contribution of progenitor cells. Am J Physiol Lung Cell Mol Physiol 286:668–678. https://doi.org/10.1152/ajplung.00108.2003

De Palma M, Venneri MA, Galli R et al (2005) Tie2 identifies a hematopoietic lineage of proangiogenic monocytes required for tumor vessel formation and a mesenchymal population of pericyte progenitors. Cancer Cell 8:211–226. https://doi.org/10.1016/j.ccr.2005.08.002

Deasy BM, Schugar RC, Chirieleison SM et al (2009) High harvest yield, high expansion, and phenotype stability of CD146 mesenchymal stromal cells from whole primitive human umbilical cord tissue. J Biomed Biotechnol 2009. https://doi.org/10.1155/2009/789526

Dellavalle A, Sampaolesi M, Tonlorenzi R et al (2007) Pericytes of human skeletal muscle are myogenic precursors distinct from satellite cells. Nat Cell Biol. https://doi.org/10.1038/ncb1542

Dellavalle A, Maroli G, Covarello D et al (2011) Pericytes resident in postnatal skeletal muscle differentiate into muscle fibres and generate satellite cells. Nat Commun. https://doi.org/10.1038/ncomms1508

Devana SK, Kelley BV, McBride OJ et al (2018) Adipose-derived human perivascular stem cells may improve achilles tendon healing in rats. Clin Orthop Relat Res. https://doi.org/10.1097/CORR.0000000000000461

Díaz-Flores L, Gutiérrez R, Madrid JF et al (2009) Pericytes. Morphofunction, interactions and pathology in a quiescent and activated mesenchymal cell niche. Histol Histopathol 24(7):909–969

Doherty MJ, Ashton BA, Walsh S et al (1998) Vascular pericytes express osteogenic potential in vitro and in vivo. J Bone Miner Res 13:828–838. https://doi.org/10.1359/jbmr.1998.13.5.828

Dulmovits BM, Herman IM (2012) Microvascular remodeling and wound healing: a role for pericytes. Int J Biochem Cell Biol 44(11):1800–1812

Faglia E, Clerici G, Clerissi J et al (2006) Early and five-year amputation and survival rate of diabetic patients with critical limb ischemia: data of a cohort study of 564 patients. Eur J Vasc Endovasc Surg. https://doi.org/10.1016/j.ejvs.2006.03.006

Fang S, Wei J, Pentinmikko N et al (2012) Generation of functional blood vessels from a single c-kit+ adult vascular endothelial stem cell. PLoS Biol 10. https://doi.org/10.1371/journal.pbio.1001407

Farrington-Rock C, Crofts NJ, Doherty MJ et al (2004) Chondrogenic and adipogenic potential of microvascular pericytes. Circulation. https://doi.org/10.1161/01.CIR.0000144457.55518.E5

Faulkner A, Tamiato A, Cathery W, Rampin A, Garcia EJ, Allen S, Mellor H, Lisa Heather GS, PM (2020) 2-oxoglutarate improves redox balance and mitochondrial function in muscle pericytes of type-2 diabetic patients. Diabetologia. https://doi.org/10.1007/s00125-020-05230-4

Fuoco C, Sangalli E, Vono R et al (2014) 3D hydrogel environment rejuvenates aged pericytes for skeletal muscle tissue engineering. Front Physiol. https://doi.org/10.3389/fphys.2014.00203

Garoffolo G, Ruiter MS, Piola M et al (2020) Coronary artery mechanics induces human saphenous vein remodelling via recruitment of adventitial myofibroblast-like cells mediated by thrombospondin-1. Theranostics 10:2597–2611. https://doi.org/10.7150/thno.40595

Geraldes P, Hiraoka-Yamamoto J, Matsumoto M et al (2009) Activation of PKC-and SHP-1 by hyperglycemia causes vascular cell apoptosis and diabetic retinopathy. Nat Med. https://doi.org/10.1038/nm.2052

Gitiaux C, Kostallari E, Lafuste P et al (2013) Whole microvascular unit deletions in dermatomyositis. Ann Rheum Dis. https://doi.org/10.1136/annrheumdis-2012-201822

Hirsch AT, Haskal ZJ, Hertzer NR et al (2006) ACC/AHA 2005 practice guidelines for the management of patients with peripheral arterial disease (lower extremity, renal, mesenteric, and abdominal aortic). Circulation 113(11):e463–e654

Hoehn K, Marieb EN (2012) Human anatomy & physiology. Benjamin-Cummings Publ Co, San Francisco. https://doi.org/10.1007/BF00845519

Hoong SL, Blann AD, Aun YC et al (2004) Plasma vascular endothelial growth factor, angiopoietin-1, and angiopoietin-2 in diabetes: implications for cardiovascular risk and effects of multifactorial intervention. Diabetes Care. https://doi.org/10.2337/diacare.27.12.2918

Howson KM, Aplin AC, Gelati M et al (2005) The postnatal rat aorta contains pericyte progenitor cells that form spheroidal colonies in suspension culture. Am J Physiol Cell Physiol. https://doi.org/10.1152/ajpcell.00168.2005

Hu Y, Zhang Z, Torsney E et al (2004) Abundant progenitor cells in the adventitia contribute to atheroscleroses of vein grafts in ApoE-deficient mice. J Clin Invest 113:1258–1265. https://doi.org/10.1172/JCI19628

Ieronimakis N, Hays AL, Janebodin K et al (2013) Coronary adventitial cells are linked to perivascular cardiac fibrosis via TGF$β$1 signaling in the mdx mouse model of Duchenne muscular dystrophy. J Mol Cell Cardiol 63:122–134. https://doi.org/10.1016/j.yjmcc.2013.07.014

Invernici G, Emanueli C, Madeddu P et al (2007) Human fetal aorta contains vascular progenitor cells capable of inducing vasculogenesis, angiogenesis, and myogenesis in vitro and in a murine model of peripheral ischemia. Am J Pathol 170:1879–1892. https://doi.org/10.2353/ajpath.2007.060646

James AW, Zara JN, Corselli M et al (2012a) An abundant perivascular source of stem cells for bone tissue engineering. Stem Cells Transl Med. https://doi.org/10.5966/sctm.2012-0053

James AW, Zara JN, Zhang X et al (2012b) Perivascular stem cells: a prospectively purified mesenchymal stem cell population for bone tissue engineering. Stem Cells Transl Med 1:510–519. https://doi.org/10.5966/sctm.2012-0002

Katare R, Riu F, Mitchell K et al (2011) Transplantation of human pericyte progenitor cells improves the repair of infarcted heart through activation of an angiogenic program involving micro-RNA-132. Circ Res 109:894–906. https://doi.org/10.1161/CIRCRESAHA.111.251546

Khan WS, Tew SR, Adesida AB, Hardingham TE (2008) Human infrapatellar fat pad-derived stem cells express the pericyte marker 3G5 and show enhanced chondrogenesis after expansion in fibroblast growth factor-2. Arthritis Res Ther 10:1–11. https://doi.org/10.1186/ar2448

Kim MS, Dean LS (2011) In-stent restenosis. Cardiovasc Ther 29(3):190–198

Kim SJ, Kim SA, Choi YA et al (2020) Alpha-smooth muscle actin-positive perivascular cells in diabetic retina and choroid. Int J Mol Sci 21. https://doi.org/10.3390/ijms21062158

Klatte-Schulz F, Pauly S, Scheibel M et al (2012) Influence of age on the cell biological characteristics and the stimulation potential of male human tenocyte-like cells. Eur Cells Mater. https://doi.org/10.22203/eCM.v024a06

Klein D, Weißhardt P, Kleff V et al (2011) Vascular wall-resident CD44+ multipotent stem cells give rise to pericytes and smooth muscle cells and contribute to new vessel maturation. PLoS One. https://doi.org/10.1371/journal.pone.0020540

Kostallari E, Baba-Amer Y, Alonso-Martin S et al (2015) Pericytes in the myovascular niche promote post-natal myofiber growth and satellite cell quiescence. Development. https://doi.org/10.1242/dev.115386

Kramann R, Schneider RK, Dirocco DP et al (2015) Perivascular Gli1+ progenitors are key contributors to injury-induced organ fibrosis. Cell Stem Cell 16:51–66. https://doi.org/10.1016/j.stem.2014.11.004

Kramann R, Goettsch C, Wongboonsin J et al (2016) Adventitial MSC-like cells are progenitors of vascular smooth muscle cells and drive vascular calcification in chronic kidney disease. Cell Stem Cell 19:628–642. https://doi.org/10.1016/j.stem.2016.08.001

Kumar A, D'Souza SS, Moskvin OV et al (2017) Specification and diversification of pericytes and smooth muscle cells from mesenchymoangioblasts. Cell Rep. https://doi.org/10.1016/j.celrep.2017.05.019
Larson DM, Carson MP, Haudenschild CC (1987) Junctional transfer of small molecules in cultured bovine brain microvascular endothelial cells and pericytes. Microvasc Res. https://doi.org/10.1016/0026-2862(87)90052-5
Lavasani M, Thompson SD, Pollett JB et al (2014) Human muscle-derived stem/progenitor cells promote functional murine peripheral nerve regeneration. J Clin Invest. https://doi.org/10.1172/JCI44071
Lepper C, Partridge TA, Fan CM (2011) An absolute requirement for pax7-positive satellite cells in acute injury-induced skeletal muscle regeneration. Development. https://doi.org/10.1242/dev.067595
Leszczynska A, O'Doherty A, Farrell E et al (2016) Differentiation of vascular stem cells contributes to ectopic calcification of atherosclerotic plaque. Stem Cells. https://doi.org/10.1002/stem.2315
Li Z, Yan G, Diao Q et al (2019) Transplantation of human endometrial perivascular cells with elevated CYR61 expression induces angiogenesis and promotes repair of a full-thickness uterine injury in rat. Stem Cell Res Ther 10:1–16. https://doi.org/10.1186/s13287-019-1272-3
Lusis AJ (2000) Atherosclerosis. Nature 407(6801):233–234
Maier CL, Shepherd BR, Yi T, Pober JS (2010) Explant outgrowth, propagation and characterization of human pericytes. Microcirculation 17:367–380. https://doi.org/10.1111/j.1549-8719.2010.00038.x
Majesky MW, Horita H, Ostriker A et al (2017) Differentiated smooth muscle cells generate a subpopulation of resident vascular progenitor cells in the adventitia regulated by Klf4. Circ Res 120:296–311. https://doi.org/10.1161/CIRCRESAHA.116.309322
Mazurek R, Dave JM, Chandran RR et al (2017) Vascular cells in blood vessel wall development and disease. In: Advances in pharmacology, APHA. Academic, London
Mekala SR, Wörsdörfer P, Bauer J et al (2018) Generation of cardiomyocytes from vascular adventitia-resident stem cells. Circ Res 123:686–699. https://doi.org/10.1161/CIRCRESAHA.117.312526
Michelis KC, Nomura-Kitabayashi A, Lecce L et al (2018) CD90 identifies adventitial mesenchymal progenitor cells in adult human medium- and large-sized arteries. Stem Cell Reports 11:242–257. https://doi.org/10.1016/j.stemcr.2018.06.001
Montani D, Perros F, Gambaryan N et al (2011) C-kit-positive cells accumulate in remodeled vessels of idiopathic pulmonary arterial hypertension. Am J Respir Crit Care Med 184:116–123. https://doi.org/10.1164/rccm.201006-0905OC
Montemurro T, Andriolo G, Montelatici E et al (2011) Differentiation and migration properties of human foetal umbilical cord perivascular cells: potential for lung repair. J Cell Mol Med 15:796–808. https://doi.org/10.1111/j.1582-4934.2010.01047.x
Morgan JE, Pagel CN, Sherrratt T, Partridge TA (1993) Long-term persistence and migration of myogenic cells injected into pre-irradiated muscles of mdx mice. J Neurol Sci. https://doi.org/10.1016/0022-510X(93)90224-M
Nehls V, Drenckhahn D (1991) Heterogeneity of microvascular pericytes for smooth muscle type alpha-actin. J Cell Biol. https://doi.org/10.1083/jcb.113.1.147
Nehls V, Drenckhahn D (1993) The versatility of microvascular pericytes: from mesenchyme to smooth muscle? Histochemistry 99:1–12
Nehls V, Denzer K, Drenckhahn D (1992) Pericyte involvement in capillary sprouting during angiogenesis in situ. Cell Tissue Res. https://doi.org/10.1007/BF00645048
Ni Z, Deng J, Potter CMF et al (2019) Recipient c-kit lineage cells repopulate smooth muscle cells of transplant arteriosclerosis in mouse models. Circ Res 125:223–241. https://doi.org/10.1161/CIRCRESAHA.119.314855
Ozerdem U, Grako KA, Dahlin-Huppe K et al (2001) NG2 proteoglycan is expressed exclusively by mural cells during vascular morphogenesis. Dev Dyn. https://doi.org/10.1002/dvdy.1200

Pasquinelli G, Tazzari PL, Vaselli C et al (2007) Thoracic aortas from multiorgan donors are suitable for obtaining resident angiogenic mesenchymal stromal cells. Stem Cells. https://doi.org/10.1634/stemcells.2006-0731

Passman JN, Dong XR, Wu SP et al (2008) A sonic hedgehog signaling domain in the arterial adventitia supports resident Sca1+ smooth muscle progenitor cells. Proc Natl Acad Sci U S A 105:9349–9354. https://doi.org/10.1073/pnas.0711382105

Paulson M (1992) Basement membrane proteins: structure, assembly, and cellular interactions. Crit Rev Biochem Mol Biol. https://doi.org/10.3109/10409239209082560

Prattichizzo F, Giuliani A, De Nigris V et al (2016) Extracellular microRNAs and endothelial hyperglycaemic memory: a therapeutic opportunity? Diabetes Obes Metab 18(9):855–867

Psaltis PJ, Puranik AS, Spoon DB et al (2014) Characterization of a resident population of adventitial macrophage progenitor cells in postnatal vasculature. Circ Res 115:364–375. https://doi.org/10.1161/CIRCRESAHA.115.303299

Robin C, Bollerot K, Mendes S et al (2009) Human placenta is a potent hematopoietic niche containing hematopoietic stem and progenitor cells throughout development. Cell Stem Cell 5:385–395. https://doi.org/10.1016/j.stem.2009.08.020

Romeo G, Liu WH, Asnaghi V et al (2002) Activation of nuclear factor-κB induced by diabetes and high glucose regulates a proapoptotic program in retinal pericytes. Diabetes. https://doi.org/10.2337/diabetes.51.7.2241

Rouget C (1873) Memoire sur le developpement, la structure et les proprietes physiologiques des capillaries sanguins et lymphatiques. Arch Physiol Norm Pathol 5:603–663

Russell JS, Brown JM (2014) Circulating mouse Flk1+/c-kit+/CD45- cells function as endothelial progenitors cells (EPCs) and stimulate the growth of human tumor xenografts. Mol Cancer 13:1–16. https://doi.org/10.1186/1476-4598-13-177

Saeedi P, Petersohn I, Salpea P et al (2019) Global and regional diabetes prevalence estimates for 2019 and projections for 2030 and 2045: results from the International Diabetes Federation Diabetes Atlas, 9th edition. Diabetes Res Clin Pract. https://doi.org/10.1016/j.diabres.2019.107843

Sainz J, Zen AAH, Caligiuri G et al (2006) Isolation of "side population" progenitor cells from healthy arteries of adult mice. Arterioscler Thromb Vasc Biol 26:281–286. https://doi.org/10.1161/01.ATV.0000197793.83391.91

Sallum AME, Varsani H, Holton JL et al (2013) Morphometric analyses of normal pediatric brachial biceps and quadriceps muscle tissue. Histol Histopathol. https://doi.org/10.14670/HH-28.525

Sambasivan R, Yao R, Kissenpfennig A et al (2011) Pax7-expressing satellite cells are indispensable for adult skeletal muscle regeneration. Development. https://doi.org/10.1242/dev.067587

Sampaolesi M, Torrente Y, Innocenzi A et al (2003) Cell therapy of α-sarcoglycan null dystrophic mice through intra-arterial delivery of mesoangioblasts. Science (80-). https://doi.org/10.1126/science.1082254

Sarugaser R, Lickorish D, Baksh D et al (2005) Human umbilical cord perivascular (HUCPV) cells: a source of mesenchymal progenitors. Stem Cells 23:220–229. https://doi.org/10.1634/stemcells.2004-0166

Schwab KE, Gargett CE (2007) Co-expression of two perivascular cell markers isolates mesenchymal stem-like cells from human endometrium. Hum Reprod 22:2903–2911. https://doi.org/10.1093/humrep/dem265

Shepro D, Morel NM (1993) Pericyte physiology. FASEB J. https://doi.org/10.1096/fasebj.7.11.8370472

Shi Y, O'Brien JE, Fard A et al (1996) Adventitial myofibroblasts contribute to neointimal formation in injured porcine coronary arteries. Circulation. https://doi.org/10.1161/01.CIR.94.7.1655

Spinetti G, Fortunato O, Caporali A et al (2013) MicroRNA-15a and MicroRNA-16 impair human circulating proangiogenic cell functions and are increased in the proangiogenic cells and serum of patients with critical limb ischemia. Circ Res. https://doi.org/10.1161/CIRCRESAHA.111.300418

Spinetti G, Sangalli E, Tagliabue E et al (2020) microRNA-21/PDCD4 proapoptotic signaling from circulating CD34 + cells to vascular endothelial cells: a potential contributor to adverse cardiovascular outcomes in patients with critical limb ischemia. Diabetes Care. https://doi.org/10.2337/dc19-2227

Tedesco FS, Gerli MFM, Perani L et al (2012) Transplantation of genetically corrected human iPSC-derived progenitors in mice with limb-girdle muscular dystrophy. Sci Transl Med. https://doi.org/10.1126/scitranslmed.3003541

Tigges U, Komatsu M, Stallcup WB (2013) Adventitial pericyte progenitor/mesenchymal stem cells participate in the restenotic response to arterial injury. J Vasc Res 50:134–144. https://doi.org/10.1159/000345524

Tilton RG, Kilo C, Williamson JR (1979) Pericyte-endothelial relationships in cardiac and skeletal muscle capillaries. Microvasc Res. https://doi.org/10.1016/0026-2862(79)90041-4

Tilton RG, Faller AM, Burkhardt JK et al (1985) Pericyte degeneration and acellular capillaries are increased in the feet of human diabetic patients. Diabetologia. https://doi.org/10.1007/BF00703132

Toledo-Flores D, Williamson A, Schwarz N et al (2019) Vasculogenic properties of adventitial Sca-1 + CD45 + progenitor cells in mice: a potential source of vasa vasorum in atherosclerosis. Sci Rep 9:1–15. https://doi.org/10.1038/s41598-019-43765-8

Torsney E, Mandal K, Halliday A et al (2007) Characterisation of progenitor cells in human atherosclerotic vessels. Atherosclerosis 191:259–264. https://doi.org/10.1016/j.atherosclerosis.2006.05.033

Tsai TN, Kirton JP, Campagnolo P et al (2012) Contribution of stem cells to neointimal formation of decellularized vessel grafts in a novel mouse model. Am J Pathol 181:362–373. https://doi.org/10.1016/j.ajpath.2012.03.021

Tsang WP, Shu Y, Kwok PL et al (2013) CD146+ human umbilical cord perivascular cells maintain stemness under hypoxia and as a cell source for skeletal regeneration. PLoS One 8:1–13. https://doi.org/10.1371/journal.pone.0076153

Tsukada T, Tippens D, Gordon D et al (1987) HHF35, a muscle-actin-specific monoclonal antibody. I. Immunocytochemial and biochemical characterization. Am J Pathol 126(1):51–60

Van Dijk CGM, Nieuweboer FE, Pei JY et al (2015) The complex mural cell: pericyte function in health and disease. Int J Cardiol 190:75–89

Varu VN, Hogg ME, Kibbe MR (2010) Critical limb ischemia. J Vasc Surg 51(1):230–241

Vono R, Fuoco C, Testa S et al (2016) Activation of the pro-oxidant PKCβIIp66Shc signaling pathway contributes to pericyte dysfunction in skeletal muscles of patients with diabetes with critical limb ischemia. Diabetes. https://doi.org/10.2337/db16-0248

Wang G, Jacquet L, Karamariti E, Xu Q (2015) Origin and differentiation of vascular smooth muscle cells. J Physiol. https://doi.org/10.1113/JP270033

Wang Y, Xu J, Chang L et al (2019) Relative contributions of adipose-resident CD146+ pericytes and CD34+ adventitial progenitor cells in bone tissue engineering. NPJ Regen Med 4. https://doi.org/10.1038/s41536-018-0063-2

White RB, Biérinx AS, Gnocchi VF, Zammit PS (2010) Dynamics of muscle fibre growth during postnatal mouse development. BMC Dev Biol. https://doi.org/10.1186/1471-213X-10-21

Wilkinson FL, Liu Y, Rucka AK et al (2007) Contribution of VCAF-positive cells to neovascularization and calcification in atherosclerotic plaque development. J Pathol. https://doi.org/10.1002/path.2114

Wong MM, Winkler B, Karamariti E et al (2013) Sirolimus stimulates vascular stem/progenitor cell migration and differentiation into smooth muscle cells via epidermal growth factor receptor/extracellular signal-regulated kinase/$β$-catenin signaling pathway. Arterioscler Thromb Vasc Biol 33:2397–2406. https://doi.org/10.1161/ATVBAHA.113.301595

Yokota T, Ma RC, Park JY et al (2003) Role of protein kinase C on the expression of platelet-derived growth factor and endothelin-1 in the retina of diabetic rats and cultured retinal capillary pericytes. Diabetes. https://doi.org/10.2337/diabetes.52.3.838

Zengin E, Chalajour F, Gehling UM et al (2006) Vascular wall resident progenitor cells: a source for postnatal vasculogenesis. Development. https://doi.org/10.1242/dev.02315

Zhang S, Ba K, Wu L et al (2015) Adventitial cells and perictyes support chondrogenesis through different mechanisms in 3-dimensional cultures with or without nanoscaffolds. J Biomed Nanotechnol. https://doi.org/10.1166/jbn.2015.2112

Zimmerlin L, Donnenberg VS, Pfeifer ME et al (2010) Stromal vascular progenitors in adult human adipose tissue. Cytom Part A 77:22–30. https://doi.org/10.1002/cyto.a.20813

Zimmermann KW (1923) Der feinere Bau der Blutcapillaren. Z Anat Entwicklungsgesch. https://doi.org/10.1007/bf02593544

Index

A. Birbrair (ed.), *Biology of Pericytes – Recent Advances*, Stem Cell Biology and Regenerative Medicine 68, https://doi.org/10.1007/978-3-030-62129-2

The manufacturer's authorised representative in the EU is Springer Nature Customer Service Centre GmbH, Europaplatz 3, 69115 Heidelberg, Germany. If you have any concerns regarding our products, please contact ProductSafety@springernature.com

Printed and bound by CPI Group (UK) Ltd, Croydon, CR0 4YY
28/09/2026
02224092-0002